# STUDENT'S SOLUTIONS MANUAL

# INTRODUCTORY ALGEBRA

## SEVENTH EDITION

# STUDENT'S SOLUTIONS MANUAL

# INTRODUCTORY ALGEBRA
## SEVENTH EDITION

**Marvin L. Bittinger**
Indiana University—Purdue University at Indianapolis

**Mervin L. Keedy**
Purdue University

# Judith A. Penna

**ADDISON-WESLEY PUBLISHING COMPANY**
Reading, Massachusetts • Menlo Park, California • New York
Don Mills, Ontario • Wokingham, England • Amsterdam • Bonn
Sydney • Singapore • Tokyo • Madrid • San Juan • Milan • Paris

*Reprinted with corrections, September 1996.*

Reproduced by Addison-Wesley from camera-ready copy supplied by the authors.

ISBN  0-201-58962-1

5 6 7 8 9 10        9796

# Table of Contents

The author thanks Patsy Hammond for her
excellent typing and Pam Smith for her careful
proofreading. Very special thanks are extended to
Mike Penna for sharing his expertise and for
his support.

# STUDENT'S SOLUTIONS MANUAL

# INTRODUCTORY ALGEBRA
## SEVENTH EDITION

# Chapter R
# Prealgebra Review

## Exercise Set R.1

**1.** We first find some factorizations:

$$20 = 1 \cdot 20, \ 20 = 2 \cdot 10, \ 20 = 4 \cdot 5$$

The factors of 20 are 1, 2, 4, 5, 10, and 20.

**3.** We first find some factorizations:

$$72 = 1 \cdot 72, \ 72 = 2 \cdot 36, \ 72 = 3 \cdot 24,$$
$$72 = 4 \cdot 18, \ 72 = 6 \cdot 12, \ 72 = 8 \cdot 9$$

The factors of 72 are 1, 2, 3, 4, 6, 8, 9, 12, 18, 24, 36 and 72.

**5.** $15 = 3 \cdot 5$

Both factors are prime, so we have the prime factorization of 15.

**7.** $22 = 2 \cdot 11$

Both factors are prime, so we have the prime factorization of 22.

**9.** $9 = 3 \cdot 3$

Both factors are prime, so we have the prime factorization of 9.

**11.** $49 = 7 \cdot 7$

Both factors are prime, so we have the prime factorization of 49.

**13.** We begin by factoring 18 in any way that we can and continue factoring until each factor is prime.

$$18 = 2 \cdot 9 = 2 \cdot 3 \cdot 3$$

**15.** We begin by factoring 40 in any way that we can and continue factoring until each factor is prime.

$$40 = 4 \cdot 10 = 2 \cdot 2 \cdot 2 \cdot 5$$

**17.** We begin by factoring 90 in any way that we can and continue factoring until each factor is prime.

$$90 = 2 \cdot 45 = 2 \cdot 9 \cdot 5 = 2 \cdot 3 \cdot 3 \cdot 5$$

**19.** We go through the table of primes until we find a prime that divides 210. We continue dividing by that prime until it is not possible to do so any longer. We continue this process until each factor is prime.

$$210 = 2 \cdot 105 = 2 \cdot 3 \cdot 35 = 2 \cdot 3 \cdot 5 \cdot 7$$

**21.** We go through the table of primes until we find a prime that divides 91. The first such prime is 7.

$$91 = 7 \cdot 13$$

Both factors are prime, so this is the prime factorization.

**23.** We go through the table of primes until we find a prime that divides 119. The first such prime is 7.

$$119 = 7 \cdot 17$$

Both factors are prime, so this is the prime factorization.

**25.** a) We find the prime factorizations:

$$4 = 2 \cdot 2$$
$$5 = 5 \quad (\text{5 is prime})$$

b) We write 2 as a factor two times (the greatest number of times it occurs in any one factorization). We write 5 as a factor one time (the greatest number of times it occurs in any one factorization). The LCM is $2 \cdot 2 \cdot 5$, or 20.

**27.** a) We find the prime factorizations:

$$24 = 2 \cdot 2 \cdot 2 \cdot 3$$
$$36 = 2 \cdot 2 \cdot 3 \cdot 3$$

b) We write 2 as a factor three times (the greatest number of times it occurs in any one factorization). We write 3 as a factor two times (the greatest number of times it occurs in any one factorization). The LCM is $2 \cdot 2 \cdot 2 \cdot 3 \cdot 3$, or 72.

**29.** $3 = 3 \quad (\text{3 is prime})$
$15 = 3 \cdot 5$
The LCM is $3 \cdot 5$, or 15.

**31.** $30 = 2 \cdot 3 \cdot 5$
$40 = 2 \cdot 2 \cdot 2 \cdot 5$
The LCM is $2 \cdot 2 \cdot 2 \cdot 3 \cdot 5$, or 120.

**33.** 13 and 23 are both prime. The LCM is $13 \cdot 23$, or 299.

**35.** $18 = 2 \cdot 3 \cdot 3$
$30 = 2 \cdot 3 \cdot 5$
The LCM is $2 \cdot 3 \cdot 3 \cdot 5$, or 90.

**37.** $30 = 2 \cdot 3 \cdot 5$
$36 = 2 \cdot 2 \cdot 3 \cdot 3$
The LCM is $2 \cdot 2 \cdot 3 \cdot 3 \cdot 5$, or 180.

**39.** $24 = 2 \cdot 2 \cdot 2 \cdot 3$
$30 = 2 \cdot 3 \cdot 5$
The LCM is $2 \cdot 2 \cdot 2 \cdot 3 \cdot 5$, or 120.

**41.** 17 and 29 are both prime. The LCM is $17 \cdot 29$, or 493.

**43.** $12 = 2 \cdot 2 \cdot 3$

$28 = 2 \cdot 2 \cdot 7$

The LCM is $2 \cdot 2 \cdot 3 \cdot 7$, or 84.

**45.** 2, 3, and 5 are all prime. The LCM is $2 \cdot 3 \cdot 5$, or 30.

**47.** $24 = 2 \cdot 2 \cdot 2 \cdot 3$

$36 = 2 \cdot 2 \cdot 3 \cdot 3$

$12 = 2 \cdot 2 \cdot 3$

The LCM is $2 \cdot 2 \cdot 2 \cdot 3 \cdot 3$, or 72.

**49.** $5 = 5$    (5 is prime)

$12 = 2 \cdot 2 \cdot 3$

$15 = 3 \cdot 5$

The LCM is $2 \cdot 2 \cdot 3 \cdot 5$, or 60.

**51.** $6 = 2 \cdot 3$

$12 = 2 \cdot 2 \cdot 3$

$18 = 2 \cdot 3 \cdot 3$

The LCM is $2 \cdot 2 \cdot 3 \cdot 3$, or 36.

**53.** $8 = 2 \cdot 2 \cdot 2$

$12 = 2 \cdot 2 \cdot 3$

The greatest number of times 2 occurs in the factorizations of 8 and 12 is three times. The greatest number of times 3 occurs is once. Thus, the LCM must contain exactly three factors of 2 and one factor of 3. The LCM is $2 \cdot 2 \cdot 2 \cdot 3$.

a) No; $2 \cdot 2 \cdot 3 \cdot 3$ is not a multiple of 8.

b) No; $2 \cdot 2 \cdot 3$ is not a multiple of 8.

c) No; $2 \cdot 3 \cdot 3$ is not a multiple of 8 or 12.

d) Yes; $2 \cdot 2 \cdot 2 \cdot 3$ is a multiple of both 8 and 12, and it is the smallest such multiple.

**55.** Use a calculator to find multiples of 7800. Divide each one by 2700 to determine if it is also a multiple of 2700. The first such number, 70,200, is the LCM. ($70,200 = 7800 \cdot 9 = 2700 \cdot 26$)

**57.** Saturn: $30 = 2 \cdot 3 \cdot 5$

Uranus: $84 = 2 \cdot 2 \cdot 3 \cdot 7$

The LCM is $2 \cdot 2 \cdot 3 \cdot 5 \cdot 7$, or 420.

Saturn and Uranus will appear in the same position every 420 years.

## Exercise Set R.2

**1.** $\dfrac{3}{4} = \dfrac{3}{4} \cdot 1$    Identity property of 1

$= \dfrac{3}{4} \cdot \dfrac{3}{3}$    Using $\dfrac{3}{3}$ for 1

$= \dfrac{9}{12}$    Multiplying numerators and denominators

**3.** $\dfrac{3}{5} = \dfrac{3}{5} \cdot 1$    Identity property of 1

$= \dfrac{3}{5} \cdot \dfrac{20}{20}$    Using $\dfrac{20}{20}$ for 1

$= \dfrac{60}{100}$    Multiplying numerators and denominators

**5.** $\dfrac{13}{20} = \dfrac{13}{20} \cdot 1$    Identity property of 1

$= \dfrac{13}{20} \cdot \dfrac{8}{8}$    Using $\dfrac{8}{8}$ for 1

$= \dfrac{104}{160}$

**7.** We will use $\dfrac{3}{3}$ for 1 since $24 = 8 \cdot 3$.

$\dfrac{7}{8} = \dfrac{7}{8} \cdot 1 = \dfrac{7}{8} \cdot \dfrac{3}{3} = \dfrac{21}{24}$

**9.** We will use $\dfrac{4}{4}$ for 1 since $16 = 4 \cdot 4$.

$\dfrac{5}{4} = \dfrac{5}{4} \cdot 1 = \dfrac{5}{4} \cdot \dfrac{4}{4} = \dfrac{20}{16}$

**11.** $\dfrac{18}{27} = \dfrac{2 \cdot 9}{3 \cdot 9}$    Factoring numerator and denominator

$= \dfrac{2}{3} \cdot \dfrac{9}{9}$    Factoring the fractional expression

$= \dfrac{2}{3} \cdot 1$

$= \dfrac{2}{3}$    Identity property of 1

**13.** $\dfrac{56}{14} = \dfrac{4 \cdot 14}{1 \cdot 14}$    Factoring and inserting a factor of 1 in the numerator

$= \dfrac{4}{1} \cdot \dfrac{14}{14}$    Factoring the fractional expression

$= \dfrac{4}{1} \cdot 1$

$= 4$    Identity property of 1

**15.** $\dfrac{6}{42} = \dfrac{1 \cdot 6}{7 \cdot 6}$    Factoring and inserting a factor of 1 in the numerator

$= \dfrac{1}{7} \cdot \dfrac{6}{6}$

$= \dfrac{1}{7} \cdot 1$

$= \dfrac{1}{7}$

**17.** $\dfrac{56}{7} = \dfrac{8 \cdot 7}{1 \cdot 7} = \dfrac{8}{1} \cdot \dfrac{7}{7} = \dfrac{8}{1} \cdot 1 = 8$

**19.** $\dfrac{19}{76} = \dfrac{1 \cdot 19}{4 \cdot 19}$    Factoring and inserting a factor of 1 in the numerator

$= \dfrac{1 \cdot \cancel{19}}{4 \cdot \cancel{19}}$    Removing a factor of 1: $\dfrac{19}{19} = 1$

$= \dfrac{1}{4}$

**21.** $\dfrac{100}{20} = \dfrac{5 \cdot 20}{1 \cdot 20}$    Factoring and inserting a factor of 1 in the denominator

$= \dfrac{5 \cdot \cancel{20}}{1 \cdot \cancel{20}}$    Removing a factor of 1: $\dfrac{20}{20} = 1$

$= \dfrac{5}{1}$

$= 5$    Simplifying

**23.** $\dfrac{425}{525} = \dfrac{17 \cdot 25}{21 \cdot 25}$    Factoring the numerator and the denominator

$= \dfrac{17 \cdot \cancel{25}}{21 \cdot \cancel{25}}$    Removing a factor of 1: $\dfrac{25}{25} = 1$

$= \dfrac{17}{21}$

**25.** $\dfrac{2600}{1400} = \dfrac{2 \cdot 13 \cdot 100}{2 \cdot 7 \cdot 100}$    Factoring

$= \dfrac{13 \cdot \cancel{2} \cdot \cancel{100}}{7 \cdot \cancel{2} \cdot \cancel{100}}$    Removing a factor of 1:

$\dfrac{2 \cdot 100}{2 \cdot 100} = 1$

$= \dfrac{13}{7}$

**27.** $\dfrac{8 \cdot x}{6 \cdot x} = \dfrac{2 \cdot 4 \cdot x}{2 \cdot 3 \cdot x}$    Factoring

$= \dfrac{4 \cdot \cancel{2} \cdot \cancel{x}}{3 \cdot \cancel{2} \cdot \cancel{x}}$    Removing a factor of 1: $\dfrac{2 \cdot x}{2 \cdot x} = 1$

$= \dfrac{4}{3}$

**29.** $\dfrac{1}{3} \cdot \dfrac{1}{4} = \dfrac{1 \cdot 1}{3 \cdot 4}$    Multiplying numerators and denominators

$= \dfrac{1}{12}$

**31.** $\dfrac{15}{4} \cdot \dfrac{3}{4} = \dfrac{15 \cdot 3}{4 \cdot 4} = \dfrac{45}{16}$

**33.** $\dfrac{1}{3} + \dfrac{1}{3} = \dfrac{1 + 1}{3}$    Adding numerators; keeping the same denominator

$= \dfrac{2}{3}$

**35.** $\dfrac{4}{9} + \dfrac{13}{18} = \dfrac{4}{9} \cdot \dfrac{2}{2} + \dfrac{13}{18}$    LCD is 18.

$= \dfrac{8}{18} + \dfrac{13}{18}$

$= \dfrac{21}{18}$

$= \dfrac{7 \cdot \cancel{3}}{6 \cdot \cancel{3}} = \dfrac{7}{6}$    Simplifying

**37.** $\dfrac{3}{10} + \dfrac{8}{15} = \dfrac{3}{10} \cdot \dfrac{3}{3} + \dfrac{8}{15} \cdot \dfrac{2}{2}$    LCD is 30.

$= \dfrac{9}{30} + \dfrac{16}{30}$

$= \dfrac{25}{30}$

$= \dfrac{5 \cdot \cancel{5}}{6 \cdot \cancel{5}} = \dfrac{5}{6}$    Simplifying

**39.** $\dfrac{5}{4} - \dfrac{3}{4} = \dfrac{2}{4}$

$= \dfrac{1 \cdot \cancel{2}}{2 \cdot \cancel{2}} = \dfrac{1}{2}$

**41.** $\dfrac{11}{12} - \dfrac{3}{8} = \dfrac{11}{12} \cdot \dfrac{2}{2} - \dfrac{3}{8} \cdot \dfrac{3}{3}$    LCD is 24.

$= \dfrac{22}{24} - \dfrac{9}{24}$

$= \dfrac{13}{24}$

**43.** $\dfrac{11}{12} - \dfrac{2}{5} = \dfrac{11}{12} \cdot \dfrac{5}{5} - \dfrac{2}{5} \cdot \dfrac{12}{12}$    LCD is 60.

$= \dfrac{55}{60} - \dfrac{24}{60}$

$= \dfrac{31}{60}$

**45.** $\dfrac{7}{6} \div \dfrac{3}{5} = \dfrac{7}{6} \cdot \dfrac{5}{3}$    $\left( \dfrac{5}{3} \text{ is the reciprocal of } \dfrac{3}{5}. \right)$

$= \dfrac{35}{18}$    Multiplying

**47.** $\dfrac{8}{9} \div \dfrac{4}{15} = \dfrac{8}{9} \cdot \dfrac{15}{4} = \dfrac{8 \cdot 15}{9 \cdot 4} = \dfrac{2 \cdot \cancel{4} \cdot \cancel{3} \cdot 5}{\cancel{3} \cdot 3 \cdot \cancel{4}} = \dfrac{10}{3}$

**49.** $\dfrac{1}{8} \div \dfrac{1}{4} = \dfrac{1}{8} \cdot \dfrac{4}{1} = \dfrac{1 \cdot 4}{8 \cdot 1} = \dfrac{1 \cdot \cancel{4}}{2 \cdot \cancel{4} \cdot 1} = \dfrac{1}{2}$

**51.** $\dfrac{\frac{13}{12}}{\frac{39}{5}} = \dfrac{13}{12} \div \dfrac{39}{5} = \dfrac{13}{12} \cdot \dfrac{5}{39} = \dfrac{13 \cdot 5}{12 \cdot 39} = \dfrac{\cancel{13} \cdot 5}{12 \cdot 3 \cdot \cancel{13}} = \dfrac{5}{36}$

**53.** $100 \div \dfrac{1}{5} = \dfrac{100}{1} \div \dfrac{1}{5} = \dfrac{100}{1} \cdot \dfrac{5}{1} = \dfrac{100 \cdot 5}{1 \cdot 1} = \dfrac{500}{1} = 500$

**55.** $\dfrac{3}{4} \div 10 = \dfrac{3}{4} \cdot \dfrac{1}{10} = \dfrac{3 \cdot 1}{4 \cdot 10} = \dfrac{3}{40}$

**57.** We begin by factoring 28 in any way that we can and continue factoring until each factor is prime.

$$28 = 4 \cdot 7 = 2 \cdot 2 \cdot 7$$

**59.** We begin by factoring 1000 in any way that we can and continue factoring until each factor is prime.

$$1000 = 4 \cdot 250 = 2 \cdot 2 \cdot 2 \cdot 125 = 2 \cdot 2 \cdot 2 \cdot 5 \cdot 25 =$$
$$2 \cdot 2 \cdot 2 \cdot 5 \cdot 5 \cdot 5$$

**61.** $16 = 2 \cdot 2 \cdot 2 \cdot 2$
$24 = 2 \cdot 2 \cdot 2 \cdot 3$
The LCM $= 2 \cdot 2 \cdot 2 \cdot 2 \cdot 3$, or 48.

**63.** $\dfrac{192}{256} = \dfrac{3 \cdot 64}{4 \cdot 64} = \dfrac{3}{4}$

**65.** $\dfrac{64 \cdot a \cdot b}{16 \cdot a \cdot b} = \dfrac{4 \cdot 16 \cdot a \cdot b}{1 \cdot 16 \cdot a \cdot b} = \dfrac{4}{1} = 4$

**67.** $\dfrac{36 \cdot (2 \cdot h)}{8 \cdot (9 \cdot h)} = \dfrac{9 \cdot 4 \cdot 2 \cdot h}{2 \cdot 4 \cdot 9 \cdot h} = 1$

## Exercise Set R.3

**1.** 5.3.
1 place $\qquad$ $5.3 = \dfrac{53}{10}$ — 1 zero

**3.** 0.67.
2 places $\qquad$ $0.67 = \dfrac{67}{100}$ — 2 zeros

**5.** 2.0007,
4 places $\qquad$ $2.0007 = \dfrac{20,007}{10,000}$ — 4 zeros

**7.** 7889.8.
1 place $\qquad$ $7889.8 = \dfrac{78,898}{10}$ — 1 zero

**9.** $\dfrac{1}{10}$ — 1 zero $\quad$ 0.1. — 1 place $\quad$ $\dfrac{1}{10} = 0.1$

**11.** $\dfrac{1}{10,000}$ — 4 zeros $\quad$ 0.0001. — 4 places $\quad$ $\dfrac{1}{10,000} = 0.0001$

**13.** $\dfrac{9999}{1000}$ — 3 zeros $\quad$ 9.999. — 3 places $\quad$ $\dfrac{9999}{1000} = 9.999$

**15.** $\dfrac{4578}{10,000}$ — 4 zeros $\quad$ 0.4578. — 4 places $\quad$ $\dfrac{4578}{10,000} = 0.4578$

**17.**
$$\begin{array}{r} \overset{1\phantom{0}\overset{1}{0}\phantom{0}}{4\,1\,5.\,7\,8} \\ +\quad 2\,9.\,1\,6 \\ \hline 4\,4\,4.\,9\,4 \end{array}$$

**19.**
$$\begin{array}{r} \overset{\phantom{00}1\phantom{0000}}{2\,3\,4.\,0\,0\,0} \\ +\,1\,5\,6.\,6\,1\,7 \\ \hline 3\,9\,0.\,6\,1\,7 \end{array}$$

**21.** $85 + 67.95 + 2.774$

We have:
$$\begin{array}{r} \overset{\phantom{00}1\ 1\ 1\phantom{00}}{8\,5.} \\ 6\,7.\,9\,5 \\ +\quad 2.\,7\,7\,4 \\ \hline 1\,5\,5.\,7\,2\,4 \end{array}$$

**23.** $17.95 + 16.99 + 28.85$

We have:
$$\begin{array}{r} \overset{2\ 2\ 1\phantom{00}}{1\,7.\,9\,5} \\ 1\,6.\,9\,9 \\ +\,2\,8.\,8\,5 \\ \hline 6\,3.\,7\,9 \end{array}$$

**25.**
$$\begin{array}{r} \overset{\phantom{0}7\ \cancel{8}\ \cancel{1}\ \cancel{1}\ 10}{7\,8.\,1\,1\,0} \\ -\,4\,5.\,8\,7\,6 \\ \hline 3\,2.\,2\,3\,4 \end{array}$$

**27.**
$$\begin{array}{r} \overset{7\ \cancel{8}\ 9\ 10}{3\,8.\,7\,0\,0} \\ -\,1\,1.\,8\,6\,5 \\ \hline 2\,6.\,8\,3\,5 \end{array}$$

**29.**
$$\begin{array}{r} \overset{4\ \cancel{6}\ 18}{5\,7.\,8\,6} \\ -\quad 9.\,9\,5 \\ \hline 4\,7.\,9\,1 \end{array}$$

**31.**
$$\begin{array}{r} \overset{2\ 9\ 9\ 9\ 10}{3.\,0\,0\,0\,0} \\ -\,1.\,0\,8\,0\,7 \\ \hline 1.\,9\,1\,9\,3 \end{array}$$

**33.**
$$\begin{array}{r} 7.\,3\,4 \longleftarrow \text{2 decimal places} \\ \times \quad 1.\,8 \longleftarrow \text{1 decimal place} \\ \hline 5\,8\,7\,2 \\ 7\,3\,4\,0 \\ \hline 1\,3.\,2\,1\,2 \longleftarrow \text{3 decimal places} \end{array}$$

**35.**
$$
\begin{array}{r}
\overset{5}{\phantom{0}}\overset{1}{\phantom{0}}\phantom{0} \\
0.\,8\,6 \longleftarrow \text{2 decimal places} \\
\times\quad 0.\,9\,3 \longleftarrow \text{2 decimal places} \\
\hline
2\,5\,8 \\
7\,7\,4\,0 \\
\hline
0.\,7\,9\,9\,8 \longleftarrow \text{4 decimal places}
\end{array}
$$

**37.**
$$
\begin{array}{r}
1\,7.\,9\,5 \longleftarrow \text{2 decimal places} \\
\times\qquad 1\,0 \longleftarrow \text{0 decimal places} \\
\hline
1\,7\,9.\,5\,0 \longleftarrow \text{2 decimal places}
\end{array}
$$

**39.**
$$
\begin{array}{r}
1\,8.\,9\,4 \longleftarrow \text{2 decimal places} \\
\times\quad 0.\,0\,1 \longleftarrow \text{2 decimal places} \\
\hline
0.\,1\,8\,9\,4 \longleftarrow \text{4 decimal places}
\end{array}
$$

**41.**
$$
\begin{array}{r}
\overset{1}{\phantom{0}}\ \overset{2}{\phantom{0}}\phantom{00} \\
\overset{4}{\phantom{0}}\ \overset{5}{\phantom{0}}\phantom{00} \\
0.\,4\,5\,7 \longleftarrow \text{3 decimal places} \\
\times\quad 3.\,0\,8 \longleftarrow \text{2 decimal places} \\
\hline
3\,6\,5\,6 \\
1\,3\,7\,1\,0\,0 \\
\hline
1.\,4\,0\,7\,5\,6 \longleftarrow \text{5 decimal places}
\end{array}
$$

**43.**
$$
\begin{array}{r}
\overset{5}{\phantom{0}}\ \overset{3}{\phantom{0}}\ \overset{1}{\phantom{0}} \\
\overset{5}{\phantom{0}}\ \overset{3}{\phantom{0}}\ \overset{1}{\phantom{0}} \\
3.\,6\,4\,2 \longleftarrow \text{3 decimal places} \\
\times\quad 0.\,9\,9 \longleftarrow \text{2 decimal places} \\
\hline
3\,2\,7\,7\,8 \\
3\,2\,7\,7\,8\,0 \\
\hline
3.\,6\,0\,5\,5\,8 \longleftarrow \text{5 decimal places}
\end{array}
$$

**45.** Place the decimal point in the quotient directly above the decimal point in the dividend. Then divide as whole numbers.

$$
\begin{array}{r}
2.\,3\phantom{0} \\
72\,\overline{)\,1\,6\,5.\,6} \\
1\,4\,4\phantom{.0} \\
\hline
2\,1\,6\phantom{.} \\
2\,1\,6\phantom{.} \\
\hline
0
\end{array}
$$

**47.**
$$
\begin{array}{r}
5.\,2\phantom{0} \\
8.5_\wedge\,\overline{)\,4\,4.\,2_\wedge 0} \\
4\,2\,5\phantom{0} \\
\hline
1\,7\,0 \\
1\,7\,0 \\
\hline
0
\end{array}
$$

**49.**
$$
\begin{array}{r}
0.\,0\,2\,3 \\
9.9_\wedge\,\overline{)\,0.\,2_\wedge 2\,7\,7} \\
1\,9\,8\phantom{0} \\
\hline
2\,9\,7 \\
2\,9\,7 \\
\hline
0
\end{array}
$$

**51.**
$$
\begin{array}{r}
1\,8.\,7\,5 \\
0.6\,4_\wedge\,\overline{)\,1\,2.\,0\,0_\wedge 0\,0} \\
6\,4\phantom{.0000} \\
\hline
5\,6\,0\phantom{000} \\
5\,1\,2\phantom{000} \\
\hline
4\,8\,0\phantom{00} \\
4\,4\,8\phantom{00} \\
\hline
3\,2\,0\phantom{0} \\
3\,2\,0\phantom{0} \\
\hline
0
\end{array}
$$

**53.**
$$
\begin{array}{r}
6\,6\,0.\phantom{00} \\
1.0\,5_\wedge\,\overline{)\,6\,9\,3.\,0\,0_\wedge} \\
6\,3\,0\phantom{.00} \\
\hline
6\,3\,0\phantom{0} \\
6\,3\,0\phantom{0} \\
\hline
0\phantom{0} \\
0\phantom{0} \\
\hline
0
\end{array}
$$

**55.**
$$
\begin{array}{r}
0.\,6\,8 \\
8.6_\wedge\,\overline{)\,5.\,8_\wedge 4\,8} \\
5\,1\,6\phantom{0} \\
\hline
6\,8\,8 \\
6\,8\,8 \\
\hline
0
\end{array}
$$

**57.**
$$
\begin{array}{r}
0.\,3\,4\,3\,7\,5 \\
32\,\overline{)\,1\,1.\,0\,0\,0\,0\,0} \\
9\,6\phantom{.00000} \\
\hline
1\,4\,0\phantom{0000} \\
1\,2\,8\phantom{0000} \\
\hline
1\,2\,0\phantom{000} \\
9\,6\phantom{000} \\
\hline
2\,4\,0\phantom{00} \\
2\,2\,4\phantom{00} \\
\hline
1\,6\,0\phantom{0} \\
1\,6\,0\phantom{0} \\
\hline
0
\end{array}
$$

Decimal notation for $\dfrac{11}{32}$ is 0.34375.

**59.**
$$
\begin{array}{r}
1.\,1\,8\,1\,8 \\
11\,\overline{)\,1\,3.\,0\,0\,0\,0} \\
1\,1\phantom{.0000} \\
\hline
2\,0\phantom{000} \\
1\,1\phantom{000} \\
\hline
9\,0\phantom{00} \\
8\,8\phantom{00} \\
\hline
2\,0\phantom{0} \\
1\,1\phantom{0} \\
\hline
9\,0 \\
8\,8 \\
\hline
2
\end{array}
$$

Since 2 and 9 alternate as remainders, the sequence of digits following the decimal point in the quotient repeats. Thus, decimal notation for $\dfrac{13}{11}$ is 1.1818..., or $1.\overline{18}$.

**61.**

```
     0.5 5
9 ) 5.0 0
     4 5
     ‾‾‾‾
       5 0
       4 5
       ‾‾‾
         5
```

The number 5 repeats as a remainder, so the digit 5 will repeat in the quotient. Thus, decimal notation for $\frac{5}{9}$ is $0.55\ldots$, or $0.\overline{5}$.

**63.**

```
       2.1 1
9 ) 1 9.0 0
     1 8
     ‾‾‾‾
       1 0
         9
       ‾‾‾
         1
```

The number 1 repeats as a remainder, so the digit 1 will repeat in the quotient. Thus, decimal notation for $\frac{19}{9}$ is $2.11\ldots$, or $2.\overline{1}$.

**65.** 745.06534

Round to the nearest hundredth: The digit in the hundredths place is 6. The next digit to the right, 5, is 5 or higher, so we round up: 745.07

Round to the nearest tenth: The digit in the tenths place is 0. The next digit to the right, 6, is 5 or higher, so we round up: 745.1

Round to the nearest one: The digit in the ones place is 5. The next digit to the right, 0, is less than 5, so we round down: 745

Round to the nearest ten: The digit in the tens place is 4. The next digit to the right, 5, is 5 or higher, so we round up: 750

Round to the nearest hundred: The digit in the hundreds place is 7. The next digit to the right, 4, is less than 5, so we round down: 700

**67.** 6780.50568

Round to the nearest hundredth: The digit in the hundredths place is 0. The next digit to the right, 5, is 5 or higher, so we round up: 6780.51

Round to the nearest tenth: The digit in the tenths place is 5. The next digit to the right, 0, is less than 5, so we round down: 6780.5

Round to the nearest one: The digit in the ones place is 0. The next digit to the right, 5, is 5 or higher, so we round up: 6781

Round to the nearest ten: The digit in the tens place is 8. The next digit to the right, 0, is less than 5, so we round down: 6780

Round to the nearest hundred: The digit in the hundreds place is 7. The next digit to the right, 8, is 5 or higher, so we round up: 6800

**69.** $17.988

Round to the nearest cent (nearest hundredth): The digit in the hundredths place is 8. The next digit to the right, 8, is 5 or higher, so we round up: $17.99

Round to the nearest dollar (nearest one): The digit in the ones place is 7. The next digit to the right, 9, is 5 or higher, so we round up: $18

**71.** $346.075

Round to the nearest cent (nearest hundredth): The digit in the hundredths place is 7. The next digit to the right, 5, is 5 or higher, so we round up: $346.08

Round to the nearest dollar (nearest one): The digit in the ones place is 6. The next digit to the right, 0, is less than 5, so we round down: $346

**73.** $16.95

The digit in the ones place is 6. The next digit to the right, 9, is 5 or higher, so we round up: $17

**75.** $189.50

The digit in the ones place is 9. The next digit to the right, 5, is 5 or higher, so we round up: $190

**77.**

```
            1 2.3 4 5 6 7
8 1 ) 1 0 0 0.0 0 0 0 0
        8 1
        ‾‾‾
        1 9 0
        1 6 2
        ‾‾‾‾‾
          2 8 0
          2 4 3
          ‾‾‾‾‾
            3 7 0
            3 2 4
            ‾‾‾‾‾
              4 6 0
              4 0 5
              ‾‾‾‾‾
                5 5 0
                4 8 6
                ‾‾‾‾‾
                  6 4 0
                  5 6 7
                  ‾‾‾‾‾
                    7 3
```

$\frac{1000}{81} \approx 12.34567$

We round to the nearest:

ten-thousandth: 12.3457

thousandth: 12.346

hundredth: 12.35

tenth: 12.3

one: 12

## Exercise Set R.4

**1.** 63%    0.63.
              ↑__|

Move the decimal point 2 places to the left.

$63\% = 0.63$

**3.** 94.1%    0. 94.1
          ⌐⌐↑

Move the decimal point 2 places to the left.
94.1% = 0.941

**5.** 1%    0 .01.
          ⌐⌐↑

Move the decimal point 2 places to the left.
1% = 0.01

**7.** 0.61%    0 .00.61
            ⌐⌐↑

Move the decimal point 2 places to the left.
0.61% = 0.0061

**9.** 240%    2 .40.
            ⌐⌐↑

Move the decimal point 2 places to the left.
240% = 2.4

**11.** 3.25%    0 .03.25
              ⌐⌐↑

Move the decimal point 2 places to the left.
3.25% = 0.0325

**13.** $60\% = 60 \times \dfrac{1}{100}$    Replacing % by $\times \dfrac{1}{100}$

  $= \dfrac{60}{100}$

**15.** $28.9\% = 28.9 \times \dfrac{1}{100}$    Replacing % by $\times \dfrac{1}{100}$

  $= \dfrac{28.9}{100}$

  $= \dfrac{28.9}{100} \cdot \dfrac{10}{10}$    Multiplying by 1 to get a whole number in the numerator

  $= \dfrac{289}{1000}$

**17.** $110\% = 110 \times \dfrac{1}{100} = \dfrac{110}{100}$

**19.** $0.042\% = 0.042 \times \dfrac{1}{100} = \dfrac{0.042}{100} = \dfrac{0.042}{100} \times \dfrac{1000}{1000} =$

  $\dfrac{42}{100,000}$

**21.** $250\% = 250 \times \dfrac{1}{100} = \dfrac{250}{100}$

**23.** $3.47\% = 3.47 \times \dfrac{1}{100} = \dfrac{3.47}{100} = \dfrac{3.47}{100} \times \dfrac{100}{100} = \dfrac{347}{10,000}$

**25.** 1    1.00.
            ⌐⌐↑

Move the decimal point 2 places to the right.
1 = 100%

**27.** 0.996    0.99.6
                ⌐⌐↑

Move the decimal point 2 places to the right.
0.996 = 99.6%

**29.** 0.0047    0.00. 47
                  ⌐⌐↑

Move the decimal point 2 places to the right.
0.0047 = 0.47%

**31.** 0.072    0.07. 2
                ⌐⌐↑

Move the decimal point 2 places to the right.
0.072 = 7.2%

**33.** 9.2    9.20.
              ⌐⌐↑

Move the decimal point 2 places to the right.
9.2 = 920%

**35.** 0.0068    0.00. 68
                  ⌐⌐↑

Move the decimal point 2 places to the right.
0.0068 = 0.68%

**37.** $\dfrac{1}{6} = 0.16\overline{6} = 16.\overline{6}\%$, or $16\dfrac{2}{3}\%$

**39.** $\dfrac{13}{20} = 0.65 = 65\%$

**41.** $\dfrac{29}{100} = 0.29 = 29\%$

**43.** $\dfrac{8}{10} = 0.8 = 80\%$

**45.** $\dfrac{3}{5} = 0.6 = 60\%$

**47.** $\dfrac{2}{3} = 0.66\overline{6} = 66.\overline{6}\%$, or $66\dfrac{2}{3}\%$

**49.** $\dfrac{7}{4} = 1.75 = 175\%$

**51.** $\dfrac{3}{4} = 0.75 = 75\%$

**53.**
```
      2.2 5
  4 ) 9.0 0
      8
      ‾‾‾
      1 0
        8
      ‾‾‾
        2 0
        2 0
      ‾‾‾
         0
```

Decimal notation for $\dfrac{9}{4}$ is 2.25.

**55.**

$$
\begin{array}{r}
1.4\,1\,6\,6 \\
1\,2\,\overline{)1\,7.0\,0\,0\,0} \\
\underline{1\,2\phantom{.0000}} \\
5\,0 \\
\underline{4\,8} \\
2\,0 \\
\underline{1\,2} \\
8\,0 \\
\underline{7\,2} \\
8\,0 \\
\underline{7\,2} \\
8
\end{array}
$$

The number 8 repeats as a remainder, so the digit 6 will repeat in the quotient. Thus, $\frac{17}{12} = 1.41\overline{6}$.

**57.**

$$
\begin{array}{r}
0.9\,0\,9\,0\,9 \\
1\,1\,\overline{)1\,0.0\,0\,0\,0\,0} \\
\underline{9\,9\phantom{.00000}} \\
1\,0\,0 \\
\underline{9\,9} \\
1\,0\,0 \\
\underline{9\,9} \\
1
\end{array}
$$

The pattern of the remainders repeats, so the sequence of digits in the quotient repeats. Thus, $\frac{10}{11} = 0.\overline{90}$.

**59.** $18\% + 14\% = 0.18 + 0.14 = 0.32 = 32\%$

**61.** $1 - 30\% = 1 - 0.3 = 0.7 = 70\%$

**63.** $27 \times 100\% = 27 \times 1 = 27 = 2700\%$

**65.** $3(1 + 15\%) = 3(1 + 0.15) = 3(1.15) = 3.45 = 345\%$

**67.** $\frac{100\%}{40} = \frac{1}{40} = 0.025 = 2.5\%$

## Exercise Set R.5

**1.** $\underbrace{5 \times 5 \times 5 \times 5}_{4 \text{ factors}} = 5^4$

**3.** $\underbrace{10 \times 10 \times 10}_{3 \text{ factors}} = 10^3$

**5.** $\underbrace{1 \times 1 \times 1}_{3 \text{ factors}} = 1^3$

**7.** $7^2 = 7 \cdot 7 = 49$

**9.** $9^5 = 9 \cdot 9 \cdot 9 \cdot 9 \cdot 9 = 59,049$

**11.** $10^2 = 10 \cdot 10 = 100$

**13.** $1^4 = 1 \cdot 1 \cdot 1 \cdot 1 = 1$

**15.** $(2.3)^2 = (2.3)(2.3) = 5.29$

**17.** $(0.2)^3 = (0.2)(0.2)(0.2) = 0.008$

**19.** $(20.4)^2 = (20.4)(20.4) = 416.16$

**21.** $\left(\frac{3}{8}\right)^2 = \left(\frac{3}{8}\right)\left(\frac{3}{8}\right) = \frac{9}{64}$

**23.** $5^3 = 5 \cdot 5 \cdot 5 = 125$

**25.** $1000 \times (1.02)^3 = 1000 \times (1.02 \times 1.02 \times 1.02) =$
$1000 \times 1.061208 = 1061.208$

**27.** $9 + 2 \times 8 = 9 + 16$    Multiplying
$\phantom{9 + 2 \times 8} = 25$    Adding

**29.** $9 \times 8 + 7 \times 6 = 72 + 42$    Multiplying
$\phantom{9 \times 8 + 7 \times 6} = 114$    Adding

**31.** $39 - 4 \times 2 + 2 = 39 - 8 + 2$    Multiplying
$\phantom{39 - 4 \times 2 + 2} = 31 + 2$    Subtracting
$\phantom{39 - 4 \times 2 + 2} = 33$    Adding

**33.** $9 \div 3 + 16 \div 8 = 3 + 2$    Dividing
$\phantom{9 \div 3 + 16 \div 8} = 5$    Adding

**35.** $7 + 10 - 10 \div 2 = 7 + 10 - 5$    Dividing
$\phantom{7 + 10 - 10 \div 2} = 17 - 5$    Adding and subtracting
$\phantom{7 + 10 - 10 \div 2} = 12$     from left to right

**37.** $(6 \cdot 3)^2 = 18^2$    Multiplying within parentheses
$\phantom{(6 \cdot 3)^2} = 324$

**39.** $4 \cdot 5^2 = 4 \cdot 25$    Evaluating the exponential expression
$\phantom{4 \cdot 5^2} = 100$

**41.** $(8 + 2)^3 = 10^3$    Adding within parentheses
$\phantom{(8 + 2)^3} = 1000$    Evaluating the exponential expression

**43.** $6 + 4^2 = 6 + 16$    Evaluating the exponential expression
$\phantom{6 + 4^2} = 22$    Adding

**45.** $(3 - 2)^2 = 1^2$    Subtracting within parentheses
$\phantom{(3 - 2)^2} = 1$    Evaluating the exponential expression

**47.** $\phantom{=} 4^3 \div 8 - 4$
$= 64 \div 8 - 4$    Evaluating the exponential expression
$= 8 - 4$    Dividing
$= 4$    Subtracting

**49.** $\phantom{=} 120 - 3^3 \cdot 4 \div 6$
$= 120 - 27 \cdot 4 \div 6$    Evaluating the exponential expression
$= 120 - 108 \div 6$    Multiplying
$= 120 - 18$    Dividing
$= 102$    Subtracting

**51.** $6[9 + (3 + 4)] = 6[9 + 7]$    Adding inside the parentheses
$\phantom{6[9 + (3 + 4)]} = 6[16]$    Adding inside the brackets
$\phantom{6[9 + (3 + 4)]} = 96$    Multiplying

**53.** $8 + (7 + 9) = 8 + 16$    Adding inside the parentheses

$\qquad = 24$    Adding

**55.** $15(4 + 2) = 15(6)$    Adding inside the parentheses

$\qquad = 90$    Multiplying

**57.** $12 - (8 - 4) = 12 - 4$    Subtracting inside the parentheses

$\qquad = 8$    Subtracting

**59.** $1000 \div 100 \div 10 = 10 \div 10$    Dividing in order from left to right

$\qquad = 1$    Doing the second division

**61.** $2000 \div \dfrac{3}{50} \cdot \dfrac{3}{2}$

$= \dfrac{2000}{1} \cdot \dfrac{50}{3} \cdot \dfrac{3}{2}$    Dividing

$= \dfrac{100,000}{3} \cdot \dfrac{3}{2}$    Completing the division

$= \dfrac{100,000 \cdot 3}{3 \cdot 2}$    Multiplying

$= \dfrac{\cancel{2} \cdot 50,000 \cdot \cancel{3}}{1 \cdot \cancel{3} \cdot \cancel{2}}$

$= 50,000$    Simplifying

**63.** We will do the calculations in the numerator and in the denominator and then divide the results.

$\dfrac{80 - 6^2}{9^2 + 3^2} = \dfrac{80 - 36}{81 + 9}$

$= \dfrac{44}{90}$

$= \dfrac{22 \cdot \cancel{2}}{45 \cdot \cancel{2}}$

$= \dfrac{22}{45}$

**65.** $\dfrac{3(6 + 7) - 5 \cdot 4}{6 \cdot 7 + 8(4 - 1)} = \dfrac{3 \cdot 13 - 5 \cdot 4}{6 \cdot 7 + 8 \cdot 3}$

$= \dfrac{39 - 20}{42 + 24}$

$= \dfrac{19}{66}$

**67.** $\dfrac{5}{16} = 0.3125 \qquad 0.31.25$

Move the decimal point two places to the right.

$\dfrac{5}{16} = 0.3125 = 31.25\%$

**69.** We begin by factoring 48 in any way that we can and continue factoring until each factor is prime.

$48 = 2 \cdot 24 = 2 \cdot 4 \cdot 6 = 2 \cdot 2 \cdot 2 \cdot 2 \cdot 3$

**71.** $\dfrac{64}{96} = \dfrac{2 \cdot \cancel{32}}{3 \cdot \cancel{32}} = \dfrac{2}{3}$

**73.** $\dfrac{10^5}{10^3} = \dfrac{\cancel{10} \cdot \cancel{10} \cdot \cancel{10} \cdot 10 \cdot 10}{\cancel{10} \cdot \cancel{10} \cdot \cancel{10} \cdot 1} = 10 \cdot 10 = 10^2$

**75.** $\dfrac{5^4}{5^2} = \dfrac{\cancel{5} \cdot \cancel{5} \cdot 5 \cdot 5}{\cancel{5} \cdot \cancel{5} \cdot 1} = 5 \cdot 5 = 5^2$

**77.** See the answer section in the text.

# Chapter 1

# Introduction to Real Numbers and Algebraic Expressions

## Exercise Set 1.1

1. Substitute 29 for $x$: $29 - 6 = 23$
   Substitute 34 for $x$: $34 - 6 = 28$
   Substitute 47 for $x$: $47 - 6 = 41$

3. Substitute 45 m for $b$ and 86 m for $h$, and carry out the multiplication:

$$A = \frac{1}{2}bh = \frac{1}{2}(45 \text{ m})(86 \text{ m})$$
$$= \frac{1}{2}(45)(86)(\text{m})(\text{m})$$
$$= 1935 \text{ m}^2$$

5. Substitute 65 for $r$ and 4 for $t$, and carry out the multiplication:
$$d = rt = 65 \cdot 4 = 260 \text{ mi}$$

7. $8x = 8 \cdot 7 = 56$

9. $\dfrac{a}{b} = \dfrac{24}{3} = 8$

11. $\dfrac{3p}{q} = \dfrac{3 \cdot 2}{6} = \dfrac{6}{6} = 1$

13. $\dfrac{x + y}{5} = \dfrac{10 + 20}{5} = \dfrac{30}{5} = 6$

15. $\dfrac{x - y}{8} = \dfrac{20 - 4}{8} = \dfrac{16}{8} = 2$

17. $b + 7$, or $7 + b$

19. $c - 12$

21. $4 + q$, or $q + 4$

23. $a + b$, or $b + a$

25. $y - x$

27. $x + w$, or $w + x$

29. $n - m$

31. $s + r$, or $r + s$

33. $2z$

35. $3m$

37. Let $x$ represent the number. Then we have $89\%x$, or $0.89x$.

39. The distance traveled is the product of the speed and the time. Thus the driver traveled $55t$ miles.

41. We use a factor tree.

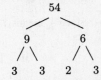

The prime factorization is $2 \cdot 3 \cdot 3 \cdot 3$.

43. We use the list of primes. The first prime that is a factor of 108 is 2.
$$108 = 2 \cdot 54$$

We keep dividing by 2 until it is no longer possible to do so.
$$108 = 2 \cdot 2 \cdot 27$$

Now we do the same thing for the next prime, 3.
$$108 = 2 \cdot 2 \cdot 3 \cdot 3 \cdot 3$$

This is the prime factorization of 108.

45. $6 = 2 \cdot 3$
    $18 = 2 \cdot 3 \cdot 3$
    The LCM is $2 \cdot 3 \cdot 3$, or 18.

47. $10 = 2 \cdot 5$
    $20 = 2 \cdot 2 \cdot 5$
    $30 = 2 \cdot 3 \cdot 5$
    The LCM is $2 \cdot 2 \cdot 3 \cdot 5$, or 60.

49. $x + 3y$

51. $2x - 3$

## Exercise Set 1.2

1. The integer $-1286$ corresponds to 1286 ft below sea level; the integer 13,804 corresponds to 13,804 ft above sea level.

3. The integer $-8,000,000$ corresponds to a deficit of $8 million.

5. The integer $-3$ corresponds to 3 sec before liftoff; the integer 128 corresponds to 128 sec after liftoff.

**7.** The number $\frac{10}{3}$ can be named $3\frac{1}{3}$, or $3.3\overline{3}$. The graph is $\frac{1}{3}$ of the way from 3 to 4.

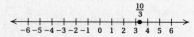

**9.** The graph of $-5.2$ is $\frac{2}{10}$ of the way from $-5$ to $-6$.

**11.** We first find decimal notation for $\frac{7}{8}$. Since $\frac{7}{8}$ means $7 \div 8$, we divide.

$$
\begin{array}{r}
0.8\,7\,5 \\
8\,\overline{)7.0\,0\,0} \\
6\,4 \\
\hline
6\,0 \\
5\,6 \\
\hline
4\,0 \\
4\,0 \\
\hline
0
\end{array}
$$

Thus $\frac{7}{8} = 0.875$, so $-\frac{7}{8} = -0.875$.

**13.** $\frac{5}{6}$ means $5 \div 6$, so we divide.

$$
\begin{array}{r}
0.8\,3\,3\,\ldots \\
6\,\overline{)5.0\,0\,0} \\
4\,8 \\
\hline
2\,0 \\
1\,8 \\
\hline
2\,0 \\
1\,8 \\
\hline
0
\end{array}
$$

We have $\frac{5}{6} = 0.8\overline{3}$.

**15.** $\frac{7}{6}$ means $7 \div 6$, so we divide.

$$
\begin{array}{r}
1.1\,6\,6\,\ldots \\
6\,\overline{)7.0\,0\,0} \\
6 \\
\hline
1\,0 \\
6 \\
\hline
4\,0 \\
3\,6 \\
\hline
4\,0 \\
3\,6 \\
\hline
4
\end{array}
$$

We have $\frac{7}{6} = 1.1\overline{6}$.

**17.** $\frac{2}{3}$ means $2 \div 3$, so we divide.

$$
\begin{array}{r}
0.6\,6\,6\,\ldots \\
3\,\overline{)2.0\,0\,0} \\
1\,8 \\
\hline
2\,0 \\
1\,8 \\
\hline
2\,0 \\
1\,8 \\
\hline
2
\end{array}
$$

We have $\frac{2}{3} = 0.\overline{6}$.

**19.** We first find decimal notation for $\frac{1}{2}$. Since $\frac{1}{2}$ means $1 \div 2$, we divide.

$$
\begin{array}{r}
0.5 \\
2\,\overline{)1.0} \\
1\,0 \\
\hline
0
\end{array}
$$

Thus $\frac{1}{2} = 0.5$, so $-\frac{1}{2} = -0.5$

**21.** $\frac{1}{10}$ means $1 \div 10$, so we divide.

$$
\begin{array}{r}
0.1 \\
1\,0\,\overline{)1.0} \\
1\,0 \\
\hline
0
\end{array}
$$

We have $\frac{1}{10} = 0.1$

**23.** Since 8 is to the right of 0, we have $8 > 0$.

**25.** Since $-8$ is to the left of 3, we have $-8 < 3$.

**27.** Since $-8$ is to the left of 8, we have $-8 < 8$.

**29.** Since $-8$ is to the left of $-5$, we have $-8 < -5$.

**31.** Since $-5$ is to the right of $-11$, we have $-5 > -11$.

**33.** Since $-6$ is to the left of $-5$, we have $-6 < -5$.

**35.** Since 2.14 is to the right of 1.24, we have $2.14 > 1.24$.

**37.** Since $-14.5$ is to the left of 0.011, we have $-14.5 < 0.011$.

**39.** Since $-12.88$ is to the left of $-6.45$, we have $-12.88 < -6.45$.

**41.** Convert to decimal notation $\frac{5}{12} = 0.4166\ldots$ and $\frac{11}{25} = 0.44$. Since $0.4166\ldots$ is to the left of 0.44, $\frac{5}{12} < \frac{11}{25}$.

**43.** $-3 \geq -11$ is true since $-3 > -11$ is true.

**45.** $0 \geq 8$ is false since neither $0 > 8$ nor $0 = 8$ is true.

**47.** $x < -6$ has the same meaning as $-6 > x$.

**49.** $y \geq -10$ has the same meaning as $-10 \leq y$.

**51.** The distance of $-3$ from 0 is 3, so $|-3| = 3$.

**53.** The distance of 10 from 0 is 10, so $|10| = 10$.

**55.** The distance of 0 from 0 is 0, so $|0| = 0$.

**57.** The distance of $-24$ from 0 is 24, so $|-24| = 24$.

**59.** The distance of $-\frac{2}{3}$ from 0 is $\frac{2}{3}$, so $\left|-\frac{2}{3}\right| = \frac{2}{3}$.

**61.** The distance of $\frac{0}{4}$ from 0 is $\frac{0}{4}$, or 0, so $\left|\frac{0}{4}\right| = 0$.

**63.** $-\frac{2}{3}, \frac{1}{2}, -\frac{3}{4}, -\frac{5}{6}, \frac{3}{8}, \frac{1}{6}$ can be written in decimal notation as $-0.\overline{6}, 0.5, -0.75, -0.8\overline{3}, 0.375, 0.1\overline{6}$, respectively. Listing from least to greatest, we have
$$-\frac{5}{6}, -\frac{3}{4}, -\frac{2}{3}, \frac{1}{6}, \frac{3}{8}, \frac{1}{2}.$$

## Exercise Set 1.3

**1.** $2 + (-9)$ The absolute values are 2 and 9. The difference is $9 - 2$, or 7. The negative number has the larger absolute value, so the answer is negative. $2 + (-9) = -7$

**3.** $-11 + 5$ The absolute values are 11 and 5. The difference is $11 - 5$, or 6. The negative number has the larger absolute value, so the answer is negative. $-11 + 5 = -6$

**5.** $-6 + 6$ A negative and a positive number. The numbers have the same absolute value. The sum is 0. $-6 + 6 = 0$

**7.** $-3 + (-5)$ Two negatives. Add the absolute values, getting 8. Make the answer negative. $-3 + (-5) = -8$

**9.** $-7 + 0$ One number is 0. The answer is the other number. $-7 + 0 = -7$

**11.** $0 + (-27)$ One number is 0. The answer is the other number. $0 + (-27) = -27$

**13.** $17 + (-17)$ A negative and a positive number. The numbers have the same absolute value. The sum is 0. $17 + (-17) = 0$

**15.** $-17 + (-25)$ Two negatives. Add the absolute values, getting 42. Make the answer negative. $-17 + (-25) = -42$

**17.** $18 + (-18)$ A positive and a negative number. The numbers have the same absolute value. The sum is 0. $18 + (-18) = 0$

**19.** $-28 + 28$ A negative and a positive number. The numbers have the same absolute value. The sum is 0. $-28 + 28 = 0$

**21.** $8 + (-5)$ The absolute values are 8 and 5. The difference is $8 - 5$, or 3. The positive number has the larger absolute value, so the answer is positive. $8 + (-5) = 3$

**23.** $-4 + (-5)$ Two negatives. Add the absolute values, getting 9. Make the answer negative. $-4 + (-5) = -9$

**25.** $13 + (-6)$ The absolute values are 13 and 6. The difference is $13 - 6$, or 7. The positive number has the larger absolute value, so the answer is positive. $13 + (-6) = 7$

**27.** $-25 + 25$ A negative and a positive number. The numbers have the same absolute value. The sum is 0. $-25 + 25 = 0$

**29.** $53 + (-18)$ The absolute values are 53 and 18. The difference is $53 - 18$, or 35. The positive number has the larger absolute value, so the answer is positive. $53 + (-18) = 35$

**31.** $-8.5 + 4.7$ The absolute values are 8.5 and 4.7. The difference is $8.5 - 4.7$, or 3.8. The negative number has the larger absolute value, so the answer is negative. $-8.5 + 4.7 = -3.8$

**33.** $-2.8 + (-5.3)$ Two negatives. Add the absolute values, getting 8.1. Make the answer negative. $-2.8 + (-5.3) = -8.1$

**35.** $-\frac{3}{5} + \frac{2}{5}$ The absolute values are $\frac{3}{5}$ and $\frac{2}{5}$. The difference is $\frac{3}{5} - \frac{2}{5}$, or $\frac{1}{5}$. The negative number has the larger absolute value, so the answer is negative. $-\frac{3}{5} + \frac{2}{5} = -\frac{1}{5}$

**37.** $-\frac{3}{9} + \left(-\frac{5}{9}\right)$ Two negatives. Add the absolute values, getting $\frac{8}{9}$. Make the answer negative. $-\frac{3}{9} + \left(-\frac{5}{9}\right) = -\frac{8}{9}$

**39.** $-\frac{5}{8} + \frac{1}{4}$ The absolute values are $\frac{5}{8}$ and $\frac{1}{4}$. The difference is $\frac{5}{8} - \frac{2}{8}$, or $\frac{3}{8}$. The negative number has the larger absolute value, so the answer is negative. $-\frac{5}{8} + \frac{1}{4} = -\frac{3}{8}$

**41.** $-\frac{5}{8} + \left(-\frac{1}{6}\right)$ Two negatives. Add the absolute values, getting $\frac{15}{24} + \frac{4}{24}$, or $\frac{19}{24}$. Make the answer negative.
$$-\frac{5}{8} + \left(-\frac{1}{6}\right) = -\frac{19}{24}$$

**43.** $-\frac{3}{8} + \frac{5}{12}$ The absolute values are $\frac{3}{8}$ and $\frac{5}{12}$. The difference is $\frac{10}{24} - \frac{9}{24}$, or $\frac{1}{24}$. The positive number has the larger absolute value, so the answer is positive. $-\frac{3}{8} + \frac{5}{12} = \frac{1}{24}$

**45.** $76 + (-15) + (-18) + (-6)$
a) Add the negative numbers: $-15 + (-18) + (-6) = -39$
b) Add the results: $76 + (-39) = 37$

**47.** $-44 + \left(-\frac{3}{8}\right) + 95 + \left(-\frac{5}{8}\right)$

a) Add the negative numbers: $-44 + \left(-\frac{3}{8}\right) + \left(-\frac{5}{8}\right) = -45$
b) Add the results: $-45 + 95 = 50$

**49.** We add from left to right.

$$
\begin{array}{rl}
& 98 + (-54) + 113 + (-998) + 44 + (-612) \\
= & \phantom{00}44 \phantom{00} + 113 + (-998) + 44 + (-612) \\
= & \phantom{0000000}157 \phantom{0} + (-998) + 44 + (-612) \\
= & \phantom{00000000000000}-841 \phantom{0} + 44 + (-612) \\
= & \phantom{000000000000000000000}-797 + (-612) \\
= & \phantom{0000000000000000000000000000}-1409
\end{array}
$$

**51.** The additive inverse of 24 is $-24$ because $24 + (-24) = 0$.

**53.** The additive inverse of $-26.9$ is $26.9$ because $-26.9 + 26.9 = 0$.

**55.** If $x = 8$, then $-x = -8$. (The opposite of 8 is $-8$.)

**57.** If $x = -\dfrac{13}{8}$ then $-x = -\left(-\dfrac{13}{8}\right) = \dfrac{13}{8}$. (The opposite of $-\dfrac{13}{8}$ is $\dfrac{13}{8}$.)

**59.** If $x = -43$ then $-(-x) = -(-(-43)) = -43$. (The opposite of the opposite of $-43$ is $-43$.)

**61.** If $x = \dfrac{4}{3}$ then $-(-x) = -\left(-\dfrac{4}{3}\right) = \dfrac{4}{3}$. (The opposite of the opposite of $\dfrac{4}{3}$ is $\dfrac{4}{3}$.)

**63.** $-(-24) = 24$ (The opposite of $-24$ is 24.)

**65.** $-\left(-\dfrac{3}{8}\right) = \dfrac{3}{8}$ (The opposite of $-\dfrac{3}{8}$ is $\dfrac{3}{8}$.)

**67.** 57%     0.57.

Move the decimal point two places to the left.
57% = 0.57.

**69.** 52.9%     0.52.9

Move the decimal point two places to the left.
52.9% = 0.529.

**71.** $\dfrac{5}{4} = 1.25 = 125\%$

**73.** $\dfrac{13}{25} = 0.52 = 52\%$

**75.** When $x$ is positive, the opposite of $x$, $-x$, is negative.

**77.** If $a$ is positive, $-a$ is negative. Thus $-a + b$, the sum of two negatives, is negative.

## Exercise Set 1.4

**1.** $2 - 9 = 2 + (-9) = -7$

**3.** $0 - 4 = 0 + (-4) = -4$

**5.** $-8 - (-2) = -8 + 2 = -6$

**7.** $-11 - (-11) = -11 + 11 = 0$

**9.** $12 - 16 = 12 + (-16) = -4$

**11.** $20 - 27 = 20 + (-27) = -7$

**13.** $-9 - (-3) = -9 + 3 = -6$

**15.** $-40 - (-40) = -40 + 40 = 0$

**17.** $7 - 7 = 7 + (-7) = 0$

**19.** $7 - (-7) = 7 + 7 = 14$

**21.** $8 - (-3) = 8 + 3 = 11$

**23.** $-6 - 8 = -6 + (-8) = -14$

**25.** $-4 - (-9) = -4 + 9 = 5$

**27.** $1 - 8 = 1 + (-8) = -7$

**29.** $-6 - (-5) = -6 + 5 = -1$

**31.** $8 - (-10) = 8 + 10 = 18$

**33.** $0 - 10 = 0 + (-10) = -10$

**35.** $-5 - (-2) = -5 + 2 = -3$

**37.** $-7 - 14 = -7 + (-14) = -21$

**39.** $0 - (-5) = 0 + 5 = 5$

**41.** $-8 - 0 = -8 + 0 = -8$

**43.** $7 - (-5) = 7 + 5 = 12$

**45.** $2 - 25 = 2 + (-25) = -23$

**47.** $-42 - 26 = -42 + (-26) = -68$

**49.** $-71 - 2 = -71 + (-2) = -73$

**51.** $24 - (-92) = 24 + 92 = 116$

**53.** $-50 - (-50) = -50 + 50 = 0$

**55.** $-\dfrac{3}{8} - \dfrac{5}{8} = -\dfrac{3}{8} + \left(-\dfrac{5}{8}\right) = -\dfrac{8}{8} = -1$

**57.** $\dfrac{3}{4} - \dfrac{2}{3} = \dfrac{3}{4} + \left(-\dfrac{2}{3}\right) = \dfrac{9}{12} + \left(-\dfrac{8}{12}\right) = \dfrac{1}{12}$

**59.** $-\dfrac{3}{4} - \dfrac{2}{3} = -\dfrac{3}{4} + \left(-\dfrac{2}{3}\right) = -\dfrac{9}{12} + \left(-\dfrac{8}{12}\right) = -\dfrac{17}{12}$

**61.** $-\dfrac{5}{8} - \left(-\dfrac{3}{4}\right) = -\dfrac{5}{8} + \dfrac{3}{4} = -\dfrac{5}{8} + \dfrac{6}{8} = \dfrac{1}{8}$

**63.** $6.1 - (-13.8) = 6.1 + 13.8 = 19.9$

**65.** $-2.7 - 5.9 = -2.7 + (-5.9) = -8.6$

**67.** $0.99 - 1 = 0.99 + (-1) = -0.01$

**69.** $-79 - 114 = -79 + (-114) = -193$

**71.** $0 - (-500) = 0 + 500 = 500$

**73.** $-2.8 - 0 = -2.8 + 0 = -2.8$

**75.** $7 - 10.53 = 7 + (-10.53) = -3.53$

**77.** $\dfrac{1}{6} - \dfrac{2}{3} = \dfrac{1}{6} + \left(-\dfrac{2}{3}\right) = \dfrac{1}{6} + \left(-\dfrac{4}{6}\right) = -\dfrac{3}{6}$, or $-\dfrac{1}{2}$

**79.** $-\dfrac{4}{7} - \left(-\dfrac{10}{7}\right) = -\dfrac{4}{7} + \dfrac{10}{7} = \dfrac{6}{7}$

**81.** $-\dfrac{7}{10} - \dfrac{10}{15} = -\dfrac{7}{10} + \left(-\dfrac{10}{15}\right) = -\dfrac{21}{30} + \left(-\dfrac{20}{30}\right) = -\dfrac{41}{30}$

**83.** $\dfrac{1}{5} - \dfrac{1}{3} = \dfrac{1}{5} + \left(-\dfrac{1}{3}\right) = \dfrac{3}{15} + \left(-\dfrac{5}{15}\right) = -\dfrac{2}{15}$

**85.** $18 - (-15) - 3 - (-5) + 2 = 18 + 15 + (-3) + 5 + 2 = 37$

**87.** $-31 + (-28) - (-14) - 17 = (-31) + (-28) + 14 + (-17) = -62$

**89.** $-34 - 28 + (-33) - 44 = (-34) + (-28) + (-33) + (-44) = -139$

**91.** $-93 - (-84) - 41 - (-56) = (-93) + 84 + (-41) + 56 = 6$

**93.** $-5 - (-30) + 30 + 40 - (-12) = (-5) + 30 + 30 + 40 + 12 = 107$

**95.** $132 - (-21) + 45 - (-21) = 132 + 21 + 45 + 21 = 219$

**97.** The number $-\$530$ represents the original debt, and $-\$156$ represents the amount of the debt that is canceled. We subtract to find how much is still owed: $-\$530 - (-\$156) = -\$530 + \$156 = -\$374$. Jose owes $\$374$ now.

**99.** We subtract the lower temperature from the higher temperature: $-5 - (-12) = -5 + 12 = 7$. The temperature dropped 7°F.

**101.** We draw a picture of the situation.

Sea level
-28,538 ft
Puerto Rico Trench
To find
-34,370 ft
Marianas Trench

We subtract the lower altitude from the higher altitude: $-28,538 - (-34,370) = -28,538 + 34,370 = 5832$. The Puerto Rico Trench is 5832 ft higher than the Marianas Trench.

**103.** $5^3 = 5 \times 5 \times 5 = 125$

**105.** $256 \div 64 \div 2^3 + 100 = 256 \div 64 \div 8 + 100$
$= 4 \div 8 + 100$
$= \dfrac{1}{2} + 100$
$= 100\dfrac{1}{2}$, or $100.5$

**107.** $58.3\%$     $0.58.3$

Move the decimal point two places to the left.
$58.3\% = 0.583$

**109.** Use a calculator to do this exercise.
$123,907 - 433,789 = -309,882$

**111.** False. $3 - 0 = 3, 0 - 3 = -3, 3 - 0 \neq 0 - 3$

**113.** True

**115.** True by definition of opposites.

**117.** The changes during weeks 1 to 5 are represented by the integers $-13, -16, 36, -11$, and 19, respectively. We add to find the total rise or fall:

$$-13 + (-16) + 36 + (-11) + 19 = 15$$

The market rose 15 points during the 5 week period.

## Exercise Set 1.5

**1.** $-8$

**3.** $-48$

**5.** $-24$

**7.** $-72$

**9.** $16$

**11.** $42$

**13.** $-120$

**15.** $-238$

**17.** $1200$

**19.** $98$

**21.** $-72$

**23.** $-12.4$

**25.** $30$

**27.** $21.7$

**29.** $\dfrac{2}{3} \cdot \left(-\dfrac{3}{5}\right) = -\left(\dfrac{2 \cdot 3}{3 \cdot 5}\right) = -\left(\dfrac{2}{5} \cdot \dfrac{3}{3}\right) = -\dfrac{2}{5}$

**31.** $-\dfrac{3}{8} \cdot \left(-\dfrac{2}{9}\right) = \dfrac{3 \cdot 2}{8 \cdot 9} = \dfrac{3 \cdot 2 \cdot 1}{4 \cdot 2 \cdot 3 \cdot 3} = \dfrac{3 \cdot 2}{3 \cdot 2} \cdot \dfrac{1}{4 \cdot 3} = \dfrac{1}{12}$

**33.** $-17.01$

**35.** $-\dfrac{5}{9} \cdot \dfrac{3}{4} = -\left(\dfrac{5 \cdot 3}{9 \cdot 4}\right) = -\dfrac{5 \cdot 3}{3 \cdot 3 \cdot 4} = -\dfrac{5}{3 \cdot 4} \cdot \dfrac{3}{3} = -\dfrac{5}{12}$

**37.** $7 \cdot (-4) \cdot (-3) \cdot 5 = 7 \cdot 12 \cdot 5 = 7 \cdot 60 = 420$

**39.** $-\dfrac{2}{3} \cdot \dfrac{1}{2} \cdot \left(-\dfrac{6}{7}\right) = -\dfrac{2}{6} \cdot \left(-\dfrac{6}{7}\right) = \dfrac{2 \cdot 6}{7 \cdot 6} = \dfrac{2}{7} \cdot \dfrac{6}{6} = \dfrac{2}{7}$

**41.** $-3 \cdot (-4) \cdot (-5) = 12 \cdot (-5) = -60$

**43.** $-2 \cdot (-5) \cdot (-3) \cdot (-5) = 10 \cdot 15 = 150$

**45.** $-\dfrac{2}{45}$

**47.** $-7 \cdot (-21) \cdot 13 = 147 \cdot 13 = 1911$

**49.** $-4 \cdot (-1.8) \cdot 7 = (7.2) \cdot 7 = 50.4$

**51.** $-\dfrac{1}{9} \cdot \left(-\dfrac{2}{3}\right) \cdot \left(\dfrac{5}{7}\right) = \dfrac{2}{27} \cdot \dfrac{5}{7} = \dfrac{10}{189}$

**53.** $4 \cdot (-4) \cdot (-5) \cdot (-12) = -16 \cdot (60) = -960$

**55.** $0.07 \cdot (-7) \cdot 6 \cdot (-6) = 0.07 \cdot 6 \cdot (-7) \cdot (-6) = 0.42 \cdot (42) = 17.64$

**57.** $\left(-\dfrac{5}{6}\right)\left(\dfrac{1}{8}\right)\left(-\dfrac{3}{7}\right)\left(-\dfrac{1}{7}\right) = \left(-\dfrac{5}{48}\right)\left(\dfrac{3}{49}\right) = -\dfrac{5 \cdot 3}{16 \cdot 3 \cdot 49} = -\dfrac{5}{16 \cdot 49} \cdot \dfrac{3}{3} = -\dfrac{5}{784}$

**59.** 0, The product of 0 and any real number is 0.

**61.** $(-8)(-9)(-10) = 72(-10) = -720$

**63.** $(-6)(-7)(-8)(-9)(-10) = 42 \cdot 72 \cdot (-10) = 3024 \cdot (-10) = -30,240$

**65.**
$$(-3x)^2 = (-3 \cdot 7)^2 \quad \text{Substituting}$$
$$= (-21)^2 \quad \text{Multiplying inside the parentheses}$$
$$= (-21)(-21) \quad \text{Evaluating the power}$$
$$= 441$$
$$-3x^2 = -3(7)^2 \quad \text{Substituting}$$
$$= -3 \cdot 49 \quad \text{Evaluating the power}$$
$$= -147$$

**67.** When $x = 2$:
$$5x^2 = 5(2)^2 \quad \text{Substituting}$$
$$= 5 \cdot 4 \quad \text{Evaluating the power}$$
$$= 20$$
When $x = -2$:
$$5x^2 = 5(-2)^2 \quad \text{Substituting}$$
$$= 5 \cdot 4 \quad \text{Evaluating the power}$$
$$= 20$$

**69.** $-6[(-5) + (-7)] = -6[-12] = 72$

**71.** $-(3^5) \cdot [-(2^3)] = -243[-8] = 1944$

**73.** $|(-2)^3 + 4^2| - (2 - 7)^2 = |(-2)^3 + 4^2| - (-5)^2 = |-8 + 16| - 25 = |8| - 25 = 8 - 25 = -17$

**75.** a) $a$ and $b$ have different signs;
b) either $a$ or $b$ or both must be zero;
c) $a$ and $b$ have the same sign

## Exercise Set 1.6

**1.** $48 \div (-6) = -8$  Check: $-8(-6) = 48$

**3.** $\dfrac{28}{-2} = -14$  Check: $-14(-2) = 28$

**5.** $\dfrac{-24}{8} = -3$  Check: $-3 \cdot 8 = -24$

**7.** $\dfrac{-36}{-12} = 3$  Check: $3(-12) = -36$

**9.** $\dfrac{-72}{9} = -8$  Check: $-8 \cdot 9 = -72$

**11.** $-100 \div (-50) = 2$  Check: $2(-50) = -100$

**13.** $-108 \div 9 = -12$  Check: $9(-12) = -108$

**15.** $\dfrac{200}{-25} = -8$  Check: $-8(-25) = 200$

**17.** Undefined

**19.** $\dfrac{20}{-7} = -\dfrac{20}{7}$  Check: $-\dfrac{20}{7}(-7) = 20$

**21.** The reciprocal of $\dfrac{15}{7}$ is $\dfrac{7}{15}$ because $\dfrac{15}{7} \cdot \dfrac{7}{15} = 1$.

**23.** The reciprocal of $-\dfrac{47}{13}$ is $-\dfrac{13}{47}$ because $\left(-\dfrac{47}{13}\right) \cdot \left(-\dfrac{13}{47}\right) = 1$.

**25.** The reciprocal of 13 is $\dfrac{1}{13}$ because $13 \cdot \dfrac{1}{13} = 1$.

**27.** The reciprocal of 4.3 is $\dfrac{1}{4.3}$ because $4.3 \cdot \dfrac{1}{4.3} = 1$.

**29.** The reciprocal of $-\dfrac{1}{7.1}$ is $-7.1$ because $\left(-\dfrac{1}{7.1}\right)(-7.1) = 1$.

**31.** The reciprocal of $\dfrac{p}{q}$ is $\dfrac{q}{p}$ because $\dfrac{p}{q} \cdot \dfrac{q}{p} = 1$.

**33.** The reciprocal of $\dfrac{1}{4y}$ is $4y$ because $\dfrac{1}{4y} \cdot 4y = 1$.

**35.** The reciprocal of $\dfrac{2a}{3b}$ is $\dfrac{3b}{2a}$ because $\dfrac{2a}{3b} \cdot \dfrac{3b}{2a} = 1$.

**37.** $4 \cdot \dfrac{1}{17}$

**39.** $8\left(-\dfrac{1}{13}\right)$

**41.** $13.9\left(-\dfrac{1}{1.5}\right)$

**43.** $x \cdot y$

**45.** $(3x + 4)\dfrac{1}{5}$

**47.** $(5a - b)\left(\dfrac{1}{5a + b}\right)$

**49.** $\dfrac{3}{4} \div \left(-\dfrac{2}{3}\right) = \dfrac{3}{4} \cdot \left(-\dfrac{3}{2}\right) = -\dfrac{9}{8}$

**51.** $-\dfrac{5}{4} \div \left(-\dfrac{3}{4}\right) = -\dfrac{5}{4} \cdot \left(-\dfrac{4}{3}\right) = \dfrac{20}{12} = \dfrac{5 \cdot 4}{3 \cdot 4} = \dfrac{5}{3}$

**53.** $-\dfrac{2}{7} \div \left(-\dfrac{4}{9}\right) = -\dfrac{2}{7} \cdot \left(-\dfrac{9}{4}\right) = \dfrac{18}{28} = \dfrac{9 \cdot 2}{14 \cdot 2} = \dfrac{9}{14}$

**55.** $-\dfrac{3}{8} \div \left(-\dfrac{8}{3}\right) = -\dfrac{3}{8} \cdot \left(-\dfrac{3}{8}\right) = \dfrac{9}{64}$

**57.** $-6.6 \div 3.3 = -2$  Do the long division. Make the answer negative.

**59.** $\dfrac{-11}{-13} = \dfrac{11}{13}$  The opposite of a number divided by the opposite of another number is the quotient of the two numbers.

**61.** $\dfrac{48.6}{-3} = -16.2$  Do the long division. Make the answer negative.

**63.** $\dfrac{-9}{17-17} = \dfrac{-9}{0}$  Division by 0 is undefined.

**65.** $\dfrac{264}{468} = \dfrac{4 \cdot 66}{4 \cdot 117} = \dfrac{\cancel{4} \cdot \cancel{3} \cdot 22}{\cancel{4} \cdot \cancel{3} \cdot 39} = \dfrac{22}{39}$

**67.** $2^3 - 5 \cdot 3 + 8 \cdot 10 \div 2$

$\begin{aligned}
&= 8 - 5 \cdot 3 + 8 \cdot 10 \div 2 && \text{Evaluating the power} \\
&= 8 - 15 + 80 \div 2 && \text{Multiplying and dividing} \\
&= 8 - 15 + 40 && \text{in order from left to right} \\
&= -7 + 40 && \text{Adding and subtracting} \\
&= 33 && \text{in order from left to right}
\end{aligned}$

**69.** $\dfrac{7}{8} = 0.875 = 87.5\%$

**71.** We find $\dfrac{1}{-10.5}$ using a calculator. If a reciprocal key is available, enter $-10.5$ and then use the reciprocal key. If a reciprocal key is not available, find $\dfrac{1}{-10.5}$, or $1 \div (-10.5)$. In either case, the result is $-0.\overline{095238}$.

**73.** There are none. For $a \neq 0$, $-a$ and $a$ have opposite signs but $a$ and $\dfrac{1}{a}$ have the same sign. For $a = 0$, $-a = a$, but $\dfrac{1}{a}$ is undefined.

**75.** $-n$ is positive and $m$ is negative, so $\dfrac{-n}{m}$ is the quotient of a positive and a negative number and, thus, is negative.

**77.** $\dfrac{-n}{m}$ is negative (see Exercise 75), so $-\left(\dfrac{-n}{m}\right)$ is the opposite of a negative number and, thus, is positive.

**79.** $-n$ and $-m$ are both positive, so $\dfrac{-n}{-m}$ is the quotient of two positive numbers and, thus, is positive. Then, $-\left(\dfrac{-n}{-m}\right)$ is the opposite of a positive number and, thus, is negative.

## Exercise Set 1.7

**1.** Note that $5y = 5 \cdot y$. We multiply by 1, using $y/y$ as an equivalent expression for 1:

$$\frac{3}{5} = \frac{3}{5} \cdot 1 = \frac{3}{5} \cdot \frac{y}{y} = \frac{3y}{5y}$$

**3.** Note that $15x = 3 \cdot 5x$. We multiply by 1, using $5x/5x$ as an equivalent expression for 1:

$$\frac{2}{3} = \frac{2}{3} \cdot 1 = \frac{2}{3} \cdot \frac{5x}{5x} = \frac{10x}{15x}$$

**5.** $\begin{aligned}
-\frac{24a}{16a} &= -\frac{3 \cdot 8a}{2 \cdot 8a} \\
&= -\frac{3}{2} \cdot \frac{8a}{8a} \\
&= -\frac{3}{2} \cdot 1 \qquad \left(\frac{8a}{8a} = 1\right) \\
&= -\frac{3}{2} \qquad \text{Identity property of 1}
\end{aligned}$

**7.** $\begin{aligned}
-\frac{42ab}{36ab} &= -\frac{7 \cdot 6ab}{6 \cdot 6ab} \\
&= -\frac{7}{6} \cdot \frac{6ab}{6ab} \\
&= -\frac{7}{6} \cdot 1 \qquad \left(\frac{6ab}{6ab} = 1\right) \\
&= -\frac{7}{6} \qquad \text{Identity property of 1}
\end{aligned}$

**9.** $8 + y$, commutative law of addition

**11.** $nm$, commutative law of multiplication

**13.** $xy + 9$, commutative law of addition

$9 + yx$, commutative law of multiplication

**15.** $c + ab$, commutative law of addition

$ba + c$, commutative law of multiplication

**17.** $(a + b) + 2$, associative law of addition

**19.** $8(xy)$, associative law of multiplication

**21.** $a + (b + 3)$, associative law of addition

**23.** $(3a)b$, associative law of multiplication

**25.** a) $(a + b) + 2 = a + (b + 2)$, associative law of addition
b) $(a + b) + 2 = (b + a) + 2$, commutative law of addition
c) $(a + b) + 2 = (b + a) + 2$ Using the commutative law first,
$\quad = b + (a + 2)$ then the associative law
There are other correct answers.

**27.** a) $5 + (v + w) = (5 + v) + w$, associative law of addition
b) $5 + (v + w) = 5 + (w + v)$, commutative law of addition
c) $5 + (v + w) = 5 + (w + v)$ Using the commutative law first,
$\quad = (5 + w) + v$ then the associative law
There are other correct answers.

**29.** a) $(xy)3 = x(y3)$, associative law of multiplication

b) $(xy)3 = (yx)3$, commutative law of multiplication

c) $(xy)3 = (yx)3$ Using the commutative
    law first,
    $= y(x3)$ then the associative law

There are other correct answers.

**31.** a) $7(ab) = (7a)b$

b) $7(ab) = (7a)b = b(7a)$

c) $7(ab) = 7(ba) = (7b)a$

There are other correct answers.

**33.** $2(b + 5) = 2 \cdot b + 2 \cdot 5 = 2b + 10$

**35.** $7(1 + t) = 7 \cdot 1 + 7 \cdot t = 7 + 7t$

**37.** $6(5x + 2) = 6 \cdot 5x + 6 \cdot 2 = 30x + 12$

**39.** $7(x + 4 + 6y) = 7 \cdot x + 7 \cdot 4 + 7 \cdot 6y = 7x + 28 + 42y$

**41.** $7(4 - 3) = 7 \cdot 4 - 7 \cdot 3 = 28 - 21 = 7$

**43.** $-3(3 - 7) = -3 \cdot 3 - (-3) \cdot 7 = -9 - (-21) = -9 + 21 = 12$

**45.** $4.1(6.3 - 9.4) = 4.1(6.3) - 4.1(9.4) = 25.83 - 38.54 = -12.71$

**47.** $7(x - 2) = 7 \cdot x - 7 \cdot 2 = 7x - 14$

**49.** $-7(y - 2) = -7 \cdot y - (-7) \cdot 2 = -7y - (-14) = -7y + 14$

**51.** $-9(-5x - 6y + 8) = -9(-5x) - (-9)6y + (-9)8$
$= 45x - (-54y) + (-72) = 45x + 54y - 72$

**53.** $-4(x - 3y - 2z) = -4 \cdot x - (-4)3y - (-4)2z$
$= -4x - (-12y) - (-8z) = -4x + 12y + 8z$

**55.** $3.1(-1.2x + 3.2y - 1.1) = 3.1(-1.2x) + (3.1)3.2y - 3.1(1.1)$
$= -3.72x + 9.92y - 3.41$

**57.** $4x + 3z$     Parts are separated by plus signs. The terms
are $4x$ and $3z$.

**59.** $7x + 8y - 9z = 7x + 8y + (-9z)$ Separating parts
                                   with plus signs

The terms are $7x$, $8y$, and $-9z$.

**61.** $2x + 4 = 2 \cdot x + 2 \cdot 2 = 2(x + 2)$

**63.** $30 + 5y = 5 \cdot 6 + 5 \cdot y = 5(6 + y)$

**65.** $14x + 21y = 7 \cdot 2x + 7 \cdot 3y = 7(2x + 3y)$

**67.** $5x + 10 + 15y = 5 \cdot x + 5 \cdot 2 + 5 \cdot 3y = 5(x + 2 + 3y)$

**69.** $8x - 24 = 8 \cdot x - 8 \cdot 3 = 8(x - 3)$

**71.** $32 - 4y = 4 \cdot 8 - 4 \cdot y = 4(8 - y)$

**73.** $8x + 10y - 22 = 2 \cdot 4x + 2 \cdot 5y - 2 \cdot 11 = 2(4x + 5y - 11)$

**75.** $ax - a = a \cdot x - a \cdot 1 = a(x - 1)$

**77.** $ax - ay - az = a \cdot x - a \cdot y - a \cdot z = a(x - y - z)$

**79.** $18x - 12y + 6 = 6 \cdot 3x - 6 \cdot 2y + 6 \cdot 1 = 6(3x - 2y + 1)$

**81.** $3ax - 2ay + a = a \cdot 3x - a \cdot 2y + a \cdot 1 = a(3x - 2y + 1)$

**83.** $9a + 10a = (9 + 10)a = 19a$

**85.** $10a - a = 10a - 1 \cdot a = (10 - 1)a = 9a$

**87.** $2x + 9z + 6x = 2x + 6x + 9z = (2 + 6)x + 9z = 8x + 9z$

**89.** $7x + 6y^2 + 9y^2 = 7x + (6 + 9)y^2 = 7x + 15y^2$

**91.** $41a + 90 - 60a - 2 = 41a - 60a + 90 - 2$
$= (41 - 60)a + (90 - 2)$
$= -19a + 88$

**93.** $23 + 5t + 7y - t - y - 27$
$= 23 - 27 + 5t - 1 \cdot t + 7y - 1 \cdot y$
$= (23 - 27) + (5 - 1)t + (7 - 1)y$
$= -4 + 4t + 6y$, or $4t + 6y - 4$

**95.** $\frac{1}{2}b + \frac{1}{2}b = \left(\frac{1}{2} + \frac{1}{2}\right)b = 1b = b$

**97.** $2y + \frac{1}{4}y + y = 2y + \frac{1}{4}y + 1 \cdot y = \left(2 + \frac{1}{4} + 1\right)y = 3\frac{1}{4}y$, or
$\frac{13}{4}y$

**99.** $11x - 3x = (11 - 3)x = 8x$

**101.** $6n - n = (6 - 1)n = 5n$

**103.** $y - 17y = (1 - 17)y = -16y$

**105.** $-8 + 11a - 5b + 6a - 7b + 7$
$= 11a + 6a - 5b - 7b - 8 + 7$
$= (11 + 6)a + (-5 - 7)b + (-8 + 7)$
$= 17a - 12b - 1$

**107.** $9x + 2y - 5x = (9 - 5)x + 2y = 4x + 2y$

**109.** $11x + 2y - 4x - y = (11 - 4)x + (2 - 1)y = 7x + y$

**111.** $2.7x + 2.3y - 1.9x - 1.8y = (2.7 - 1.9)x + (2.3 - 1.8)y = $
$0.8x + 0.5y$

**113.** $\frac{1}{5}x + \frac{4}{5}y + \frac{2}{5}x - \frac{1}{5}y = \left(\frac{1}{5} + \frac{2}{5}\right)x + \left(\frac{4}{5} - \frac{1}{5}\right)y = \frac{3}{5}x + \frac{3}{5}y$

**115.** $\frac{11}{12} + \frac{15}{16} = \frac{11}{12} \cdot \frac{4}{4} + \frac{15}{16} \cdot \frac{3}{3}$   LCD is 48
$= \frac{44}{48} + \frac{45}{48}$
$= \frac{89}{48}$

**117.** $16 = 2 \cdot 2 \cdot 2 \cdot 2$
$18 = 2 \cdot 3 \cdot 3$
$24 = 2 \cdot 2 \cdot 2 \cdot 3$
The LCM is $2 \cdot 2 \cdot 2 \cdot 2 \cdot 3 \cdot 3$, or 144.

**119.** $\frac{1}{8} - \frac{1}{3} = \frac{1}{8} + \left(-\frac{1}{3}\right) = \frac{3}{24} + \left(-\frac{8}{24}\right) = -\frac{5}{24}$

**121.** No; for any replacement other than 5 the two expressions do not have the same value. For example, let $t = 2$. Then $3 \cdot 2 + 5 = 6 + 5 = 11$, but $3 \cdot 5 + 2 = 15 + 2 = 17$.

**123.** Yes; commutative law of addition

**125.**  $q + qr + qrs + qrst$                 There are no like terms.

$= q \cdot 1 + q \cdot r + q \cdot rs + q \cdot rst$

$= q(1 + r + rs + rst)$                 Factoring

---

## Exercise Set 1.8

**1.** $-(2x + 7) = -2x - 7$     Changing the sign of each term

**3.** $-(5x - 8) = -5x + 8$     Changing the sign of each term

**5.** $-4a + 3b - 7c$

**7.** $-6x + 8y - 5$

**9.** $-3x + 5y + 6$

**11.** $8x + 6y + 43$

**13.**  $9x - (4x + 3) = 9x - 4x - 3$   Removing parentheses by changing the sign of every term

$= 5x - 3$     Collecting like terms

**15.** $2a - (5a - 9) = 2a - 5a + 9 = -3a + 9$

**17.** $2x + 7x - (4x + 6) = 2x + 7x - 4x - 6 = 5x - 6$

**19.** $2x - 4y - 3(7x - 2y) = 2x - 4y - 21x + 6y = -19x + 2y$

**21.**  $15x - y - 5(3x - 2y + 5z)$

$= 15x - y - 15x + 10y - 25z$   Multiplying each term in parentheses by $-5$

$= 9y - 25z$

**23.** $(3x + 2y) - 2(5x - 4y) = 3x + 2y - 10x + 8y = -7x + 10y$

**25.**  $(12a - 3b + 5c) - 5(-5a + 4b - 6c)$

$= 12a - 3b + 5c + 25a - 20b + 30c$

$= 37a - 23b + 35c$

**27.**  $[9 - 2(5 - 4)] = [9 - 2 \cdot 1]$   Computing $5 - 4$

$= [9 - 2]$   Computing $2 \cdot 1$

$= 7$

**29.** $8[7 - 6(4 - 2)] = 8[7 - 6(2)] = 8[7 - 12] = 8[-5] = -40$

**31.**  $[4(9 - 6) + 11] - [14 - (6 + 4)]$

$= [4(3) + 11] - [14 - 10]$

$= [12 + 11] - [14 - 10]$

$= 23 - 4$

$= 19$

**33.**  $[10(x + 3) - 4] + [2(x - 1) + 6]$

$= [10x + 30 - 4] + [2x - 2 + 6]$

$= [10x + 26] + [2x + 4]$

$= 10x + 26 + 2x + 4$

$= 12x + 30$

**35.**  $[7(x + 5) - 19] - [4(x - 6) + 10]$

$= [7x + 35 - 19] - [4x - 24 + 10]$

$= [7x + 16] - [4x - 14]$

$= 7x + 16 - 4x + 14$

$= 3x + 30$

**37.**  $3\{[7(x - 2) + 4] - [2(2x - 5) + 6]\}$

$= 3\{[7x - 14 + 4] - [4x - 10 + 6]\}$

$= 3\{[7x - 10] - [4x - 4]\}$

$= 3\{7x - 10 - 4x + 4\}$

$= 3\{3x - 6\}$

$= 9x - 18$

**39.**  $4\{[5(x - 3) + 2] - 3[2(x + 5) - 9]\}$

$= 4\{[5x - 15 + 2] - 3[2x + 10 - 9]\}$

$= 4\{[5x - 13] - 3[2x + 1]\}$

$= 4\{5x - 13 - 6x - 3\}$

$= 4\{-x - 16\}$

$= -4x - 64$

**41.**  $8 - 2 \cdot 3 - 9 = 8 - 6 - 9$   Multiplying

$= 2 - 9$   Doing all additions and subtractions in order from

$= -7$   left to right

**43.**  $(8 - 2 \cdot 3) - 9 = (8 - 6) - 9$   Multiplying inside the parentheses

$= 2 - 9$   Subtracting inside the parentheses

$= -7$

**45.** $[(-24) \div (-3)] \div \left(-\dfrac{1}{2}\right) = 8 \div \left(-\dfrac{1}{2}\right) = 8 \cdot (-2) = -16$

**47.** $16 \cdot (-24) + 50 = -384 + 50 = -334$

**49.** $2^4 + 2^3 - 10 = 16 + 8 - 10 = 24 - 10 = 14$

**51.** $5^3 + 26 \cdot 71 - (16 + 25 \cdot 3) = 5^3 + 26 \cdot 71 - (16 + 75) = 5^3 + 26 \cdot 71 - 91 = 125 + 26 \cdot 71 - 91 = 125 + 1846 - 91 = 1971 - 91 = 1880$

**53.** $4 \cdot 5 - 2 \cdot 6 + 4 = 20 - 12 + 4 = 8 + 4 = 12$

**55.** $4^3/8 = 64/8 = 8$

**57.** $8(-7) + 6(-5) = -56 - 30 = -86$

**59.** $19 - 5(-3) + 3 = 19 + 15 + 3 = 34 + 3 = 37$

**61.** $9 \div (-3) + 16 \div 8 = -3 + 2 = -1$

**63.** $6 - 4^2 = 6 - 16 = -10$

**65.** $(3 - 8)^2 = (-5)^2 = 25$

**67.** $12 - 20^3 = 12 - 8000 = -7988$

**69.** $2 \cdot 10^3 - 5000 = 2 \cdot 1000 - 5000 = 2000 - 5000 = -3000$

**71.** $6[9 - (3 - 4)] = 6[9 - (-1)] = 6[9 + 1] = 6[10] = 60$

**73.** $-1000 \div (-100) \div 10 = 10 \div 10 = 1$

**75.** $8 - (7 - 9) = 8 - (-2) = 8 + 2 = 10$

**77.** $\dfrac{10 - 6^2}{9^2 + 3^2} = \dfrac{10 - 36}{81 + 9} = \dfrac{-26}{90} = -\dfrac{13}{45}$

**79.** $\dfrac{3(6 - 7) - 5 \cdot 4}{6 \cdot 7 - 8(4 - 1)} = \dfrac{3(-1) - 5 \cdot 4}{42 - 8 \cdot 3} = \dfrac{-3 - 20}{42 - 24} = -\dfrac{23}{18}$

**81.** $\dfrac{2^3 - 3^2 + 12 \cdot 5}{-32 \div (-16) \div (-4)} = \dfrac{8 - 9 + 12 \cdot 5}{-32 \div (-16) \div (-4)} =$

$\dfrac{8 - 9 + 60}{2 \div (-4)} = \dfrac{8 - 9 + 60}{-\frac{1}{2}} = \dfrac{-1 + 60}{-\frac{1}{2}} = \dfrac{59}{-\frac{1}{2}} =$

$59(-2) = -118$

**83.** We divide by the first prime number, 2, until it is no longer possible to do so.

$$236 = 2 \cdot 118$$
$$236 = 2 \cdot 2 \cdot 59$$

Each factor in $2 \cdot 2 \cdot 59$ is a prime number, so this is the prime factorization.

**85.** $\dfrac{2}{3} \div \dfrac{5}{12} = \dfrac{2}{3} \cdot \dfrac{12}{5} = \dfrac{2 \cdot 12}{3 \cdot 5} = \dfrac{2 \cdot \cancel{3} \cdot 4}{\cancel{3} \cdot 5} = \dfrac{8}{5}$

**87.** $10^2 = 10 \cdot 10 = 100$

**89.** $6y + 2x - 3a + c = 6y - (-2x) - 3a - (-c) = 6y - (-2x + 3a - c)$

**91.** $6m + 3n - 5m + 4b = 6m - (-3n) - 5m - (-4b) =$
$6m - (-3n + 5m - 4b)$

**93.** $\quad \{x - [f - (f - x)] + [x - f]\} - 3x$
$= \{x - [f - f + x] + [x - f]\} - 3x$
$= \{x - [x] + [x - f]\} - 3x$
$= \{x - x + x - f\} - 3x = x - f - 3x = -2x - f$

# Chapter 2
# Solving Equations and Inequalities

## Exercise Set 2.1

**1.** $x + 17 = 32$    Writing the equation

$$\frac{15 + 17 \mid 32}{32 \mid} \quad \text{Substituting 15 for } x$$
$\phantom{xx}32 \mid$    TRUE

Since the left-hand and right-hand sides are the same, 15 is a solution of the equation.

**3.** $x - 7 = 12$    Writing the equation

$\dfrac{21 - 7 \mid 12}{\phantom{xx}14 \mid}$   Substituting 21 for $x$
        FALSE

Since the left-hand and right-hand sides are not the same, 21 is not a solution of the equation.

**5.** $6x = 54$    Writing the equation

$\dfrac{6(-7) \mid 54}{\phantom{xx}-42 \mid}$   Substituting
        FALSE

$-7$ is not a solution of the equation.

**7.** $\dfrac{x}{6} = 5$    Writing the equation

$\dfrac{\frac{30}{6} \mid 5}{\phantom{xx}5 \mid}$   Substituting
        TRUE

5 is a solution of the equation.

**9.** $5x + 7 = 107$

$\dfrac{5 \cdot 19 + 7 \mid 107}{95 + 7}$   Substituting
$\phantom{xxx}102 \mid$    FALSE

19 is not a solution of the equation.

**11.** $7(y - 1) = 63$

$\dfrac{7(-11 - 1) \mid 63}{7(-12)}$   Substituting
$\phantom{xxx}-84 \mid$    FALSE

$-11$ is not a solution of the equation.

**13.** $x + 2 = 6$

$x + 2 - 2 = 6 - 2$    Subtracting 2 on both sides
$\phantom{xxxx}x = 4$    Simplifying

Check:   $x + 2 = 6$

$\dfrac{4 + 2 \mid 6}{\phantom{xx}6 \mid}$    TRUE

The solution is 4.

**15.** $x + 15 = -5$

$x + 15 - 15 = -5 - 15$    Subtracting 15 on both sides
$\phantom{xxxx}x = -20$

Check:   $x + 15 = -5$

$\dfrac{-20 + 15 \mid -5}{\phantom{xx}-5 \mid}$    TRUE

The solution is $-20$.

**17.** $x + 6 = -8$

$x + 6 - 6 = -8 - 6$
$\phantom{xxxx}x = -14$

Check:   $x + 6 = -8$

$\dfrac{-14 + 6 \mid -8}{\phantom{xx}-8 \mid}$    TRUE

The solution is $-14$.

**19.** $x + 16 = -2$

$x + 16 - 16 = -2 - 16$
$\phantom{xxxx}x = -18$

Check:   $x + 16 = -2$

$\dfrac{-18 + 16 \mid -2}{\phantom{xx}-2 \mid}$    TRUE

The solution is $-18$.

**21.** $x - 9 = 6$

$x - 9 + 9 = 6 + 9$
$\phantom{xxxx}x = 15$

Check:   $x - 9 = 6$

$\dfrac{15 - 9 \mid 6}{\phantom{xx}6 \mid}$    TRUE

The solution is 15.

**23.** $x - 7 = -21$

$x - 7 + 7 = -21 + 7$
$\phantom{xxxx}x = -14$

Check:   $x - 7 = -21$

$\dfrac{-14 - 7 \mid -21}{\phantom{xx}-21 \mid}$    TRUE

The solution is $-14$.

**25.** $5 + t = 7$

$-5 + 5 + t = -5 + 7$
$\phantom{xxxx}t = 2$

Check:   $5 + t = 7$

$\dfrac{5 + 2 \mid 7}{\phantom{xx}7 \mid}$    TRUE

The solution is 2.

**27.** $-7 + y = 13$

$7 + (-7) + y = 7 + 13$
$\phantom{xxxx}y = 20$

Check:   $-7 + y = 13$

$\dfrac{-7 + 20 \mid 13}{\phantom{xx}13 \mid}$    TRUE

The solution is 20.

**29.**
$$-3 + t = -9$$
$$3 + (-3) + t = 3 + (-9)$$
$$t = -6$$

Check:
$$
\begin{array}{c|c}
-3 + t = -9 \\
\hline
-3 + (-6) & -9 \\
-9 &
\end{array}
$$  TRUE

The solution is $-6$.

**31.**
$$r + \frac{1}{3} = \frac{8}{3}$$
$$r + \frac{1}{3} - \frac{1}{3} = \frac{8}{3} - \frac{1}{3}$$
$$r = \frac{7}{3}$$

Check:
$$
\begin{array}{c|c}
r + \frac{1}{3} = \frac{8}{3} \\
\hline
\frac{7}{3} + \frac{1}{3} & \frac{8}{3} \\
\frac{8}{3} &
\end{array}
$$  TRUE

The solution is $\frac{7}{3}$.

**33.**
$$m + \frac{5}{6} = -\frac{11}{12}$$
$$m + \frac{5}{6} - \frac{5}{6} = -\frac{11}{12} - \frac{5}{6}$$
$$m = -\frac{11}{12} - \frac{5}{6} \cdot \frac{2}{2}$$
$$m = -\frac{11}{12} - \frac{10}{12}$$
$$m = -\frac{21}{12} = -\frac{\cancel{3} \cdot 7}{\cancel{3} \cdot 4}$$
$$m = -\frac{7}{4}$$

Check:
$$
\begin{array}{c|c}
m + \frac{5}{6} = -\frac{11}{12} \\
\hline
-\frac{7}{4} + \frac{5}{6} & -\frac{11}{12} \\
-\frac{21}{12} + \frac{10}{12} & \\
-\frac{11}{12} &
\end{array}
$$  TRUE

The solution is $-\frac{7}{4}$.

**35.**
$$x - \frac{5}{6} = \frac{7}{8}$$
$$x - \frac{5}{6} + \frac{5}{6} = \frac{7}{8} + \frac{5}{6}$$
$$x = \frac{7}{8} \cdot \frac{3}{3} + \frac{5}{6} \cdot \frac{4}{4}$$
$$x = \frac{21}{24} + \frac{20}{24}$$
$$x = \frac{41}{24}$$

Check:
$$
\begin{array}{c|c}
x - \frac{5}{6} = \frac{7}{8} \\
\hline
\frac{41}{24} - \frac{5}{6} & \frac{7}{8} \\
\frac{41}{24} - \frac{20}{24} & \frac{21}{24} \\
\frac{21}{24} &
\end{array}
$$  TRUE

The solution is $\frac{41}{24}$.

**37.**
$$-\frac{1}{5} + z = -\frac{1}{4}$$
$$\frac{1}{5} - \frac{1}{5} + z = \frac{1}{5} - \frac{1}{4}$$
$$z = \frac{1}{5} \cdot \frac{4}{4} - \frac{1}{4} \cdot \frac{5}{5}$$
$$z = \frac{4}{20} - \frac{5}{20}$$
$$z = -\frac{1}{20}$$

Check:
$$
\begin{array}{c|c}
-\frac{1}{5} + z = -\frac{1}{4} \\
\hline
-\frac{1}{5} + \left(-\frac{1}{20}\right) & -\frac{1}{4} \\
-\frac{4}{20} + \left(-\frac{1}{20}\right) & -\frac{5}{20} \\
-\frac{5}{20} &
\end{array}
$$  TRUE

The solution is $-\frac{1}{20}$.

**39.**
$$x + 2.3 = 7.4$$
$$x + 2.3 - 2.3 = 7.4 - 2.3$$
$$x = 5.1$$

Check:
$$
\begin{array}{c|c}
x + 2.3 = 7.4 \\
\hline
5.1 + 2.3 & 7.4 \\
7.4 &
\end{array}
$$  TRUE

The solution is $5.1$.

**41.**
$$7.6 = x - 4.8$$
$$7.6 + 4.8 = x - 4.8 + 4.8$$
$$12.4 = x$$

Check:
$$
\begin{array}{c|c}
7.6 = x - 4.8 \\
\hline
7.6 & 12.4 - 4.8 \\
& 7.6
\end{array}
$$  TRUE

The solution is $12.4$.

**43.**
$$-9.7 = -4.7 + y$$
$$4.7 + (-9.7) = 4.7 + (-4.7) + y$$
$$-5 = y$$

Check:
$$
\begin{array}{c|c}
-9.7 = -4.7 + y \\
\hline
-9.7 & -4.7 + (-5) \\
& -9.7
\end{array}
$$  TRUE

The solution is $-5$.

**45.**
$$5\frac{1}{6} + x = 7$$
$$-5\frac{1}{6} + 5\frac{1}{6} + x = -5\frac{1}{6} + 7$$
$$x = -\frac{31}{6} + \frac{42}{6}$$
$$x = \frac{11}{6}, \text{ or } 1\frac{5}{6}$$

Check: 
$$\begin{array}{c|c} 5\frac{1}{6} + x = 7 \\ \hline 5\frac{1}{6} + 1\frac{5}{6} & 7 \\ 7 & \end{array} \text{ TRUE}$$

The solution is $\frac{11}{6}$, or $1\frac{5}{6}$.

**47.**
$$q + \frac{1}{3} = -\frac{1}{7}$$
$$q + \frac{1}{3} - \frac{1}{3} = -\frac{1}{7} - \frac{1}{3}$$
$$q = -\frac{1}{7} \cdot \frac{3}{3} - \frac{1}{3} \cdot \frac{7}{7}$$
$$q = -\frac{3}{21} - \frac{7}{21}$$
$$q = -\frac{10}{21}$$

Check: 
$$\begin{array}{c|c} q + \frac{1}{3} = -\frac{1}{7} \\ \hline -\frac{10}{21} + \frac{1}{3} & -\frac{1}{7} \\ -\frac{10}{21} + \frac{7}{21} & -\frac{3}{21} \\ -\frac{3}{21} & \end{array} \text{ TRUE}$$

The solution is $-\frac{10}{21}$.

**49.** $-3 + (-8)$ Two negative numbers. We add the absolute values, getting 11, and make the answer negative.
$$-3 + (-8) = -11$$

**51.** $-\frac{2}{3} \cdot \frac{5}{8} = -\frac{2 \cdot 5}{3 \cdot 8} = -\frac{\not{2} \cdot 5}{3 \cdot \not{2} \cdot 4} = -\frac{5}{12}$

**53.** $\frac{2}{3} \div \left(-\frac{4}{9}\right) = \frac{2}{3} \cdot \left(-\frac{9}{4}\right) = -\frac{2 \cdot 9}{3 \cdot 4} = -\frac{\not{2} \cdot \not{3} \cdot 3}{\not{3} \cdot \not{2} \cdot 2} = -\frac{3}{2}$

**55.**
$$-356.788 = -699.034 + t$$
$$699.034 + (-356.788) = 699.034 + (-699.034) + t$$
$$342.246 = t$$

The solution is 342.246.

**57.**
$$x + \frac{4}{5} = -\frac{2}{3} - \frac{4}{15}$$
$$x + \frac{4}{5} = -\frac{2}{3} \cdot \frac{5}{5} - \frac{4}{15} \quad \text{Adding on the right side}$$
$$x + \frac{4}{5} = -\frac{10}{15} - \frac{4}{15}$$
$$x + \frac{4}{5} = -\frac{14}{15}$$
$$x + \frac{4}{5} - \frac{4}{5} = -\frac{14}{15} - \frac{4}{5}$$
$$x = -\frac{14}{15} - \frac{4}{5} \cdot \frac{3}{3}$$
$$x = -\frac{14}{15} - \frac{12}{15}$$
$$x = -\frac{26}{15}$$

The solution is $-\frac{26}{15}$.

**59.**
$$16 + x - 22 = -16$$
$$x - 6 = -16 \quad \text{Adding on the left side}$$
$$x - 6 + 6 = -16 + 6$$
$$x = -10$$

The solution is $-10$.

**61.**
$$x + 3 = 3 + x$$
$$x + 3 - 3 = 3 + x - 3$$
$$x = x$$

$x = x$ is true for all real numbers. Thus the solution is all real numbers.

**63.**
$$-\frac{3}{2} + x = -\frac{5}{17} - \frac{3}{2}$$
$$\frac{3}{2} - \frac{3}{2} + x = \frac{3}{2} - \frac{5}{17} - \frac{3}{2}$$
$$x = \left(\frac{3}{2} - \frac{3}{2}\right) - \frac{5}{17}$$
$$x = -\frac{5}{17}$$

The solution is $-\frac{5}{17}$.

**65.**
$$|x| + 6 = 19$$
$$|x| + 6 - 6 = 19 - 6$$
$$|x| = 13$$

$x$ represents a number whose distance from 0 is 13. Thus $x = -13$ or $x = 13$.

The solutions are $-13$ and 13.

---

## Exercise Set 2.2

**1.**
$$6x = 36$$
$$\frac{6x}{6} = \frac{36}{6} \quad \text{Dividing by 6 on both sides}$$
$$1 \cdot x = 6 \quad \text{Simplifying}$$
$$x = 6 \quad \text{Identity property of 1}$$

Check: 
$$\begin{array}{c|c} 6x = 36 \\ \hline 6 \cdot 6 & 36 \\ 36 & \end{array} \text{ TRUE}$$

The solution is 6.

**3.** $5x = 45$

$\dfrac{5x}{5} = \dfrac{45}{5}$    Dividing by 5 on both sides

$1 \cdot x = 9$    Simplifying

$x = 9$    Identity property of 1

Check:  $\begin{array}{c|c} 5x = 45 \\ \hline 5 \cdot 9 & 45 \\ 45 & \end{array}$    TRUE

The solution is 9.

**5.** $84 = 7x$

$\dfrac{84}{7} = \dfrac{7x}{7}$    Dividing by 7 on both sides

$12 = 1 \cdot x$

$12 = x$

Check:  $\begin{array}{c|c} 84 = 7x \\ \hline 84 & 7 \cdot 12 \\ & 84 \end{array}$    TRUE

The solution is 12.

**7.** $-x = 40$

$-1 \cdot x = 40$

$-1 \cdot (-1 \cdot x) = -1 \cdot 40$

$1 \cdot x = -40$

$x = -40$

Check:  $\begin{array}{c|c} -x = 40 \\ \hline -(-40) & 40 \\ 40 & \end{array}$    TRUE

The solution is $-40$.

**9.** $-x = -1$

$-1 \cdot x = -1$

$-1 \cdot (-1 \cdot x) = -1 \cdot (-1)$

$1 \cdot x = 1$

$x = 1$

Check:  $\begin{array}{c|c} -x = -1 \\ \hline -(1) & -1 \\ -1 & \end{array}$    TRUE

The solution is 1.

**11.** $7x = -49$

$\dfrac{7x}{7} = \dfrac{-49}{7}$

$1 \cdot x = -7$

$x = -7$

Check:  $\begin{array}{c|c} 7x = -49 \\ \hline 7(-7) & -49 \\ -49 & \end{array}$    TRUE

The solution is $-7$.

**13.** $-12x = 72$

$\dfrac{-12x}{-12} = \dfrac{72}{-12}$

$1 \cdot x = -6$

$x = -6$

Check:  $\begin{array}{c|c} -12x = 72 \\ \hline -12(-6) & 72 \\ 72 & \end{array}$    TRUE

The solution is $-6$.

**15.** $-21x = -126$

$\dfrac{-21x}{-21} = \dfrac{-126}{-21}$

$1 \cdot x = 6$

$x = 6$

Check:  $\begin{array}{c|c} -21x = -126 \\ \hline -21 \cdot 6 & -126 \\ -126 & \end{array}$    TRUE

The solution is 6.

**17.** $\dfrac{t}{7} = -9$

$7 \cdot \dfrac{1}{7}t = 7 \cdot (-9)$

$1 \cdot t = -63$

$t = -63$

Check:  $\begin{array}{c|c} \dfrac{t}{7} = -9 \\ \hline \dfrac{-63}{7} & -9 \\ -9 & \end{array}$    TRUE

The solution is $-63$.

**19.** $\dfrac{3}{4}x = 27$

$\dfrac{4}{3} \cdot \dfrac{3}{4}x = \dfrac{4}{3} \cdot 27$

$1 \cdot x = \dfrac{4 \cdot \cancel{3} \cdot 3 \cdot 3}{\cancel{3} \cdot 1}$

$x = 36$

Check:  $\begin{array}{c|c} \dfrac{3}{4}x = 27 \\ \hline \dfrac{3}{4} \cdot 36 & 27 \\ 27 & \end{array}$    TRUE

The solution is 36.

**21.** $\dfrac{-t}{3} = 7$

$3 \cdot \dfrac{1}{3} \cdot (-t) = 3 \cdot 7$

$-t = 21$

$-1 \cdot (-1 \cdot t) = -1 \cdot 21$

$1 \cdot t = -21$

$t = -21$

Check:  $\begin{array}{c|c} \dfrac{-t}{3} = 7 \\ \hline \dfrac{-(-21)}{3} & 7 \\ \dfrac{21}{3} & \\ 7 & \end{array}$    TRUE

The solution is $-21$.

**23.**
$$-\frac{m}{3} = \frac{1}{5}$$

$$-\frac{1}{3} \cdot m = \frac{1}{5}$$

$$-3 \cdot \left(-\frac{1}{3} \cdot m\right) = -3 \cdot \frac{1}{5}$$

$$m = -\frac{3}{5}$$

Check:
$$\frac{-\dfrac{m}{3} = \dfrac{1}{5}}{\begin{array}{c|c} -\dfrac{-\dfrac{3}{5}}{3} & \dfrac{1}{5} \\[2mm] -\left(-\dfrac{3}{5} \div 3\right) & \\[2mm] -\left(-\dfrac{3}{5} \cdot \dfrac{1}{3}\right) & \\[2mm] -\left(-\dfrac{1}{5}\right) & \\[2mm] \dfrac{1}{5} & \text{TRUE} \end{array}}$$

The solution is $-\dfrac{3}{5}$.

**25.**
$$-\frac{3}{5}r = \frac{9}{10}$$

$$-\frac{5}{3} \cdot \left(-\frac{3}{5}r\right) = -\frac{5}{3} \cdot \frac{9}{10}$$

$$1 \cdot r = -\frac{\cancel{5} \cdot \cancel{3} \cdot 3}{\cancel{3} \cdot \cancel{5} \cdot 2}$$

$$r = -\frac{3}{2}$$

Check:
$$\frac{-\dfrac{3}{5}r = \dfrac{9}{10}}{\begin{array}{c|c} -\dfrac{3}{5} \cdot \left(-\dfrac{3}{2}\right) & \dfrac{9}{10} \\[3mm] \dfrac{9}{10} & \text{TRUE} \end{array}}$$

The solution is $-\dfrac{3}{2}$.

**27.**
$$-\frac{3}{2}r = -\frac{27}{4}$$

$$-\frac{2}{3} \cdot \left(-\frac{3}{2}r\right) = -\frac{2}{3} \cdot \left(-\frac{27}{4}\right)$$

$$1 \cdot r = \frac{\cancel{2} \cdot \cancel{3} \cdot 3 \cdot 3}{\cancel{3} \cdot \cancel{2} \cdot 2}$$

$$r = \frac{9}{2}$$

Check:
$$\frac{-\dfrac{3}{2}r = -\dfrac{27}{4}}{\begin{array}{c|c} -\dfrac{3}{2} \cdot \dfrac{9}{2} & -\dfrac{27}{4} \\[3mm] -\dfrac{27}{4} & \text{TRUE} \end{array}}$$

The solution is $\dfrac{9}{2}$.

**29.**
$$6.3x = 44.1$$

$$\frac{6.3x}{6.3} = \frac{44.1}{6.3}$$

$$1 \cdot x = 7$$

$$x = 7$$

Check:
$$\frac{6.3x = 44.1}{\begin{array}{c|c} 6.3 \cdot 7 & 44.1 \\ 44.1 & \text{TRUE} \end{array}}$$

The solution is 7.

**31.**
$$-3.1y = 21.7$$

$$\frac{-3.1y}{-3.1} = \frac{21.7}{-3.1}$$

$$1 \cdot y = -7$$

$$y = -7$$

Check:
$$\frac{3.1y = 21.7}{\begin{array}{c|c} -3.1(-7) & 21.7 \\ 21.7 & \text{TRUE} \end{array}}$$

The solution is $-7$.

**33.**
$$38.7m = 309.6$$

$$\frac{38.7m}{38.7} = \frac{309.6}{38.7}$$

$$1 \cdot m = 8$$

$$m = 8$$

Check:
$$\frac{38.7m = 309.6}{\begin{array}{c|c} 38.7 \cdot 8 & 309.6 \\ 309.6 & \text{TRUE} \end{array}}$$

The solution is 8.

**35.**
$$-\frac{2}{3}y = -10.6$$

$$-\frac{3}{2} \cdot \left(-\frac{2}{3}y\right) = -\frac{3}{2} \cdot (-10.6)$$

$$1 \cdot y = \frac{31.8}{2}$$

$$y = 15.9$$

Check:
$$\frac{-\dfrac{2}{3}y = -10.6}{\begin{array}{c|c} -\dfrac{2}{3} \cdot (15.9) & -10.6 \\[3mm] -\dfrac{31.8}{3} & \\[3mm] -10.6 & \text{TRUE} \end{array}}$$

The solution is 15.9.

**37.** $3x + 4x = (3+4)x = 7x$

**39.** $-4x + 11 - 6x + 18x = (-4 - 6 + 18)x + 11 = 8x + 11$

**41.** $3x - (4 + 2x) = 3x - 4 - 2x = x - 4$

**43.** $8y - 6(3y + 7) = 8y - 18y - 42 = -10y - 42$

**45.**
$$-0.2344m = 2028.732$$

$$\frac{-0.2344m}{-0.2344} = \frac{2028.732}{-0.2344}$$

$$1 \cdot m = -8655$$

$$m = -8655$$

The solution is $-8655$.

**47.** For all $x$, $0 \cdot x = 0$. There is no solution to $0 \cdot x = 9$.

**49.**
$$2|x| = -12$$
$$\frac{2|x|}{2} = \frac{-12}{2}$$
$$1 \cdot |x| = -6$$
$$|x| = -6$$

Absolute value cannot be negative. The equation has no solution.

**51.**
$$3x = \frac{b}{a}$$
$$\frac{1}{3} \cdot 3x = \frac{1}{3} \cdot \frac{b}{a}$$
$$x = \frac{b}{3a}$$

The solution is $\frac{b}{3a}$.

**53.**
$$\frac{a}{b}x = 4$$
$$\frac{b}{a} \cdot \frac{a}{b}x = \frac{b}{a} \cdot 4$$
$$x = \frac{4b}{a}$$

The solution is $\frac{4b}{a}$.

## Exercise Set 2.3

**1.**
$$5x + 6 = 31$$

| | |
|---|---|
| $5x + 6 - 6 = 31 - 6$ | Subtracting 6 on both sides |
| $5x = 25$ | Simplifying |
| $\frac{5x}{5} = \frac{25}{5}$ | Dividing by 5 on both sides |
| $x = 5$ | Simplifying |

Check: 
$$\begin{array}{c|c} 5x + 6 = 31 \\ \hline 5 \cdot 5 + 6 & 31 \\ 25 + 6 & \\ 31 & \text{TRUE} \end{array}$$

The solution is 5.

**3.**
$$8x + 4 = 68$$

| | |
|---|---|
| $8x + 4 - 4 = 68 - 4$ | Subtracting 4 on both sides |
| $8x = 64$ | Simplifying |
| $\frac{8x}{8} = \frac{64}{8}$ | Dividing by 8 on both sides |
| $x = 8$ | Simplifying |

Check: 
$$\begin{array}{c|c} 8x + 4 = 68 \\ \hline 8 \cdot 8 + 4 & 68 \\ 64 + 4 & \\ 68 & \text{TRUE} \end{array}$$

The solution is 8.

**5.**
$$4x - 6 = 34$$

| | |
|---|---|
| $4x - 6 + 6 = 34 + 6$ | Adding 6 on both sides |
| $4x = 40$ | |
| $\frac{4x}{4} = \frac{40}{4}$ | Dividing by 4 on both sides |
| $x = 10$ | |

Check: 
$$\begin{array}{c|c} 4x - 6 = 34 \\ \hline 4 \cdot 10 - 6 & 34 \\ 40 - 6 & \\ 34 & \text{TRUE} \end{array}$$

The solution is 10.

**7.**
$$3x - 9 = 33$$
$$3x - 9 + 9 = 33 + 9$$
$$3x = 42$$
$$\frac{3x}{3} = \frac{42}{3}$$
$$x = 14$$

Check: 
$$\begin{array}{c|c} 3x - 9 = 33 \\ \hline 3 \cdot 14 - 9 & 33 \\ 42 - 9 & \\ 33 & \text{TRUE} \end{array}$$

The solution is 14.

**9.**
$$7x + 2 = -54$$
$$7x + 2 - 2 = -54 - 2$$
$$7x = -56$$
$$\frac{7x}{7} = \frac{-56}{7}$$
$$x = -8$$

Check: 
$$\begin{array}{c|c} 7x + 2 = -54 \\ \hline 7(-8) + 2 & -54 \\ -56 + 2 & \\ -54 & \text{TRUE} \end{array}$$

The solution is $-8$.

**11.**
$$-45 = 6y + 3$$
$$-45 - 3 = 6y + 3 - 3$$
$$-48 = 6y$$
$$\frac{-48}{6} = \frac{6y}{6}$$
$$-8 = y$$

Check: 
$$\begin{array}{c|c} -45 = 6y + 3 \\ \hline -45 & 6(-8) + 3 \\ & -48 + 3 \\ & -45 \quad \text{TRUE} \end{array}$$

The solution is $-8$.

**13.**
$$-4x + 7 = 35$$
$$-4x + 7 - 7 = 35 - 7$$
$$-4x = 28$$
$$\frac{-4x}{-4} = \frac{28}{-4}$$
$$x = -7$$

Check: 
$$\begin{array}{c|c} -4x + 7 = 35 \\ \hline -4(-7) + 7 & 35 \\ 28 + 7 & \\ 35 & \text{TRUE} \end{array}$$

The solution is $-7$.

**15.**
$$-7x - 24 = -129$$
$$-7x - 24 + 24 = -129 + 24$$
$$-7x = -105$$
$$\frac{-7x}{-7} = \frac{-105}{-7}$$
$$x = 15$$

Check:  $\dfrac{-7x - 24 = -129}{-7 \cdot 15 - 24 \ \big| \ -129}$

$\qquad\qquad -105 - 24$

$\qquad\qquad\qquad -129 \ \big| \qquad$ TRUE

The solution is 15.

**17.** $5x + 7x = 72$

$\qquad 12x = 72 \qquad$ Collecting like terms

$\qquad \dfrac{12x}{12} = \dfrac{72}{12} \qquad$ Dividing by 12 on both sides

$\qquad\quad x = 6$

Check:  $\dfrac{5x + 7x = 72}{5 \cdot 6 + 7 \cdot 6 \ \big| \ 72}$

$\qquad\qquad 30 + 42$

$\qquad\qquad\quad 72 \ \big| \qquad$ TRUE

The solution is 6.

**19.** $8x + 7x = 60$

$\qquad 15x = 60 \qquad$ Collecting like terms

$\qquad \dfrac{15x}{15} = \dfrac{60}{15} \qquad$ Dividing by 15 on both sides

$\qquad\quad x = 4$

Check:  $\dfrac{8x + 7x = 60}{8 \cdot 4 + 7 \cdot 4 \ \big| \ 60}$

$\qquad\qquad 32 + 28$

$\qquad\qquad\quad 60 \ \big| \qquad$ TRUE

The solution is 4.

**21.** $4x + 3x = 42$

$\qquad 7x = 42$

$\qquad \dfrac{7x}{7} = \dfrac{42}{7}$

$\qquad\quad x = 6$

Check:  $\dfrac{4x + 3x = 42}{4 \cdot 6 + 3 \cdot 6 \ \big| \ 42}$

$\qquad\qquad 24 + 18$

$\qquad\qquad\quad 42 \ \big| \qquad$ TRUE

The solution is 6.

**23.** $-6y - 3y = 27$

$\qquad -9y = 27$

$\qquad \dfrac{-9y}{-9} = \dfrac{27}{-9}$

$\qquad\quad y = -3$

Check:  $\dfrac{-6y - 3y = 27}{-6(-3) - 3(-3) \ \big| \ 27}$

$\qquad\qquad 18 + 9$

$\qquad\qquad\quad 27 \ \big| \qquad$ TRUE

The solution is $-3$.

**25.** $-7y - 8y = -15$

$\qquad -15y = -15$

$\qquad \dfrac{-15y}{-15} = \dfrac{-15}{-15}$

$\qquad\quad y = 1$

Check:  $\dfrac{-7y - 8y = -15}{-7 \cdot 1 - 8 \cdot 1 \ \big| \ -15}$

$\qquad\qquad -7 - 8$

$\qquad\qquad\quad -15 \ \big| \qquad$ TRUE

The solution is 1.

**27.** $10.2y - 7.3y = -58$

$\qquad 2.9y = -58$

$\qquad \dfrac{2.9y}{2.9} = \dfrac{-58}{2.9}$

$\qquad\quad y = -20$

Check:  $\dfrac{10.2y - 7.3y = -58}{10.2(-20) - 7.3(-20) \ \big| \ -58}$

$\qquad\qquad -204 + 146$

$\qquad\qquad\quad -58 \ \big| \qquad$ TRUE

The solution is $-20$.

**29.** $x + \dfrac{1}{3}x = 8$

$\qquad \left(1 + \dfrac{1}{3}\right)x = 8$

$\qquad \dfrac{4}{3}x = 8$

$\qquad \dfrac{3}{4} \cdot \dfrac{4}{3}x = \dfrac{3}{4} \cdot 8$

$\qquad\quad x = 6$

Check:  $x + \dfrac{1}{3}x = 8$

$\qquad\quad \dfrac{6 + \dfrac{1}{3} \cdot 6 \ \big| \ 8}{}$

$\qquad\qquad 6 + 2$

$\qquad\qquad\quad 8 \ \big| \qquad$ TRUE

The solution is 6.

**31.** $8y - 35 = 3y$

$\qquad 8y = 3y + 35 \qquad$ Adding 35 and simplifying

$\qquad 8y - 3y = 35 \qquad$ Subtracting $3y$ and simplifying

$\qquad 5y = 35 \qquad$ Collecting like terms

$\qquad \dfrac{5y}{5} = \dfrac{35}{5} \qquad$ Dividing by 5

$\qquad\quad y = 7$

Check:  $\dfrac{8y - 35 = 3y}{8 \cdot 7 - 35 \ \big| \ 3 \cdot 7}$

$\qquad\qquad 56 - 35 \ \big| \ 21$

$\qquad\qquad\quad 21 \ \big| \qquad$ TRUE

The solution is 7.

**33.** $8x - 1 = 23 - 4x$

$\qquad 8x + 4x = 23 + 1 \qquad$ Adding 1 and $4x$ and simplifying

$\qquad 12x = 24 \qquad$ Collecting like terms

$\qquad \dfrac{12x}{12} = \dfrac{24}{12} \qquad$ Dividing by 12

$\qquad\quad x = 2$

Check:  $\dfrac{8x - 1 = 23 - 4x}{8 \cdot 2 - 1 \ \big| \ 23 - 4 \cdot 2}$

$\qquad\qquad 16 - 1 \ \big| \ 23 - 8$

$\qquad\qquad\quad 15 \ \big| \ 15 \qquad$ TRUE

The solution is 2.

**35.** $2x - 1 = 4 + x$
$2x - x = 4 + 1$　　　Adding 1 and $-x$
$x = 5$　　　　　Collecting like terms

Check:
$$\begin{array}{c|c} 2x - 1 = 4 + x \\ \hline 2 \cdot 5 - 1 & 4 + 5 \\ 10 - 1 & 9 \\ 9 & \text{TRUE} \end{array}$$

The solution is 5.

**37.** $6x + 3 = 2x + 11$
$6x - 2x = 11 - 3$
$4x = 8$
$\dfrac{4x}{4} = \dfrac{8}{4}$
$x = 2$

Check:
$$\begin{array}{c|c} 6x + 3 = 2x + 11 \\ \hline 6 \cdot 2 + 3 & 2 \cdot 2 + 11 \\ 12 + 3 & 4 + 11 \\ 15 & 15 \quad \text{TRUE} \end{array}$$

The solution is 2.

**39.** $5 - 2x = 3x - 7x + 25$
$5 - 2x = -4x + 25$
$4x - 2x = 25 - 5$
$2x = 20$
$\dfrac{2x}{2} = \dfrac{20}{2}$
$x = 10$

Check:
$$\begin{array}{c|c} 5 - 2x = 3x - 7x + 25 \\ \hline 5 - 2 \cdot 10 & 3 \cdot 10 - 7 \cdot 10 + 25 \\ 5 - 20 & 30 - 70 + 25 \\ -15 & -40 + 25 \\ & -15 \quad \text{TRUE} \end{array}$$

The solution is 10.

**41.** $4 + 3x - 6 = 3x + 2 - x$
$3x - 2 = 2x + 2$　　　Collecting like terms on
　　　　　　　　　　each side
$3x - 2x = 2 + 2$
$x = 4$

Check:
$$\begin{array}{c|c} 4 + 3x - 6 = 3x + 2 - x \\ \hline 4 + 3 \cdot 4 - 6 & 3 \cdot 4 + 2 - 4 \\ 4 + 12 - 6 & 12 + 2 - 4 \\ 16 - 6 & 14 - 4 \\ 10 & 10 \quad \text{TRUE} \end{array}$$

The solution is 4.

**43.** $4y - 4 + y + 24 = 6y + 20 - 4y$
$5y + 20 = 2y + 20$
$5y - 2y = 20 - 20$
$3y = 0$
$y = 0$

Check:
$$\begin{array}{c|c} 4y - 4 + y + 24 = 6y + 20 - 4y \\ \hline 4 \cdot 0 - 4 + 0 + 24 & 6 \cdot 0 + 20 - 4 \cdot 0 \\ 0 - 4 + 0 + 24 & 0 + 20 - 0 \\ 20 & 20 \quad \text{TRUE} \end{array}$$

The solution is 0.

**45.** $\dfrac{7}{2}x + \dfrac{1}{2}x = 3x + \dfrac{3}{2} + \dfrac{5}{2}x$

The least common multiple of all the denominators is 2. We multiply by 2 on both sides.

$$2\left(\dfrac{7}{2}x + \dfrac{1}{2}x\right) = 2\left(3x + \dfrac{3}{2} + \dfrac{5}{2}x\right)$$
$$2 \cdot \dfrac{7}{2}x + 2 \cdot \dfrac{1}{2}x = 2 \cdot 3x + 2 \cdot \dfrac{3}{2} + 2 \cdot \dfrac{5}{2}x$$
$$7x + x = 6x + 3 + 5x$$
$$8x = 11x + 3$$
$$8x - 11x = 3$$
$$-3x = 3$$
$$\dfrac{-3x}{-3} = \dfrac{3}{-3}$$
$$x = -1$$

Check:
$$\begin{array}{c|c} \dfrac{7}{2}x + \dfrac{1}{2}x = 3x + \dfrac{3}{2} + \dfrac{5}{2}x \\ \hline \dfrac{7}{2}(-1) + \dfrac{1}{2}(-1) & 3(-1) + \dfrac{3}{2} + \dfrac{5}{2}(-1) \\ -\dfrac{7}{2} - \dfrac{1}{2} & -3 + \dfrac{3}{2} - \dfrac{5}{2} \\ -4 & -\dfrac{8}{2} \\ & -4 \quad \text{TRUE} \end{array}$$

The solution is $-1$.

**47.** $\dfrac{2}{3} + \dfrac{1}{4}t = \dfrac{1}{3}$

The least common multiple of all the denominators is 12. We multiply by 12 on both sides.

$$12\left(\dfrac{2}{3} + \dfrac{1}{4}t\right) = 12 \cdot \dfrac{1}{3}$$
$$12 \cdot \dfrac{2}{3} + 12 \cdot \dfrac{1}{4}t = 12 \cdot \dfrac{1}{3}$$
$$8 + 3t = 4$$
$$3t = 4 - 8$$
$$3t = -4$$
$$\dfrac{3t}{3} = \dfrac{-4}{3}$$
$$t = -\dfrac{4}{3}$$

Check:
$$\begin{array}{c|c} \dfrac{2}{3} + \dfrac{1}{4}t = \dfrac{1}{3} \\ \hline \dfrac{2}{3} + \dfrac{1}{4}\left(-\dfrac{4}{3}\right) & \dfrac{1}{3} \\ \dfrac{2}{3} - \dfrac{1}{3} & \\ \dfrac{1}{3} & \text{TRUE} \end{array}$$

The solution is $-\dfrac{4}{3}$.

**49.**
$$\frac{2}{3} + 3y = 5y - \frac{2}{15}, \quad \text{LCM is 15}$$

$$15\left(\frac{2}{3} + 3y\right) = 15\left(5y - \frac{2}{15}\right)$$

$$15 \cdot \frac{2}{3} + 15 \cdot 3y = 15 \cdot 5y - 15 \cdot \frac{2}{15}$$

$$10 + 45y = 75y - 2$$
$$10 + 2 = 75y - 45y$$
$$12 = 30y$$
$$\frac{12}{30} = \frac{30y}{30}$$
$$\frac{2}{5} = y$$

Check:
$$\frac{2}{3} + 3y = 5y - \frac{2}{15}$$

| $\frac{2}{3} + 3 \cdot \frac{2}{5}$ | $5 \cdot \frac{2}{5} - \frac{2}{15}$ |
|---|---|
| $\frac{2}{3} + \frac{6}{5}$ | $2 - \frac{2}{15}$ |
| $\frac{10}{15} + \frac{18}{15}$ | $\frac{30}{15} - \frac{2}{15}$ |
| $\frac{28}{15}$ | $\frac{28}{15}$  TRUE |

The solution is $\frac{2}{5}$.

**51.**
$$\frac{5}{3} + \frac{2}{3}x = \frac{25}{12} + \frac{5}{4}x + \frac{3}{4}, \quad \text{LCM is 12}$$

$$12\left(\frac{5}{3} + \frac{2}{3}x\right) = 12\left(\frac{25}{12} + \frac{5}{4}x + \frac{3}{4}\right)$$

$$12 \cdot \frac{5}{3} + 12 \cdot \frac{2}{3}x = 12 \cdot \frac{25}{12} + 12 \cdot \frac{5}{4}x + 12 \cdot \frac{3}{4}$$

$$20 + 8x = 25 + 15x + 9$$
$$20 + 8x = 15x + 34$$
$$20 - 34 = 15x - 8x$$
$$-14x = 7x$$
$$\frac{-14}{7} = \frac{7x}{7}$$
$$-2 = x$$

Check:
$$\frac{5}{3} + \frac{2}{3}x = \frac{25}{12} + \frac{5}{4}x + \frac{3}{4}$$

| $\frac{5}{3} + \frac{2}{3}(-2)$ | $\frac{25}{12} + \frac{5}{4}(-2) + \frac{3}{4}$ |
|---|---|
| $\frac{5}{3} - \frac{4}{3}$ | $\frac{25}{12} - \frac{5}{2} + \frac{3}{4}$ |
| $\frac{1}{3}$ | $\frac{25}{12} - \frac{30}{12} + \frac{9}{12}$ |
| | $\frac{4}{12}$ |
| | $\frac{1}{3}$  TRUE |

The solution is $-2$.

**53.**
$$2.1x + 45.2 = 3.2 - 8.4x$$
Greatest number of decimal places is 1
$$10(2.1x + 45.2) = 10(3.2 - 8.4x)$$
Multiplying by 10 to clear decimals
$$10(2.1x) + 10(45.2) = 10(3.2) - 10(8.4x)$$
$$21x + 452 = 32 - 84x$$
$$21x + 84x = 32 - 452$$
$$105x = -420$$
$$\frac{105x}{105} = \frac{-420}{105}$$
$$x = -4$$

Check:
$$2.1x + 45.2 = 3.2 - 8.4x$$

| $2.1(-4) + 45.2$ | $3.2 - 8.4(-4)$ |
|---|---|
| $-8.4 + 45.2$ | $3.2 + 33.6$ |
| $36.8$ | $36.8$  TRUE |

The solution is $-4$.

**55.**
$$1.03 - 0.62x = 0.71 - 0.22x$$
Greatest number of decimal places is 2
$$100(1.03 - 0.62x) = 100(0.71 - 0.22x)$$
Multiplying by 100 to clear decimals
$$100(1.03) - 100(0.62x) = 100(0.71) - 100(0.22x)$$
$$103 - 62x = 71 - 22x$$
$$32 = 40x$$
$$\frac{32}{40} = \frac{40x}{40}$$
$$\frac{4}{5} = x, \text{ or}$$
$$0.8 = x$$

Check:
$$1.03 - 0.62x = 0.71 - 0.22x$$

| $1.03 - 0.62(0.8)$ | $0.71 - 0.22(0.8)$ |
|---|---|
| $1.03 - 0.496$ | $0.71 - 0.176$ |
| $0.534$ | $0.534$  TRUE |

The solution is $\frac{4}{5}$, or 0.8.

**57.**
$$\frac{2}{7}x - \frac{1}{2}x = \frac{3}{4}x + 1, \text{ LCM is 28}$$

$$28\left(\frac{2}{7}x - \frac{1}{2}x\right) = 28\left(\frac{3}{4}x + 1\right)$$

$$28 \cdot \frac{2}{7}x - 28 \cdot \frac{1}{2}x = 28 \cdot \frac{3}{4}x + 28 \cdot 1$$

$$8x - 14x = 21x + 28$$
$$-6x = 21x + 28$$
$$-6x - 21x = 28$$
$$-27x = 28$$
$$x = -\frac{28}{27}$$

Check:
$$\frac{2}{7}x - \frac{1}{2}x = \frac{3}{4}x + 1$$

| $\frac{2}{7}\left(-\frac{28}{27}\right) - \frac{1}{2}\left(-\frac{28}{27}\right)$ | $\frac{3}{4}\left(-\frac{28}{27}\right) + 1$ |
|---|---|
| $-\frac{8}{27} + \frac{14}{27}$ | $-\frac{21}{27} + 1$ |
| $\frac{6}{27}$ | $\frac{6}{27}$  TRUE |

The solution is $-\frac{28}{27}$.

**59.**  $3(2y - 3) = 27$
$\quad 6y - 9 = 27$     Using a distributive law
$\quad 6y = 27 + 9$     Adding 9
$\quad 6y = 36$
$\quad\quad y = 6$     Dividing by 6

Check: $\quad\dfrac{3(2y - 3) = 27}{}$
$\quad\dfrac{3(2 \cdot 6 - 3)}{} \Big| \; 27$
$\quad\quad 3(12 - 3)$
$\quad\quad\quad 3 \cdot 9$
$\quad\quad\quad\quad 27 \quad\Big|\quad$ TRUE

The solution is 6.

**61.**  $\quad 40 = 5(3x + 2)$
$\quad\quad 40 = 15x + 10$     Using a distributive law
$\quad 40 - 10 = 15x$
$\quad\quad 30 = 15x$
$\quad\quad 2 = x$

Check: $\quad\dfrac{40 = 5(3x + 2)}{}$
$\quad 40 \;\Big|\; 5(3 \cdot 2 + 2)$
$\quad\quad\quad\quad 5(6 + 2)$
$\quad\quad\quad\quad 5 \cdot 8$
$\quad\quad\quad\quad 40 \qquad$ TRUE

The solution is 2.

**63.**  $2(3 + 4m) - 9 = 45$
$\quad 6 + 8m - 9 = 45$     Collecting like terms
$\quad\quad 8m - 3 = 45$
$\quad\quad\quad 8m = 45 + 3$
$\quad\quad\quad 8m = 48$
$\quad\quad\quad m = 6$

Check: $\quad\dfrac{2(3 + 4m) - 9 = 45}{}$
$\quad \dfrac{2(3 + 4 \cdot 6) - 9}{} \;\Big|\; 45$
$\quad\quad 2(3 + 24) - 9$
$\quad\quad\quad 2 \cdot 27 - 9$
$\quad\quad\quad\quad 54 - 9$
$\quad\quad\quad\quad\quad 45 \;\Big|\;$ TRUE

The solution is 6.

**65.**  $5r - (2r + 8) = 16$
$\quad 5r - 2r - 8 = 16$
$\quad\quad 3r - 8 = 16$     Collecting like terms
$\quad\quad\quad 3r = 16 + 8$
$\quad\quad\quad 3r = 24$
$\quad\quad\quad r = 8$

Check: $\quad\dfrac{5r - (2r + 8) = 16}{}$
$\quad\dfrac{5 \cdot 8 - (2 \cdot 8 + 8)}{} \;\Big|\; 16$
$\quad\quad 40 - (16 + 8)$
$\quad\quad\quad 40 - 24$
$\quad\quad\quad\quad 16 \;\Big|\;$ TRUE

The solution is 8.

**67.**  $6 - 2(3x - 1) = 2$
$\quad 6 - 6x + 2 = 2$
$\quad\quad 8 - 6x = 2$
$\quad\quad 8 - 2 = 6x$
$\quad\quad\quad 6 = 6x$
$\quad\quad\quad 1 = x$

Check: $\quad\dfrac{6 - 2(3x - 1) = 2}{}$
$\quad\dfrac{6 - 2(3 \cdot 1 - 1)}{} \;\Big|\; 2$
$\quad\quad 6 - 2(3 - 1)$
$\quad\quad\quad 6 - 2 \cdot 2$
$\quad\quad\quad\quad 6 - 4$
$\quad\quad\quad\quad\quad 2 \;\Big|\;$ TRUE

The solution is 1.

**69.**  $5(d + 4) = 7(d - 2)$
$\quad 5d + 20 = 7d - 14$
$\quad 20 + 14 = 7d - 5d$
$\quad\quad 34 = 2d$
$\quad\quad 17 = d$

Check: $\quad\dfrac{5(d + 4) = 7(d - 2)}{}$
$\quad \dfrac{5(17 + 4)}{} \;\Big|\; 7(17 - 2)$
$\quad\quad 5 \cdot 21 \;\Big|\; 7 \cdot 15$
$\quad\quad 105 \;\Big|\; 105 \qquad$ TRUE

The solution is 17.

**71.**  $8(2t + 1) = 4(7t + 7)$
$\quad 16t + 8 = 28t + 28$
$\quad 16t - 28t = 28 - 8$
$\quad\quad -12t = 20$
$\quad\quad t = -\dfrac{20}{12}$
$\quad\quad t = -\dfrac{5}{3}$

Check: $\quad\dfrac{8(2t + 1) = 4(7t + 7)}{}$
$\quad 8\left(2\left(-\dfrac{5}{3}\right) + 1\right) \;\Big|\; 4\left(7\left(-\dfrac{5}{3}\right) + 7\right)$
$\quad 8\left(-\dfrac{10}{3} + 1\right) \;\Big|\; 4\left(-\dfrac{35}{3} + 7\right)$
$\quad 8\left(-\dfrac{7}{3}\right) \;\Big|\; 4\left(-\dfrac{14}{3}\right)$
$\quad -\dfrac{56}{3} \;\Big|\; -\dfrac{56}{3} \quad$ TRUE

The solution is $-\dfrac{5}{3}$.

**73.**  $3(r - 6) + 2 = 4(r + 2) - 21$
$\quad 3r - 18 + 2 = 4r + 8 - 21$
$\quad 3r - 16 = 4r - 13$
$\quad 13 - 16 = 4r - 3r$
$\quad\quad -3 = r$

Check: $\quad\dfrac{3(r - 6) + 2 = 4(r + 2) - 21}{}$
$\quad\dfrac{3(-3 - 6) + 2}{} \;\Big|\; 4(-3 + 2) - 21$
$\quad\quad 3(-9) + 2 \;\Big|\; 4(-1) - 21$
$\quad\quad -27 + 2 \;\Big|\; -4 - 21$
$\quad\quad -25 \;\Big|\; -25 \quad$ TRUE

The solution is $-3$.

**75.**  $19 - (2x + 3) = 2(x + 3) + x$
$\quad 19 - 2x - 3 = 2x + 6 + x$
$\quad 16 - 2x = 3x + 6$
$\quad 16 - 6 = 3x + 2x$
$\quad\quad 10 = 5x$
$\quad\quad 2 = x$

Check: $\dfrac{19 - (2x + 3) = 2(x + 3) + x}{\begin{array}{c|c} 19 - (2 \cdot 2 + 3) & 2(2 + 3) + 2 \\ 19 - (4 + 3) & 2 \cdot 5 + 2 \\ 19 - 7 & 10 + 2 \\ 12 & 12 \quad \text{TRUE} \end{array}}$

The solution is 2.

**77.** $2[4 - 2(3 - x)] - 1 = 4[2(4x - 3) + 7] - 25$
$2[4 - 6 + 2x] - 1 = 4[8x - 6 + 7] - 25$
$2[-2 + 2x] - 1 = 4[8x + 1] - 25$
$-4 + 4x - 1 = 32x + 4 - 25$
$4x - 5 = 32x - 21$
$-5 + 21 = 32x - 4x$
$16 = 28x$
$\dfrac{16}{28} = x$
$\dfrac{4}{7} = x$

The check is left to the student.

The solution is $\dfrac{4}{7}$.

**79.** $0.7(3x + 6) = 1.1 - (x + 2)$
$2.1x + 4.2 = 1.1 - x - 2$
$10(2.1x + 4.2) = 10(1.1 - x - 2)$    Clearing decimals
$21x + 42 = 11 - 10x - 20$
$21x + 42 = -10x - 9$
$21x + 10x = -9 - 42$
$31x = -51$
$x = -\dfrac{51}{31}$

The check is left to the student.

The solution is $-\dfrac{51}{31}$.

**81.** $a + (a - 3) = (a + 2) - (a + 1)$
$a + a - 3 = a + 2 - a - 1$
$2a - 3 = 1$
$2a = 1 + 3$
$2a = 4$
$a = 2$

Check: $\dfrac{a + (a - 3) = (a + 2) - (a + 1)}{\begin{array}{c|c} 2 + (2 - 3) & (2 + 2) - (2 + 1) \\ 2 - 1 & 4 - 3 \\ 1 & 1 \quad \text{TRUE} \end{array}}$

The solution is 2.

**83.** Do the long division. The answer is negative.

$$3.4_\wedge \overline{\smash{\big)}\, 2\,2.1_\wedge 0} \qquad \begin{array}{r} 6\,.\,5 \\ \hline 2\,0\,4 \\ \hline 1\,7\ 0 \\ 1\,7\ 0 \\ \hline 0 \end{array}$$

$-22.1 \div 3.4 = -6.5$

**85.** Since $-15$ is to the left of $-13$ on the number line, $-15$ is less than $-13$, so $-15 < -13$.

**87.** $-22.1 + 3.4$ The absolute values are 22.1 and 3.4. The difference is 18.7. The negative number has the larger absolute value, so the answer is negative.

$-22.1 + 3.4 = -18.7$

**89.** Since we are using a calculator we will not clear the decimals.
$0.008 + 9.62x - 42.8 = 0.944x + 0.0083 - x$
$9.62x - 42.792 = -0.056x + 0.0083$
$9.62x + 0.056x = 0.0083 + 42.792$
$9.676x = 42.8003$
$x = \dfrac{42.8003}{9.676}$
$x \approx 4.4233464$

The solution is approximately 4.4233464.

**91.** $0 = y - (-14) - (-3y)$
$0 = y + 14 + 3y$
$0 = 4y + 14$
$-14 = 4y$
$\dfrac{-14}{4} = y$
$-\dfrac{7}{2} = y$

The solution is approximately $-\dfrac{7}{2}$.

**93.** $\dfrac{5 + 2y}{3} = \dfrac{25}{12} + \dfrac{5y + 3}{4}$, LCM is 12
$12\left(\dfrac{5 + 2y}{3}\right) = 12\left(\dfrac{25}{12} + \dfrac{5y + 3}{4}\right)$
$4(5 + 2y) = 25 + 3(5y + 3)$
$20 + 8y = 25 + 15y + 9$
$-7y = 14$
$y = -2$

The solution is $-2$.

**95.** $-2y + 5y = 6y$
$3y = 6y$
$0 = 3y$
$0 = y$

Ths solution is 0.

**97.** $\dfrac{1}{3}(6x + 24) - 20 = -\dfrac{1}{4}(12x - 72)$
$2x + 8 - 20 = -3x + 18$     Multiplying
$2x - 12 = -3x + 18$
$5x = 30$
$x = 6$

The answer is 6.

**99.**
$$\frac{3}{4}\left(3x - \frac{1}{2}\right) - \frac{2}{3} = \frac{1}{3}$$

$$\frac{3}{4}\left(3x - \frac{1}{2}\right) = 1 \qquad \text{Adding } \frac{2}{3} \text{ on both sides}$$

$$3\left(3x - \frac{1}{2}\right) = 4 \qquad \text{Multiplying by 4}$$

$$9x - \frac{3}{2} = 4$$

$$9x = \frac{11}{2}$$

$$x = \frac{1}{9} \cdot \frac{11}{2}$$

$$x = \frac{11}{18}$$

The solution is $\frac{11}{18}$.

**101.** Addition principle:
$$4x - 8 = 32$$
$$4x = 40$$
$$x = 10$$

Multiplication principle:
$$4x - 8 = 32$$
$$\frac{1}{4}(4x - 8) = \frac{1}{4} \cdot 32$$
$$x - 2 = 8$$
$$x = 10$$

## Exercise Set 2.4

**1. Familiarize.** Let $a$ = the area of Lake Ontario. Then "four times the area of Lake Ontario" translates to $4a$.

**Translate.**

$$\underbrace{\begin{array}{c}\text{The area of}\\\text{Lake Superior}\end{array}}_{\displaystyle 30,172} \underbrace{\text{is}}_{\displaystyle =} \underbrace{\begin{array}{c}\text{four times the area}\\\text{of Lake Ontario}\end{array}}_{\displaystyle 4a}$$

**Solve.** We solve the equation.
$$30,172 = 4a$$
$$\frac{30,172}{4} = \frac{4a}{4} \qquad \text{Dividing by 4}$$
$$7543 = a$$

**Check.** Four times 7543 is 30,172. The answer checks.

**State.** The area of Lake Ontario is about 7543 mi$^2$.

**3. Familiarize.** Let $n$ = the number. Then when 268 is added to the number we have $n + 268$ (or $268 + n$).

**Translate.** We reword the problem.

$$\underbrace{268 \text{ added to a number}}_{\displaystyle n + 268} \underbrace{\text{is}}_{\displaystyle =} \underbrace{749}_{\displaystyle 749}$$

**Solve.** We solve the equation.
$$n + 268 = 749$$
$$n + 268 - 268 = 749 - 268 \qquad \text{Subtracting 268}$$
$$n = 481$$

**Check.** When 268 is added to 481, we have 749. The answer checks.

**State.** The number is 481.

**5. Familiarize.** Let $h$ = the number of hours the editor worked. Then her total income in dollars is $35h$.

**Translate.**

$$\underbrace{\text{Total income}}_{\displaystyle 35h} \underbrace{\text{is}}_{\displaystyle =} \underbrace{\$23,362.50}_{\displaystyle 23,362.50}$$

**Solve.** We solve the equation.
$$35h = 23,362.50$$
$$\frac{35h}{35} = \frac{23,362.50}{35} \qquad \text{Dividing by 35}$$
$$h = 667.5$$

**Check.** If the editor worked 667.5 hours at \$35 an hour, she earned \$35 · 667.5, or \$23,362.50. The answer checks.

**State.** The editor would have worked 667.5 hours.

**7. Familiarize.** Let $c$ = the cost of one 12-oz box of Frosted Mini-Wheats. Then four boxes cost $4c$.

**Translate.**

$$\underbrace{\text{The cost of four boxes}}_{\displaystyle 4c} \underbrace{\text{was}}_{\displaystyle =} \underbrace{\$13.20}_{\displaystyle 13.20}$$

**Solve.** We solve the equation.
$$4c = 13.20$$
$$\frac{4c}{4} = \frac{13.20}{4} \qquad \text{Dividing by 4}$$
$$c = 3.30$$

**Check.** If one box cost \$3.30, then four boxes cost 4(\$3.30), or \$13.20. The answer checks.

**State.** One box cost \$3.30.

**9. Familiarize.** Let $x$ = the number. Then six times the number is $6x$.

**Translate.** We reword the problem.

$$\underbrace{6 \text{ times a number}}_{\displaystyle 6x} \underbrace{\text{less}}_{\displaystyle -} \underbrace{18}_{\displaystyle 18} \underbrace{\text{is}}_{\displaystyle =} \underbrace{96}_{\displaystyle 96}$$

**Solve.** We solve the equation.
$$6x - 18 = 96$$
$$6x - 18 + 18 = 96 + 18 \qquad \text{Adding 18}$$
$$6x = 114$$
$$\frac{6x}{6} = \frac{114}{6} \qquad \text{Dividing by 6}$$
$$x = 19$$

**Check.** Six times 19 is 114. When 18 is subtracted from 114 we get 96. The answer checks.

**State.** The number is 19.

**11. Familiarize.** Let $y$ = the number. Then doubling the number corresponds to finding twice the number and translates to $2y$. Three-fourths of the number translates to $\frac{3}{4}y$.

**Translate.** We reword the problem.

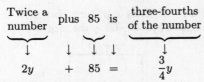

$$2y + 85 = \frac{3}{4}y$$

**Solve.** We solve the equation.

$$2y + 85 = \frac{3}{4}y$$
$$4(2y + 85) = 4 \cdot \frac{3}{4}y \qquad \text{Clearing the fraction}$$
$$8y + 340 = 3y$$
$$8y + 340 - 8y = 3y - 8y \qquad \text{Subtracting } 8y$$
$$340 = -5y$$
$$\frac{340}{-5} = \frac{-5y}{-5} \qquad \text{Dividing by } -5$$
$$-68 = y$$

**Check.** Twice $-68$ is $-136$. Adding 85 to $-136$, we get $-51$. Three-frouths of $-68$ is also $-51$. The answer checks.

**State.** The number is $-68$.

**13. Familiarize.** Let $c =$ the number of calories in a cup of low-fat milk. At breakfast Francis consumed 1 cup $+ \frac{2}{5}$ cup, or $1\frac{2}{5}$ cups, or $\frac{7}{5}$ cups of low-fat milk. The number of calories in $\frac{7}{5}$ cups is $\frac{7}{5}c$.

**Translate.** We reword the problem.

Number of calories in $\frac{7}{5}$ cups is 168

$$\frac{7}{5}c = 168$$

**Solve.** We solve the equation.

$$\frac{7}{5}c = 168$$
$$\frac{5}{7} \cdot \frac{7}{5}c = \frac{5}{7} \cdot 168 \qquad \text{Multiplying by } \frac{5}{7}$$
$$c = \frac{840}{7}$$
$$c = 120$$

**Check.** If there are 120 calories in one cup of low-fat milk, then there are $\frac{7}{5} \cdot 120$, or 168 calories in $\frac{7}{5}$ cups. The answer checks.

**State.** There are 120 calories in a cup of low-fat milk.

**15. Familiarize.** We first draw a picture. We let $x =$ the length of the first piece. Then $3x =$ the length of the second piece and $4 \cdot 3x$, or $12x =$ the length of the third piece.

| $x$ | $3x$ | $12x$ |
|---|---|---|

$\longleftarrow$ 480 m $\longrightarrow$

**Translate.** The length of the three pieces adds up to 480 m.

Length of first piece plus length of second piece plus

$$x + 3x +$$

length of third piece is 480 m

$$12x = 480$$

**Solve.** We solve the equation.

$$x + 3x + 12x = 480$$
$$16x = 480 \qquad \text{Collecting like terms}$$
$$\frac{16x}{16} = \frac{480}{16} \qquad \text{Dividing by 16}$$
$$x = 30$$

**Check.** If the shortest piece of wire is 30 m long, then the second piece is $3 \cdot 30$ m, or 90 m long and the third piece is $4 \cdot 90$ m, or 360 m long. Since 30 m + 90 m + 360 m = 480 m, the answer checks.

**State.** The first piece is 30 m long, the second piece is 90 m long, and the third piece is 360 m long.

**17. Familiarize.** Let $x =$ the smaller number. Then $x + 1 =$ the larger number.

**Translate.** We reword the problem.

First number + second number is 73

$$x + (x + 1) = 73$$

**Solve.** We solve the equation.

$$x + (x + 1) = 73$$
$$2x + 1 = 73 \qquad \text{Collecting like terms}$$
$$2x + 1 - 1 = 73 - 1 \qquad \text{Subtracting 1}$$
$$2x = 72$$
$$\frac{2x}{2} = \frac{72}{2} \qquad \text{Dividing by 2}$$
$$x = 36$$

If $x$ is 36, then $x + 1$ is 37.

**Check.** 36 and 37 are consecutive integers, and their sum is 73. The answer checks.

**State.** The page numbers are 36 and 37.

**19. Familiarize.** Let $x =$ the first even integer. Then $x + 2 =$ the next even integer.

**Translate.** We reword the problem.

First even integer + second even integer is 114

$$x + (x + 2) = 114$$

*Solve.* We solve the equation.

$$x + (x + 2) = 114$$
$$2x + 2 = 114 \qquad \text{Collecting like terms}$$
$$2x + 2 - 2 = 114 - 2$$
$$2x = 112$$
$$\frac{2x}{2} = \frac{112}{2}$$
$$x = 56$$

If $x$ is 56, then $x + 2$ is 58.

*Check.* 56 and 58 are consecutive even integers, and their sum is 114. The answer checks.

*State.* The integers are 56 and 58.

**21.** *Familiarize.* Let $a$ = Edna's age. Then $a + 1$ = Ellie's age and $(a + 1) + 1$, or $a + 2$ = Elsa's age.

*Translate.* We reword the problem.

$$
\underbrace{\text{Edna's age}}_{a} \; + \; \underbrace{\text{Ellie's age}}_{(a+1)} \; + \; \underbrace{\text{Elsa's age}}_{(a+2)} \; \text{is} \; \underset{114}{126}
$$

*Solve.* We solve the equation.

$$a + (a + 1) + (a + 2) = 126$$
$$3a + 3 = 126 \qquad \text{Collecting like terms}$$
$$3a + 3 - 3 = 126 - 3$$
$$3a = 123$$
$$\frac{3a}{3} = \frac{123}{3}$$
$$a = 41$$

If $a$ is 41, then $a + 1$ is 42 and $a + 2$ is 43.

*Check.* 41, 42, and 43 are consecutive integers, and their sum is 126. The answer checks.

*State.* Edna's, Ellie's, and Elsa's ages are 41, 42, and 43, respectively.

**23.** *Familiarize.* Let $x$ = the first odd integer. Then $x + 2$ = the next odd integer and $(x + 2) + 2$, or $x + 4$ = the third odd integer.

*Translate.* We reword the problem.

$$
\underbrace{\text{First odd integer}}_{x} \; + \; \underbrace{\text{second odd integer}}_{(x+2)} \; + \; \underbrace{\text{third odd integer}}_{(x+4)} \; \text{is} \; \underset{189}{189}
$$

*Solve.* We solve the equation.

$$x + (x + 2) + (x + 4) = 189$$
$$3x + 6 = 189 \qquad \text{Collecting like terms}$$
$$3x + 6 - 6 = 189 - 6$$
$$3x = 183$$
$$\frac{3x}{3} = \frac{183}{3}$$
$$x = 61$$

If $x$ is 61, then $x + 2$ is 63 and $x + 4$ is 65.

*Check.* 61, 63, and 65 are consecutive odd integers, and their sum is 189. The answer checks.

*State.* The integers are 61, 63, and 65.

**25.** *Familiarize.* We draw a picture. Let $w$ = the width of the rectangle. Then $w + 60$ = the length.

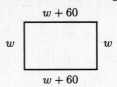

The perimeter $P$ of a rectangle is given by the formula $2l + 2w = P$, and the area $A$ of a rectangle is given by the formula $A = l \cdot w$, where $l$ = the length and $w$ = the width in each case.

*Translate.* We substitute $w + 60$ for $l$ and 520 for $P$ in the formula for perimeter.

$$2l + 2w = P$$
$$2(w + 60) + 2w = 520$$

*Solve.* We solve the equation.

$$2(w + 60) + 2w = 520$$
$$2w + 120 + 2w = 520$$
$$4w + 120 = 520$$
$$4w = 400$$
$$w = 100$$

Possible dimensions are $w = 100$ ft and $w + 60 = 160$ ft. Then the area is given by

$$A = 160 \text{ ft} \cdot 100 \text{ ft} = 16,000 \text{ ft}^2.$$

*Check.* The length is 60 ft more than the width. The perimeter is $2 \cdot 160$ ft $+ 2 \cdot 100$ ft, or 520 ft. We can redo our computation of the area also. The result checks.

*State.* The width of the rectangle is 100 ft; the length is 160 ft; and the area is 16,000 ft$^2$.

**27.** *Familiarize.* We first draw a picture. Let $w$ = the width. Then $2w$ = the length.

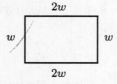

The perimeter $P$ of a rectangle is given by the formula $2l + 2w = P$, where $l$ = the length and $w$ = the width.

*Translate.* We substitute $2w$ for $l$ and $10\frac{1}{2}$ for $P$ in the formula.

$$2l + 2w = P$$
$$2 \cdot 2w + 2w = 10\frac{1}{2}$$

*Solve.* We solve the equation.

$$2 \cdot 2w + 2w = 10\tfrac{1}{2}$$

$$4w + 2w = 10\tfrac{1}{2}$$

$$6w = 10\tfrac{1}{2}$$

$$6w = \frac{21}{2} \qquad \left(10\tfrac{1}{2} = \frac{21}{2}\right)$$

$$\frac{1}{6} \cdot 6w = \frac{1}{6} \cdot \frac{21}{2}$$

$$w = \frac{21}{12}$$

$$w = \frac{7}{4}, \text{ or } 1\tfrac{3}{4}$$

If $w$ is $\frac{7}{4}$, then $2w$ is $2 \cdot \frac{7}{4}$, or $\frac{7}{2}$, or $3\tfrac{1}{2}$.

*Check.* If the width is $\frac{7}{4}$ in. and the length is $\frac{7}{2}$ in., the perimeter is $2 \cdot \frac{7}{4} + 2 \cdot \frac{7}{2} = \frac{7}{2} + 7 = \frac{21}{2}$, or $10\tfrac{1}{2}$ in. The answer checks.

*State.* The width of the cross-section is $1\tfrac{3}{4}$ in., and the length is $3\tfrac{1}{2}$ in.

**29.** *Familiarize.* We draw a picture. We let $x =$ the measure of the first angle. Then $4x =$ the measure of the second angle, and $(x + 4x) - 45$, or $5x - 45 =$ the measure of the third angle.

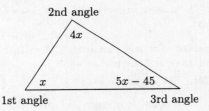

2nd angle

$4x$

$x$       $5x - 45$

1st angle       3rd angle

Recall that the measures of the angles of any triangle add up to 180°.

*Translate.*

| Measure of first angle | + | measure of second angle | + | measure of third angle | is 180. | |
|---|---|---|---|---|---|---|
| ↓ | ↓ | ↓ | ↓ | ↓ | ↓ | ↓ |
| $x$ | + | $4x$ | + | $(5x - 45)$ | = | 180 |

*Solve.* We solve the equation.

$$x + 4x + (5x - 45) = 180$$
$$10x - 45 = 180$$
$$10x = 225$$
$$x = 22.5$$

Possible andwers for the angle measures are as follows:

First angle:   $x = 22.5°$
Second angle:   $4x = 4(22.5) = 90°$
Third angle:   $5x - 45 = 5(22.5) - 45 = 112.5 - 45$
                                                    $= 67.5°$

*Check.* Consider 22.5°, 90°, and 67.5°. The second is four times the first, and the third is 45° less than five times the first. The sum is 180°. These numbers check.

*State.* The measure of the first angle is 22.5°.

**31.** *Familiarize.* The total cost is the daily charge plus the mileage charge. The mileage charge is the cost per mile times the number of miles driven. Let $m =$ the number of miles that can be driven for $80.

*Translate.* We reword the problem.

| Daily rate | plus | Cost per mile | times | Number of miles driven | is | Amount |
|---|---|---|---|---|---|---|
| ↓ | ↓ | ↓ | ↓ | ↓ | ↓ | ↓ |
| 34.95 | + | 0.10 | · | $m$ | = | 80 |

*Solve.* We solve the equation.

$$34.95 + 0.10m = 80$$
$$100(34.95 + 0.10m) = 100(80) \qquad \text{Clearing decimals}$$
$$3495 + 10m = 8000$$
$$10m = 4505$$
$$m = 450.5$$

*Check.* The mileage cost is found by multiplying 450.5 by $0.10 obtaining $45.05. Then we add $45.05 to $34.95, the daily rate, and get $80.

*State.* The businessperson can drive 450.5 mi on the car-rental allotment.

**33.** *Familiarize.* Using the labels on the drawing in the text, we let $x =$ the measure of the first angle. Then $3x =$ the measure of the second angle, and $x + 40 =$ the measure of the third angle. Recall that the sum of measures of the angles of a triangle is 180°.

*Translate.*

| Measure of first angle | + | measure of second angle | + | measure of third angle | is 180. | |
|---|---|---|---|---|---|---|
| ↓ | ↓ | ↓ | ↓ | ↓ | ↓ | ↓ |
| $x$ | + | $3x$ | + | $(x + 40)$ | = | 180 |

*Solve.* We solve the equation.

$$x + 3x + (x + 40) = 180$$
$$5x + 40 = 180$$
$$5x = 140$$
$$x = 28$$

Possible andwers for the angle measures are as follows:

First angle:   $x = 28°$
Second angle:   $3x = 3(28) = 84°$
Third angle:   $x + 40 = 28 + 40 = 68°$

*Check.* Consider 28°, 84°, and 68°. The second angle is three times the first, and the third is 40° more than the first. The sum, $28° + 84° + 68°$, is 180°. These numbers check.

*State.* The measures of the angles are 28°, 84°, and 68°.

35. a) **Familiarize.** We will use the formula

$$P = 0.3444Y - 677.$$

**Translate.** We substitute 1996 for $Y$ in the formula.

$$P = 0.3444(1996) - 677$$

**Solve.** We carry out the computation.

$$P = 0.3444(1996) - 677$$
$$P = 687.4224 - 677 \qquad \text{Multiplying}$$
$$P = 10.4224 \qquad\qquad \text{Subtracting}$$
$$P \approx 10.42 \qquad\qquad \text{Rounding to the nearest hundredth}$$

**Check.** We can repeat the computation. We can also substitute 10.4224 for $P$ in the formula and solve for $Y$. The result is 1996. The answer checks.

**State.** The average price of a ticket in 1996 will be about $10.42.

b) **Familiarize.** We will use the formula

$$P = 0.3444Y - 677.$$

**Translate.** We substitute 11.80 for $P$ in the formula.

$$11.80 = 0.3444Y - 677$$

**Solve.** We solve the equation.

$$11.80 = 0.3444Y - 677$$
$$11.80 + 677 = 0.3444Y - 677 + 677$$
$$688.80 = 0.3444Y$$
$$\frac{688.80}{0.3444} = \frac{0.3444Y}{0.3444}$$
$$2000 = Y$$

**Check.** We can go over our work. We can also substitute 2000 for $Y$ in the formula and find $P$. The result is 11.80. The answer checks.

**State.** The average price of a ticket will be $11.80 in 2000.

37. **Familiarize.** Let $a =$ the original number of apples. Then $\frac{1}{3}a$, $\frac{1}{4}a$, $\frac{1}{8}a$, and $\frac{1}{5}a$ are given to four people, respectively. The fifth and sixth people get 10 apples and 1 apple, respectively.

**Translate.** We reword the problem.

$$\underbrace{\text{The total number of apples}}_{\textstyle \frac{1}{3}a + \frac{1}{4}a + \frac{1}{8}a + \frac{1}{5}a + 10 + 1} \underset{\textstyle =}{\overset{\text{is}}{\downarrow}} \underset{\textstyle a}{\overset{a}{\downarrow}}$$

**Solve.** We solve the equation.

$$\frac{1}{3}a + \frac{1}{4}a + \frac{1}{8}a + \frac{1}{5}a + 10 + 1 = a, \text{ LCD is 120}$$

$$120\left(\frac{1}{3}a + \frac{1}{4}a + \frac{1}{8}a + \frac{1}{5}a + 11\right) = 120 \cdot a$$

$$40a + 30a + 15a + 24a + 1320 = 120a$$
$$109a + 1320 = 120a$$
$$1320 = 11a$$
$$120 = a$$

**Check.** If the original number of apples was 120, then the first four people got $\frac{1}{3} \cdot 120$, $\frac{1}{4} \cdot 120$, $\frac{1}{8} \cdot 120$, and $\frac{1}{5} \cdot 120$, or 40, 30, 15, and 24 apples, respectively. Adding all the apples we get $40 + 30 + 15 + 24 + 10 + 1$, or 120. The result checks.

**State.** There were originally 120 apples in the basket.

39. **Familiarize.** Let $m =$ the number of multiple-choice questions the student answered correctly. Then the number of points earned from the multiple-choice questions was $3m$. Since one of the 4 seven-point fill-ins was wrong, the student earned $3 \cdot 7$, or 21 points, from fill-ins.

**Translate.** We reword the problem.

$$\underbrace{\begin{array}{c}\text{Points from}\\ \text{multiple-choice}\\ \text{questions}\end{array}}_{\textstyle 3m} \underset{\textstyle +}{\overset{\text{plus}}{\downarrow}} \underbrace{\begin{array}{c}\text{points from}\\ \text{fill-in}\\ \text{questions}\end{array}}_{\textstyle 21} \underset{\textstyle =}{\overset{\text{was}}{\downarrow}} \underset{\textstyle 78}{\overset{78}{\downarrow}}$$

**Solve.** We solve the equation.

$$3m + 21 = 78$$
$$3m = 57$$
$$m = 19$$

**Check.** If 19 multiple-choice questions were answered correctly, the student earned $3 \cdot 19$, or 57 points from those questions. Since 21 points were earned from fill-ins, the total points earned were $57 + 21$, or 78. The numer checks.

**State.** The student answered 19 multiple-choice questions correctly.

41. **Familiarize.** Let $s =$ one score. Then four score $= 4s$ and four score and seven $= 4s + 7$.

**Translate.** We reword .

$$\underset{\textstyle 1776}{\overset{1776}{\downarrow}} \underset{\textstyle +}{\overset{\text{plus}}{\downarrow}} \underbrace{\text{four score and seven}}_{\textstyle (4s + 7)} \underset{\textstyle =}{\overset{\text{is}}{\downarrow}} \underset{\textstyle 1863}{\overset{1863}{\downarrow}}$$

**Solve.** We solve the equation.

$$1776 + (4s + 7) = 1863$$
$$4s + 1783 = 1863$$
$$4s = 80$$
$$s = 20$$

**Check.** If a score is 20 years, then four score and seven represents 87 years. Adding 87 to 1776 we get 1863. This checks.

**State.** A score is 20.

43. **Familiarize.** Let $x =$ the number of half dollars. Then

$$2x = \text{number of quarters,}$$
$$4x = \text{number of dimes } (2 \cdot 2x = 4x), \text{ and}$$
$$12x = \text{number of nickels } (3 \cdot 4x = 12x).$$

The value of $x$ half dollars is $0.50(x)$.

The value of $2x$ quarters is $0.25(2x)$.

The value of $4x$ dimes is $0.10(4x)$.

The value of $12x$ nickels is $0.05(12x)$.

**Translate**. The total value is $10.

$$0.50(x) + 0.25(2x) + 0.10(4x) + 0.05(12x) = 10$$

**Solve**.

$$0.50(x) + 0.25(2x) + 0.10(4x) + 0.05(12x) = 10$$
$$0.5x + 0.5x + 0.4x + 0.6x = 10$$
$$2x = 10$$
$$x = 5$$

Possible answers for the number of each coin:

Half dollars $= x = 5$

Quarters $= 2x = 2 \cdot 5 = 10$

Dimes $= 4x = 4 \cdot 5 = 20$

Nickels $= 12x = 12 \cdot 5 = 60$

**Check**. The value of

$$\begin{aligned} 5 \text{ half dollars} &= \$2.50 \\ 10 \text{ quarters} &= \$2.50 \\ 20 \text{ dimes} &= \$2.00 \\ 60 \text{ nickels} &= \$3.00 \end{aligned}$$

The total value is $10. The numbers check.

**State**. The storekeeper got 5 half dollars, 10 quarters, 20 dimes, and 60 nickels.

## Exercise Set 2.5

1. **Familiarize**. Let $x$ = the percent.

   **Translate**.

   What percent of 180 is 36?

   $$x\% \cdot 180 = 36$$

   **Solve**. We solve the equation.

   $$x\% \cdot 180 = 36$$
   $$x \times 0.01 \times 180 = 36$$
   $$x(1.8) = 36$$
   $$\frac{x(1.8)}{1.8} = \frac{36}{1.8}$$
   $$x = \frac{36}{1.8} \times \frac{10}{10} = \frac{360}{18}$$
   $$x = 20$$

   **Check**. We check by finding 20% of 180:

   $$20\% \cdot 180 = 0.2 \cdot 180 = 36$$

   **State**. The answer is 20%.

3. **Familiarize**. Let $y$ = the percent.

   **Translate**.

   What percent of 125 is 30?

   $$y\% \cdot 125 = 30$$

**Solve**. We solve the equation.

$$y\% \cdot 125 = 30$$
$$y \times 0.01 \times 125 = 30$$
$$y(1.25) = 30$$
$$\frac{y(1.25)}{1.25} = \frac{30}{1.25}$$
$$y = \frac{30}{1.25} \times \frac{100}{100} = \frac{3000}{125}$$
$$y = 24$$

**Check**. We find 24% of 125:

$$24\% \cdot 125 = 0.24 \cdot 125 = 30$$

**State**. The answer is 24%.

5. **Familiarize**. Let $y$ = the number we are taking 30% of.

   **Translate**.

   45 is 30% of what?

   $$45 = 30\% \cdot y$$

   **Solve**. We solve the equation.

   $$45 = 30\% \cdot y$$
   $$45 = 0.3y \qquad \text{Converting to decimal notation}$$
   $$\frac{45}{0.3} = \frac{y}{0.3}$$
   $$150 = y$$

   **Check**. We find 30% of 150:

   $$30\% \cdot 150 = 0.3 \times 150 = 45$$

   **State**. The answer is 150.

7. **Familiarize**. Let $y$ = the number we are taking 12% of.

   **Translate**.

   0.3 is 12% of what?

   $$0.3 = 12\% \cdot y$$

   **Solve**. We solve the equation.

   $$0.3 = 12\% \cdot y$$
   $$0.3 = 0.12y \qquad \text{Converting to decimal notation}$$
   $$\frac{0.3}{0.12} = \frac{y}{0.12}$$
   $$2.5 = y$$

   **Check**. We find 12% of 2.5:

   $$12\% \cdot 2.5 = 0.12(2.5) = 0.3$$

   **State**. The answer is 2.5.

9. **Familiarize**. Let $y$ = the unknown number.

   **Translate**.

   What is 65% of 840?

   $$y = 65\% \cdot 840$$

*Solve*. We solve the equation.

$$y = 65\% \cdot 840$$
$$y = 0.65 \times 840$$
$$y = 546$$

*Check*. The check is the computation we used to solve the equation:

$$65\% \cdot 840 = 0.65 \times 840 = 546$$

*State*. The answer is 546.

11. *Familiarize*. Let $y$ = the percent.

*Translate*.

What percent of 80 is 100?
$$y\% \quad \cdot \quad 80 \quad = \quad 100$$

*Solve*. We solve the equation.

$$y\% \cdot 80 = 100$$
$$y \times 0.01 \times 80 = 100$$
$$y(0.8) = 100$$
$$\frac{y(0.8)}{0.8} = \frac{100}{0.8}$$
$$y = 125$$

*Check*. We find 125% of 80:

$$125\% \cdot 80 = 1.25 \times 80 = 100$$

*State*. The answer is 125%.

13. *Familiarize*. Let $x$ = the unknown number.

*Translate*.

What is 2% of 40?
$$x \quad = \quad 2\% \quad \cdot \quad 40$$

*Solve*. We solve the equation.

$$x = 2\% \cdot 40$$
$$x = 0.02 \times 40$$
$$x = 0.8$$

*Check*. The check is the computation we used to solve the equation:

$$2\% \cdot 40 = 0.02 \times 40 = 0.8$$

*State*. The answer is 0.8.

15. *Familiarize*. Let $y$ = the percent.

*Translate*.

2 is what percent of 40?
$$2 \quad = \quad y\% \quad \cdot \quad 40$$

*Solve*. We solve the equation.

$$2 = y\% \cdot 40$$
$$2 = y \times 0.01 \times 40$$
$$2 = y(0.4)$$
$$\frac{2}{0.4} = \frac{y(0.4)}{0.4}$$
$$5 = y$$

*Check*. We find 5% of 40:

$$5\% \cdot 40 = 0.05 \times 40 = 2$$

*State*. The answer is 5%.

17. *Familiarize*. Write down the information.

Nonsmoker's premium: $166 per year

Smoker's premium: 170% of nonsmoker's premium

Let $p$ = the premium for a smoker.

*Translate*. We reword the problem.

What is 170% of $166?
$$p \quad = \quad 170\% \quad \cdot \quad 166$$

*Solve*. We solve the equation.

$$p = 170\% \cdot 166$$
$$p = 1.7 \cdot 166$$
$$p = 282.2$$

*Check*. The check is the computation we used to solve the equation:

$$170\% \cdot 166 = 1.7 \cdot 166 = 282.2$$

*State*. The premium for a smoker is $282.20 per year.

19. *Familiarize*. Write down the information.

Number of people: 800

Number who will catch the cold: 56

We let $p$ = the percent of people who will catch the cold.

*Translate*. We reword the problem.

What percent of 800 is 56?
$$p\% \quad \cdot \quad 800 \quad = \quad 56$$

*Solve*. We solve the equation.

$$p\% \cdot 800 = 56$$
$$p \times 0.01 \times 800 = 56$$
$$p(8) = 56$$
$$\frac{p(8)}{8} = \frac{56}{8}$$
$$p = 7$$

*Check*. We find 7% of 800:

$$7\% \cdot 800 = 0.07 \cdot 800 = 56$$

*State*. 7% will catch the cold.

21. *Familiarize*. Write down the information.

Number of items: 88

Number correct: 76

We let $y$ = the percent that were correct.

*Translate*. We reword the problem.

What percent of 88 is 76?
$$y\% \quad \cdot \quad 88 \quad = \quad 76$$

*Solve*. We solve the equation.

$$y\% \cdot 88 = 76$$
$$y \times 0.01 \times 88 = 76$$
$$y(0.88) = 76$$
$$\frac{y(0.88)}{0.88} = \frac{76}{0.88}$$
$$y \approx 86.36$$

*Check*. We find 86.36% of 88:

$$86.36\% \cdot 88 = 0.8636 \times 88 = 76 \quad \text{(Rounding)}$$

*State*. Approximately 86.36% were correct.

**23.** *Familiarize*. Write down the information.

Original cost: $24 million

Renovation cost: 2% of original cost

Let $x$ = the renovation cost.

*Translate*. We reword the problem.

What is 2% of $24 million?

$\downarrow \quad \downarrow \quad \downarrow \quad \downarrow \qquad \downarrow$

$x \quad = \quad 2\% \quad \cdot \quad 24,000,000$

*Solve*. We solve the equation.

$$x = 2\% \cdot 24,000,000$$
$$x = 0.02 \times 24,000,000$$
$$x = 480,000$$

*Check*. The check is the computation we used to solve the equation:

$$2\% \cdot 24,000,000 = 0.02 \times 24,000,000 = 480,000$$

*State*. The cost is $480,000.

**25.** *Familiarize*. Write down the information.

Sales tax rate: 8%

Purchase price: $428.86

Let $s$ = the sales tax and let $t$ = the total cost of the purchase.

*Translate*. First we find the sales tax. We reword the problem.

What is 8% of $428.86?

$\downarrow \quad \downarrow \quad \downarrow \quad \downarrow \qquad \downarrow$

$s \quad = \quad 8\% \quad \cdot \quad 428.86$

*Solve*. We solve the equation.

$$s = 8\% \cdot 428.86$$
$$s = 0.08 \cdot 428.86$$
$$s \approx 34.31$$

We add the purchase price and the sales tax to find the total cost of the purchase.

$$t = \$428.86 + \$34.31 = \$463.17$$

*Check*. We redo our computations. The results check.

*State*. The sales tax is $34.31; the total cost of the purchase is $463.17.

**27.** *Familiarize*. Write down the information.

Percent of increase in volume: 9%

Volume of water: 400 cm$^3$

Let $x$ = the amount by which the volume will increase. Let $t$ = the volume of the ice.

*Translate*. We first translate to an equation to find the amount by which the volume will increase. We reword the problem.

What is 9% of 400?

$\downarrow \quad \downarrow \quad \downarrow \quad \downarrow \quad \downarrow$

$x \quad = \quad 9\% \quad \cdot \quad 400$

*Solve*. We solve the equation.

$$x = 9\% \cdot 400$$
$$x = 0.09 \times 400$$
$$x = 36$$

We add the volume of the water and the amount of increase to find the volume of the ice.

$$t = 400 + 36 = 436$$

*Check*. We redo our computations. The results check.

*State*. The volume increases by 36 cm$^3$. The volume of the ice is 436 cm$^3$.

**29.** *Familiarize*. Write down the information.

Amount of increase: 12¢, or $0.12

Percent of increase: 12%

Let $x$ = the old price and $y$ = the new price.

*Translate*. We first find the old price. We reword the problem.

$0.12 is 8% of what?

$\downarrow \quad \downarrow \quad \downarrow \quad \downarrow \qquad \downarrow$

$0.12 \quad = \quad 8\% \quad \cdot \quad x$

*Solve*. We solve the equation.

$$0.12 = 8\% \cdot x$$
$$0.12 = 0.08x$$
$$\frac{0.12}{0.08} = \frac{0.08x}{0.08}$$
$$1.5 = x$$

We add the old price and the amount of the increase to find the new price:

$$y = \$1.50 + \$0.12 = \$1.62$$

*Check*. We redo our computations. The results check.

*State*. The old price was $1.50; the new price is $1.62.

**31.** *Familiarize*. Let $p$ = the original price. Then $40\%p$ is the amount of the reduction.

*Translate*. We reword the problem.

Original price $-$ reduction is sale price

$\downarrow \qquad\qquad \downarrow \qquad \downarrow \quad \downarrow \qquad \downarrow$

$p \qquad\quad - \quad 40\%p \quad = \quad 19.20$

**Solve.** We solve the equation.

$$p - 40\%p = 19.20$$
$$p - 0.4p = 19.20$$
$$1p - 0.4p = 19.20$$
$$0.6p = 19.20$$
$$\frac{0.6p}{0.6} = \frac{19.20}{0.6}$$
$$p = 32$$

**Check.** We find 40% of $32 and subtract it from $32:

$$40\% \times \$32 = 0.4 \times \$32 = \$12.80$$

Then $32.00 − $12.80 = $19.20, so $32 checks.

**State.** The original price was $32.

33. **Familiarize.** In each case the sum of the chance of showers and the chance that it will not rain is 100%. Let $d, t$, and $m$ represent the chances that it will not rain during the day, tonight, or tomorrow morning, respectively.

**Translate.** We reword the problem and write equations corresponding to each time period.

During the day:

| Chance of showers | plus | chance it will not rain | is | 100% |
|:---:|:---:|:---:|:---:|:---:|
| ↓ | ↓ | ↓ | ↓ | ↓ |
| 60% | + | d | = | 100% |

Tonight:

| Chance of showers | plus | chance it will not rain | is | 100% |
|:---:|:---:|:---:|:---:|:---:|
| ↓ | ↓ | ↓ | ↓ | ↓ |
| 30% | + | t | = | 100% |

Tomorrow morning:

| Chance of showers | plus | chance it will not rain | is | 100% |
|:---:|:---:|:---:|:---:|:---:|
| ↓ | ↓ | ↓ | ↓ | ↓ |
| 5% | + | m | = | 100% |

**Solve.** We solve each equation.

During the day:

$$60\% + d = 100\%$$
$$60\% + d - 60\% = 100\% - 60\%$$
$$d = 40\%$$

Tonight:

$$30\% + t = 100\%$$
$$30\% + t - 30\% = 100\% - 30\%$$
$$t = 70\%$$

Tomorrow morning:

$$5\% + m = 100\%$$
$$5\% + m - 5\% = 100\% - 5\%$$
$$m = 95\%$$

**Check.** 60% + 40% = 100%; 30% + 70% = 100%; 5% + 95% = 100%. The results check.

**State.** The chances that it will not rain during the day, tonight, and tomorrow morning are 40%, 70%, and 95%, respectively.

35. **Familiarize.** Let $p$ = the price of the gasoline as registered on the pump. Then the sales tax will be 9%$p$.

**Translate.** We reword the problem.

| Price on pump | plus | sales tax | is | $10 |
|:---:|:---:|:---:|:---:|:---:|
| ↓ | ↓ | ↓ | ↓ | ↓ |
| p | + | 9%p | = | 10 |

**Solve.** We solve the equation.

$$p + 9\%p = 10$$
$$1p + 0.09p = 10$$
$$1.09p = 10$$
$$\frac{1.09p}{1.09} = \frac{10}{1.09}$$
$$p \approx 9.17$$

**Check.** We find 9% of $9.17 and add it to $9.17:

$$9\% \times \$9.17 = 0.09 \times \$9.17 \approx \$0.83$$

Then $9.17 + $0.83 = $10, so $9.17 checks.

**State.** The attendant should have filled the tank until the pump read $9.17, not $9.10.

37. **Familiarize.** If we think of the total amount of money as 1 entire amount, then each person receives $\frac{1}{27}$ of that amount. Let $x$ = the percent.

**Translate.** We reword the problem.

| What percent | of | 1 | is | $\frac{1}{27}$? |
|:---:|:---:|:---:|:---:|:---:|
| ↓ | ↓ | ↓ | ↓ | ↓ |
| x% | · | 1 | = | $\frac{1}{27}$ |

**Solve.** We solve the equation.

$$x\% \cdot 1 = \frac{1}{27}$$
$$x \cdot \frac{1}{100} \cdot 1 = \frac{1}{27} \qquad \text{Using "times } \frac{1}{100}\text{" for \%}$$
$$\frac{x}{100} = \frac{1}{27}$$
$$x = \frac{100}{27} \qquad \text{Multiplying by 100}$$
$$x \approx 3.7$$

**Check.** If each person receives approximately 3.7% of the profit, then together they receive 27(3.7%), or 100% of the profit. The result checks.

**State.** Each person receives about 3.7% of the profit.

39. See the answer section in the text.

## Exercise Set 2.6

1.  $A = bh$

$$\frac{A}{h} = \frac{bh}{h} \qquad \text{Dividing by } h$$
$$\frac{A}{h} = b$$

**3.** $d = rt$

$\dfrac{d}{t} = \dfrac{rt}{t}$    Dividing by $t$

$\dfrac{d}{t} = r$

**5.** $I = Prt$

$\dfrac{I}{rt} = \dfrac{Prt}{rt}$    Dividing by $rt$

$\dfrac{I}{rt} = P$

**7.** $F = ma$

$\dfrac{F}{m} = \dfrac{ma}{m}$    Dividing by $m$

$\dfrac{F}{m} = a$

**9.** $P = 2l + 2w$

$P - 2l = 2l + 2w - 2l$    Subtracting $2l$

$P - 2l = 2w$

$\dfrac{P - 2l}{2} = \dfrac{2w}{2}$    Dividing by 2

$\dfrac{P - 2l}{2} = w$

**11.** $A = \pi r^2$

$\dfrac{A}{\pi} = \dfrac{\pi r^2}{\pi}$

$\dfrac{A}{\pi} = r^2$

**13.** $A = \dfrac{1}{2}bh$

$2A = 2 \cdot \dfrac{1}{2}bh$    Multiplying by 2

$2A = bh$

$\dfrac{2A}{h} = \dfrac{bh}{h}$    Dividing by $h$

$\dfrac{2A}{h} = b$

**15.** $E = mc^2$

$\dfrac{E}{c^2} = \dfrac{mc^2}{c^2}$    Dividing by $c^2$

$\dfrac{E}{c^2} = m$

**17.** $Q = \dfrac{c + d}{2}$

$2Q = 2 \cdot \dfrac{c + d}{2}$    Multiplying by 2

$2Q = c + d$

$2Q - c = c + d - c$    Subtracting $c$

$2Q - c = d$

**19.** $A = \dfrac{a + b + c}{3}$

$3A = 3 \cdot \dfrac{a + b + c}{3}$    Multiplying by 3

$3A = a + b + c$

$3A - a - c = a + b + c - a - c$    Subtracting $a$ and $c$

$3A - a - c = b$

**21.** $Ax + By = C$

$Ax + By - Ax = C - Ax$    Subtracting $Ax$

$By = C - Ax$

$\dfrac{By}{B} = \dfrac{C - Ax}{B}$    Dividing by $B$

$y = \dfrac{C - Ax}{B}$

**23.** $v = \dfrac{3k}{t}$

$tv = t \cdot \dfrac{3k}{t}$    Multiplying by $t$

$tv = 3k$

$\dfrac{tv}{v} = \dfrac{3k}{v}$    Dividing by $v$

$t = \dfrac{3k}{v}$

**25.** $H = \dfrac{D^2 N}{2.5}$

$\dfrac{2.5}{N} \cdot H = \dfrac{2.5}{N} \cdot \dfrac{D^2 N}{2.5}$    Multiplying by $\dfrac{2.5}{N}$

$\dfrac{2.5H}{N} = D^2$

**27.** $A = \dfrac{\pi r^2 S}{360}$

$\dfrac{360}{\pi r^2} \cdot A = \dfrac{360}{\pi r^2} \cdot \dfrac{\pi r^2 S}{360}$

$\dfrac{360}{\pi r^2} = S$

**29.** $R = -0.0075t + 3.85$

$R - 3.85 = -0.0075t + 3.85 - 3.85$

$R - 3.85 = -0.0075t$

$\dfrac{R - 3.85}{-0.0075} = \dfrac{-0.0075t}{-0.0075}$

$\dfrac{R - 3.85}{-0.0075} = t$

**31.** We divide:

```
        0.9 2
2 5 ) 2 3.0 0
      2 2 5
        5 0
        5 0
          0
```

Decimal notation for $\dfrac{23}{25}$ is 0.92.

**33.** $-45.8 - (-32.6) = -45.8 + 32.6 = -13.2$

**35.** $-\dfrac{2}{3} + \dfrac{5}{6} = -\dfrac{2}{3} \cdot \dfrac{2}{2} + \dfrac{5}{6}$

$= -\dfrac{4}{6} + \dfrac{5}{6}$

$= \dfrac{1}{6}$

**37.**
$$A = \frac{1}{2}ah + \frac{1}{2}bh$$
$$2A = 2\left(\frac{1}{2}ah + \frac{1}{2}bh\right) \quad \text{Clearing the fractions}$$
$$2A = ah + bh$$
$$2A - ah = bh \quad \text{Subtracting } ah$$
$$\frac{2A - ah}{h} = b \quad \text{Dividing by } h$$

$$A = \frac{1}{2}ah + \frac{1}{2}bh$$
$$2A = ah + bh \quad \text{Clearing fractions as above}$$
$$2A = h(a + b) \quad \text{Factoring}$$
$$\frac{2A}{a + b} = h \quad \text{Dividing by } a + b$$

**39.**
$$Q = 3a + 5ca$$
$$Q = a(3 + 5c) \quad \text{Factoring}$$
$$\frac{Q}{3 + 5c} = a \quad \text{Dividing by } 3 + 5c$$

**41.** $A = lw$

When $l$ and $w$ both double, we have
$$2l \cdot 2w = 4lw = 4A,$$
so $A$ quadruples.

**43.** $A = \frac{1}{2}bh$

When $b$ increases by 4 units we have
$$\frac{1}{2}(b + 4)h = \frac{1}{2}bh + 2h = A + 2h,$$
so $A$ increases by $2h$ units.

## Exercise Set 2.7

**1.** $x > -4$

a) Since $4 > -4$ is true, 4 is a solution.

b) Since $0 > -4$ is true, 0 is a solution.

c) Since $-4 > -4$ is false, $-4$ is not a solution.

d) Since $6 > -4$ is true, 6 is a solution.

e) Since $5.6 > -4$ is true, 5.6 is a solution.

**3.** $x \geq 6$

a) Since $-6 \geq 6$ is false, $-6$ is not a solution.

b) Since $0 \geq 6$ is false, 0 is not a solution.

c) Since $6 \geq 6$ is true, 6 is a solution.

d) Since $8 \geq 6$ is true, 8 is a solution.

e) Since $-3\frac{1}{2} \geq 6$ is false, $-3\frac{1}{2}$ is not a solution.

**5.** The solutions of $x > 4$ are those numbers greater than 4. They are shown on the graph by shading all points to the right of 4. The open circle at 4 indicates that 4 is not part of the graph.

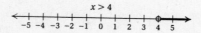

**7.** The solutions of $t < -3$ are those numbers less than $-3$. They are shown on the graph by shading all points to the left of $-3$. The open circle at $-3$ indicates that $-3$ is not part of the graph.

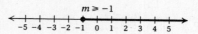

**9.** The solutions of $m \geq -1$ are are shown by shading the point for $-1$ and all points to the right of $-1$. The closed circle at $-1$ indicates that $-1$ is part of the graph.

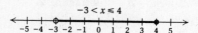

**11.** In order to be a solution of the inequality $-3 < x \leq 4$, a number must be a solution of both $-3 < x$ and $x \leq 4$. The solution set is graphed as follows:

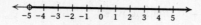

The open circle at $-3$ means that $-3$ is not part of the graph. The closed circle at 4 means that 4 is part of the graph.

**13.** In order to be a solution of the inequality $0 < x < 3$, a number must be a solution of both $0 < x$ and $x < 3$. The solution set is graphed as follows:

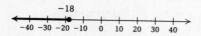

The open circles at 0 and at 3 mean that 0 and 3 are not part of the graph.

**15.**
$$x + 7 > 2$$
$$x + 7 - 7 > 2 - 7 \quad \text{Subtracting 7}$$
$$x > -5 \quad \text{Simplifying}$$

The solution set is $\{x | x > -5\}$.

The graph is as follows:

**17.**
$$x + 8 \leq -10$$
$$x + 8 - 8 \leq -10 - 8 \quad \text{Subtracting 8}$$
$$x \leq -18 \quad \text{Simplifying}$$

The solution set is $\{x | x \leq -18\}$.

The graph is as follows:

**19.**
$$y - 7 > -12$$
$$y - 7 + 7 > -12 + 7 \quad \text{Adding 7}$$
$$y > -5 \quad \text{Simplifying}$$

The solution set is $\{y | y > -5\}$.

**21.**     $2x + 3 > x + 5$
$2x + 3 - 3 > x + 5 - 3$     Subtracting 3
$2x > x + 2$     Simplifying
$2x - x > x + 2 - x$     Subtracting $x$
$x > 2$     Simplifying

The solution set is $\{x | x > 2\}$.

**23.**     $3x + 9 \leq 2x + 6$
$3x + 9 - 9 \leq 2x + 6 - 9$     Subtracting 9
$3x \leq 2x - 3$     Simplifying
$3x - 2x \leq 2x - 3 - 2x$     Subtracting $2x$
$x \leq -3$     Simplifying

The solution set is $\{x | x \leq -3\}$.

**25.**     $5x - 6 < 4x - 2$
$5x - 6 + 6 < 4x - 2 + 6$
$5x < 4x + 4$
$5x - 4x < 4x + 4 - 4x$
$x < 4$

The solution set is $\{x | x < 4\}$.

**27.**     $-9 + t > 5$
$-9 + t + 9 > 5 + 9$
$t > 14$

The solution set is $\{t | t > 14\}$.

**29.**
$$y + \frac{1}{4} \leq \frac{1}{2}$$
$$y + \frac{1}{4} - \frac{1}{4} \leq \frac{1}{2} - \frac{1}{4}$$
$$y \leq \frac{2}{4} - \frac{1}{4} \quad \text{Obtaining a common denominator}$$
$$y \leq \frac{1}{4}$$

The solution set is $\left\{y \middle| y \leq \frac{1}{4}\right\}$.

**31.**
$$x - \frac{1}{3} > \frac{1}{4}$$
$$x - \frac{1}{3} + \frac{1}{3} > \frac{1}{4} + \frac{1}{3}$$
$$x > \frac{3}{12} + \frac{4}{12} \quad \text{Obtaining a common denominator}$$
$$x > \frac{7}{12}$$

The solution set is $\left\{x \middle| x > \frac{7}{12}\right\}$.

**33.**     $5x < 35$
$$\frac{5x}{5} < \frac{35}{5} \quad \text{Dividing by 5}$$
$$x < 7$$

The solution set is $\{x | x < 7\}$. The graph is as follows:

**35.**     $-12x > -36$
$$\frac{-12x}{-12} < \frac{-36}{-12} \quad \text{Dividing by } -12$$
$$\quad \text{The symbol has to be reversed.}$$
$$x < 3 \quad \text{Simplifying}$$

The solution set is $\{x | x < 3\}$. The graph is as follows:

**37.**     $5y \geq -2$
$$\frac{5y}{5} \geq \frac{-2}{5} \quad \text{Dividing by 5}$$
$$y \geq -\frac{2}{5}$$

The solution set is $\left\{y \middle| y \geq -\frac{2}{5}\right\}$.

**39.**     $-2x \leq 12$
$$\frac{-2x}{-2} \geq \frac{12}{-2} \quad \text{Dividing by } -2$$
$$\quad \text{The symbol has to be reversed.}$$
$$x \geq -6 \quad \text{Simplifying}$$

The solution set is $\{x | x \geq 6\}$.

**41.**     $-4y \geq -16$
$$\frac{-4y}{-4} \leq \frac{-16}{-4} \quad \text{Dividing by } -4$$
$$\quad \text{The symbol has to be reversed.}$$
$$y \leq 4 \quad \text{Simplifying}$$

The solution set is $\{y | y \leq 4\}$.

**43.**     $-3x < -17$
$$\frac{-3x}{-3} > \frac{-17}{-3} \quad \text{Dividing by } -3$$
$$\quad \text{The symbol has to be reversed.}$$
$$x > \frac{17}{3} \quad \text{Simplifying}$$

The solution set is $\left\{x \middle| x > \frac{17}{3}\right\}$.

**45.**
$$-2y > \frac{1}{7}$$
$$-\frac{1}{2} \cdot (-2y) < -\frac{1}{2} \cdot \frac{1}{7}$$
$$\quad \text{The symbol has to be reversed.}$$
$$y < -\frac{1}{14}$$

The solution set is $\left\{y \middle| y < -\frac{1}{14}\right\}$.

**47.**
$$-\frac{6}{5} \leq -4x$$
$$-\frac{1}{4} \cdot \left(-\frac{6}{5}\right) \geq -\frac{1}{4} \cdot (-4x)$$
$$\frac{6}{20} \geq x$$
$$\frac{3}{10} \geq x, \text{ or } x \leq \frac{3}{10}$$

The solution set is $\left\{x \middle| \frac{3}{10} \geq x\right\}$, or $\left\{x \middle| x \leq \frac{3}{10}\right\}$.

**49.**
$$4 + 3x < 28$$
$$-4 + 4 + 3x < -4 + 28 \qquad \text{Adding } -4$$
$$3x < 24 \qquad \text{Simplifying}$$
$$\frac{3x}{3} < \frac{24}{3} \qquad \text{Dividing by 3}$$
$$x < 8$$

The solution set is $\{x | x < 8\}$.

**51.**
$$3x - 5 \le 13$$
$$3x - 5 + 5 \le 13 + 5 \qquad \text{Adding 5}$$
$$3x \le 18$$
$$\frac{3x}{3} \le \frac{18}{3} \qquad \text{Dividing by 3}$$
$$x \le 6$$

The solution set is $\{x | x \le 6\}$.

**53.**
$$13x - 7 < -46$$
$$13x - 7 + 7 < -46 + 7$$
$$13x < -39$$
$$\frac{13x}{13} < \frac{-39}{13}$$
$$x < -3$$

The solution set is $\{x | x < -3\}$.

**55.**
$$30 > 3 - 9x$$
$$30 - 3 > 3 - 9x - 3 \qquad \text{Subtracting 3}$$
$$27 > -9x$$
$$\frac{27}{-9} < \frac{-9x}{-9} \qquad \text{Dividing by } -9$$
$$\llcorner \quad \text{The symbol has to be reversed.}$$
$$-3 < x$$

The solution set is $\{x | -3 < x\}$, or $\{x | x > -3\}$.

**57.**
$$4x + 2 - 3x \le 9$$
$$x + 2 \le 9 \qquad \text{Collecting like terms}$$
$$x + 2 - 2 \le 9 - 2$$
$$x \le 7$$

The solution set is $\{x | x \le 7\}$.

**59.**
$$-3 < 8x + 7 - 7x$$
$$-3 < x + 7 \qquad \text{Collecting like terms}$$
$$-3 - 7 < x + 7 - 7$$
$$-10 < x$$

The solution set is $\{x | -10 < x\}$, or $\{x | x > -10\}$.

**61.**
$$6 - 4y > 4 - 3y$$
$$6 - 4y + 4y > 4 - 3y + 4y \qquad \text{Adding } 4y$$
$$6 > 4 + y$$
$$-4 + 6 > -4 + 4 + y \qquad \text{Adding } -4$$
$$2 > y, \text{ or } y < 2$$

The solution set is $\{y | 2 > y\}$, or $\{y | y < 2\}$.

**63.**
$$5 - 9y \le 2 - 8y$$
$$5 - 9y + 9y \le 2 - 8y + 9y$$
$$5 \le 2 + y$$
$$-2 + 5 \le -2 + 2 + y$$
$$3 \le y, \text{ or } y \ge 3$$

The solution set is $\{y | 3 \le y\}$, or $\{y | y \ge 3\}$.

**65.**
$$19 - 7y - 3y < 39$$
$$19 - 10y < 39 \qquad \text{Collecting like terms}$$
$$-19 + 19 - 10y < -19 + 39$$
$$-10y < 20$$
$$\frac{-10y}{-10} > \frac{20}{-10}$$
$$\llcorner \quad \text{The symbol has to be reversed.}$$
$$y > -2$$

The solution set is $\{y | y > -2\}$.

**67.**
$$2.1x + 45.2 > 3.2 - 8.4x$$
$$10(2.1x + 45.2) > 10(3.2 - 8.4x) \qquad \text{Multiplying by 10 to clear decimals}$$
$$21x + 452 > 32 - 84x$$
$$21x + 84x > 32 - 452 \qquad \text{Adding } 84x \text{ and subtracting 452}$$
$$105x > -420$$
$$x > -4 \qquad \text{Dividing by 105}$$

The solution set is $\{x | x > -4\}$.

**69.**
$$\frac{x}{3} - 2 \le 1$$
$$3\left(\frac{x}{3} - 2\right) \le 3 \cdot 1 \qquad \text{Multiplying by 3 to to clear the fraction}$$
$$x - 6 \le 3 \qquad \text{Simplifying}$$
$$x \le 9 \qquad \text{Adding 6}$$

The solution set is $\{x | x \le 9\}$.

**71.**
$$\frac{y}{5} + 1 \le \frac{2}{5}$$
$$5\left(\frac{y}{5} + 1\right) \le 5 \cdot \frac{2}{5} \qquad \text{Clearing fractions}$$
$$y + 5 \le 2$$
$$y \le -3 \qquad \text{Subtracting 5}$$

The solution set is $\{y | y \le -3\}$.

**73.**
$$3(2y - 3) < 27$$
$$6y - 9 < 27 \qquad \text{Removing parentheses}$$
$$6y < 36 \qquad \text{Adding 9}$$
$$y < 6 \qquad \text{Dividing by 6}$$

The solution set is $\{y | y < 6\}$.

**75.**
$$2(3 + 4m) - 9 \ge 45$$
$$6 + 8m - 9 \ge 45 \qquad \text{Removing parentheses}$$
$$8m - 3 \ge 45 \qquad \text{Collecting like terms}$$
$$8m \ge 48 \qquad \text{Adding 3}$$
$$m \ge 6 \qquad \text{Dividing by 8}$$

The solution set is $\{m | m \ge 6\}$.

**77.**
$$8(2t + 1) > 4(7t + 7)$$
$$16t + 8 > 28t + 28$$
$$16t - 28t > 28 - 8$$
$$-12t > 20$$
$$t < -\frac{20}{12} \qquad \text{Dividing by } -12 \text{ and reversing the symbol}$$
$$t < -\frac{5}{3}$$

The solution set is $\{t | t < -\frac{5}{3}\}$.

**79.** $3(r-6)+2 < 4(r+2)-21$
$3r-18+2 < 4r+8-21$
$3r-16 < 4r-13$
$-16+13 < 4r-3r$
$-3 < r$, or $r > -3$

The solution set is $\{r|r > -3\}$.

**81.** $0.8(3x+6) \geq 1.1-(x+2)$
$2.4x+4.8 \geq 1.1-x-2$
$10(2.4x+4.8) \geq 10(1.1-x-2)$  Clearing decimals
$24x+48 \geq 11-10x-20$
$24x+48 \geq -10x-9$  Collecting like terms
$24x+10x \geq -9-48$
$34x \geq -57$
$x \geq -\dfrac{57}{34}$

The solution set is $\left\{x\Big|x \geq -\dfrac{57}{34}\right\}$.

**83.** $a+(a-1) < (a+2)-(a+1)$
$2a-1 < a+2-a-1$
$2a-1 < 1$
$2a < 2$
$a < 1$

The solution set is $\{a|a < 1\}$.

**85.** $-56+(-18)$    Two negative numbers. Add the absolute values and make the answer negative.

$-56+(-18) = -74$

**87.** $-\dfrac{3}{4}+\dfrac{1}{8}$    One negative and one positive number. Find the difference of the absolute values. Then make the answer negative, since the negative number has the larger absolute value.

$-\dfrac{3}{4}+\dfrac{1}{8} = -\dfrac{6}{8}+\dfrac{1}{8} = -\dfrac{5}{8}$

**89.** $-56-(-18) = -56+18 = -38$

**91.** $-2.3-7.1 = -2.3+(-7.1) = -9.4$

**93.** Yes. If $2x-5 \geq 9$ for some value of $x$, then $2x-5 \geq 8$ for that same value of $x$ since $8 < 9$.

**95.** The solutions of $|x| < 3$ are all points whose distance from 0 is less than 3. This is equivalent to $-3 < x < 3$. The graph is as follows:

**97.** $x+3 \leq 3+x$
$x-x \leq 3-3$    Subtracting $x$ and 3
$0 \leq 0$

We get an inequality that is true for all values of $x$, so the inequality is true for all real numbers.

**99.** All integers greater than 5 are greater than or equal to 6, so we have $x \geq 6$.

## Exercise Set 2.8

**1.** $x > 8$

**3.** $y \leq -4$

**5.** $n \geq 1300$

**7.** $a \leq 500$

**9.** $3x+2 < 13$, or $2+3x < 13$

**11.** *Familiarize.* The average of the five scores is their sum divided by the number of quizzes, 5. We let $s$ represent the student's score on the last quiz.

*Translate.* The average of the five scores is given by
$$\frac{73+75+89+91+s}{5}.$$
Since this average must be at least 85, this means that it must be greater than or equal to 85. Thus, we can translate the problem to the inequality
$$\frac{73+75+89+91+s}{5} \geq 85.$$

*Solve.* We first multiply by 5 to clear the fraction.
$$5\left(\frac{73+75+89+91+s}{5}\right) \geq 5 \cdot 85$$
$$73+75+89+91+s \geq 425$$
$$328+s \geq 425$$
$$s \geq 425-328$$
$$s \geq 97$$

*Check.* Suppose $s$ is a score greater than or equal to 97. Then by successively adding 73, 75, 89, and 91 on both sides of the inequality we get
$$73+75+89+91+s \geq 425$$
so
$$\frac{73+75+89+91+s}{5} \geq \frac{425}{5}, \text{ or } 85.$$

*State.* Any score which is at least 97 will give an average quiz grade of 85. The solution set is $\{s|s \geq 97\}$.

**13.** *Familiarize.* $R = -0.075t+3.85$

In the formula $R$ represents the world record and $t$ represents the years since 1930. When $t = 0$ (1930), the record was $-0.075 \cdot 0+3.85$, or 3.85 minutes. When $t = 2$ (1932), the record was $-0.075(2)+3.85$, or 3.7 minutes. For what values of $t$ will $-0.075t+3.85$ be less than 3.5?

*Translate.* The record is to be less than 3.5. We have the inequality
$$R < 3.5.$$
To find the $t$ values which satisfy this condition we substitute $-0.075t+3.85$ for $R$.
$$-0.075t+3.85 < 3.5$$

*Solve.*
$$-0.075t+3.85 < 3.5$$
$$-0.075t < 3.5-3.85$$
$$-0.075t < -0.35$$
$$t > \frac{-0.35}{-0.075}$$
$$t > 4\frac{2}{3}$$

***Check.*** With inequalities it is impossible to check each solution. But we can check to see if the solution set we obtained seems reasonable.

When $t = 4\frac{1}{2}$, $R = -0.075(4.5) + 3.85$, or $3.5125$.

When $t = 4\frac{2}{3}$, $R = -0.075\left(\frac{14}{3}\right) + 3.85$, or $3.5$.

When $t = 4\frac{3}{4}$, $R = -0.075(4.75) + 3.85$, or $3.49375$.

Since $r = 3.5$ when $t = 4\frac{2}{3}$ and $R$ decreases as $t$ increases, $R$ will be less than $3.5$ when $t$ is greater than $4\frac{2}{3}$.

***State.*** The world record will be less than 3.5 minutes more than $4\frac{2}{3}$ years after 1930. If we let $Y =$ the year, then the solution set is $\{Y | Y \geq 1935\}$.

**15.** ***Familiarize.*** We let $d =$ the number of days after the calf's birth. Then, for the first few weeks, the calf gains $2d$ lb in $d$ days.

***Translate.***

| Birth weight | plus | weight gain | is more than | 125 lb |
|:---:|:---:|:---:|:---:|:---:|
| ↓ | ↓ | ↓ | ↓ | ↓ |
| 75 | + | 2d | > | 125 |

***Solve.***

$$75 + 2d > 125$$
$$2d > 50 \qquad \text{Subtracting 75}$$
$$d > 25 \qquad \text{Dividing by 2}$$

The solution set is $\{d | d > 25\}$.

***Check.*** We can obtain a partial check by substituting a number less than or equal to 25 in the equation. For example, when $d = 20$:

$$75 + 2d = 75 + 2 \cdot 20 = 75 + 40 = 115 < 125$$

Our result appears to be correct.

***State.*** The calf's weight is more than 125 lb more than 25 days after its birth. The solution set is $\{d | d > 25\}$.

**17.** ***Familiarize.*** Let $n$ represent the number.

***Translate.***

| The number | plus | 15 | is less than | 4 | times | the number. |
|:---:|:---:|:---:|:---:|:---:|:---:|:---:|
| ↓ | ↓ | ↓ | ↓ | ↓ | ↓ | ↓ |
| n | + | 15 | < | 4 | · | n |

***Solve.***

$$n + 15 < 4n$$
$$15 < 3n$$
$$5 < n, \text{ or } n > 5$$

***Check.*** With inequalities it is impossible to check each solution. But we can check to see if the solution set we obtained seems reasonable.

When $n = 4$, we have $4 + 15 < 4 \cdot 4$, or $19 < 16$. This is false.

When $n = 5$, we have $5 + 15 < 4 \cdot 5$, or $20 < 20$. This is false.

When $n = 6$, we have $6 + 15 < 4 \cdot 6$, or $21 < 24$. This is true.

Since the inequality is false for the numbers less than or equal to 5 that we tried and true for the number greater than 5, it would appear that $n > 5$ is correct.

***State.*** All numbers greater than 5 are solutions. The solution set is $\{n | n > 5\}$.

**19.** ***Familiarize.*** We first make a drawing. We let $l$ represent the length.

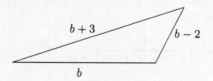

The area is the length times the width, or $4l$.

***Translate.***

| Area | is less than | 86 cm². |
|:---:|:---:|:---:|
| ↓ | ↓ | ↓ |
| 4l | < | 86 |

***Solve.***

$$4l < 86$$
$$l < 21.5$$

***Check.*** We check to see if the solution seems reasonable.

When $l = 22$, the area is $22 \cdot 4$, or $88$ cm².

When $l = 21.5$, the area is $21.5(4)$, or $86$ cm².

When $l = 21$, the area is $21 \cdot 4$, or $84$ cm².

From these calculations, it would appear that the solution is correct.

***State.*** The area will be less than 86 cm² for lengths less than 21.5 cm. The solution set is $\{l | l < 21.5 \text{ cm}\}$, or $\left\{l | l < \frac{43}{2} \text{ cm}\right\}$.

**21.** ***Familiarize.*** We first make a drawing. We let $b$ represent the length of the base. Then the lengths of the other sides are $b - 2$ and $b + 3$.

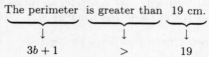

The perimeter is the sum of the lengths of the sides or $b + b - 2 + b + 3$, or $3b + 1$.

***Translate.***

| The perimeter | is greater than | 19 cm. |
|:---:|:---:|:---:|
| ↓ | ↓ | ↓ |
| 3b + 1 | > | 19 |

**Solve.**

$$3b + 1 > 19$$
$$3b > 18$$
$$b > 6$$

**Check**. We check to see if the solution seems reasonable.

When $b = 5$, the perimeter is $3 \cdot 5 + 1$, or 16 cm.

When $b = 6$, the perimeter is $3 \cdot 6 + 1$, or 19 cm.

When $b = 7$, the perimeter is $3 \cdot 7 + 1$, or 22 cm.

From these calculations, it would appear that the solution is correct.

**State**. For lengths of the base greater than 6 cm the perimeter will be greater than 19 cm. The solution set is $\{b|b > 6 \text{ cm}\}$.

**23. Familiarize**. The average number of calls per week is the sum of the calls for the three weeks divided by the number of weeks, 3. We let $c$ represent the number of calls made during the third week.

**Translate**. The average of the three weeks is given by

$$\frac{17 + 22 + c}{3}.$$

Since the average must be at least 20, this means that it must be greater than or equal to 20. Thus, we can translate the problem to the inequality

$$\frac{17 + 22 + c}{3} \geq 20.$$

**Solve**. We first multiply by 3 to clear the fraction.

$$3\left(\frac{17 + 22 + c}{3}\right) \geq 3 \cdot 20$$
$$17 + 22 + c \geq 60$$
$$39 + c \geq 60$$
$$c \geq 21$$

**Check**. Suppose $c$ is a number greater than or equal to 21. Then by adding 17 and 22 on both sides of the inequality we get

$$17 + 22 + c \geq 17 + 22 + 21$$
$$17 + 22 + c \geq 60$$

so

$$\frac{17 + 22 + c}{3} \geq \frac{60}{3}, \text{ or } 20.$$

**State**. Any number of calls which is at least 21 will maintain an average of at least 20 for the three-week period. The solution set is $\{c|c \geq 21\}$.

**25. Familiarize**. We let $t$ = the length of the service call, in hours. Then the hourly charge is $60t$. The total cost of the service call is the sum of the $70 flat fee, the hourly charge, and the $35 cost of the freon.

**Translate**.

| Flat fee | plus | hourly charge | plus | cost of freon | is at most | total charge |
|---|---|---|---|---|---|---|
| ↓ | ↓ | ↓ | ↓ | ↓ | ↓ | ↓ |
| 70 | + | 60t | + | 35 | ≤ | 150 |

**Solve.**

$$70 + 60t + 35 \leq 150$$
$$60t + 105 \leq 150 \qquad \text{Collecting like terms}$$
$$60t \leq 45 \qquad \text{Subtracting 105}$$
$$t \leq 0.75 \qquad \text{Dividing by 60}$$

**Check**. We can obtain a partial check by substituting a number greater than 0.75 in the equation. For example, when $t = 1$:

$$70 + 60t + 35 = 70 + 60 \cdot 1 + 35 = 70 + 60 + 35 = 165 > 150.$$

Our result appears to be correct.

**State**. A service call that is no more than 0.75 hr in length will allow the family to stay within its budget. The solution set is $\{t|t \leq 0.75 \text{ hr}\}$.

**27. Familiarize**. We first make a drawing. Let $b$ = the length of the base of the triangle.

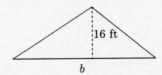

Recall that the formula for the area $A$ of a triangle is $A = \frac{1}{2}bh$. Substituting 16 for $h$, we have $\frac{1}{2}bh = \frac{1}{2} \cdot b \cdot 16 = 8b$.

**Translate**.

| Area | is at least | 200 ft$^2$ |
|---|---|---|
| ↓ | ↓ | ↓ |
| 8b | ≥ | 200 |

**Solve.**

$$8b \geq 200$$
$$b \geq 25 \qquad \text{Dividing by 8}$$

**Check**. We can obtain a partial check by substituting a number less than 25 in the equation. For example, when $b = 24$:

$$8b = 8 \cdot 24 = 192 < 200.$$

Our result appears to be correct.

**State**. The area will be at least 200 ft$^2$ when the base is 25 ft or more. The solution set is $\{b|b \geq 25 \text{ ft}\}$.

**29. Familiarize**. Let $x$ = the first odd integer. Then $x + 2$ is the next odd integer.

**Translate**.

| First odd integer | plus | next odd integer | is less than | 100 |
|---|---|---|---|---|
| ↓ | ↓ | ↓ | ↓ | ↓ |
| x | + | (x + 2) | < | 100 |

**Solve.**

$$x + (x + 2) < 100$$
$$2x + 2 < 100$$
$$2x < 98$$
$$x < 49$$

The largest odd integer less than 49 is 47. When $x = 47$, then $x + 2 = 49$.

***Check***. We can obtain a partial check by testing a pair of odd integers for which the smaller integer is greater than 47. For example, consider the pair 49 and 51. Since $49 + 51 = 100 \geq 100$, our result appears to be correct.

***State***. The largest pair of consecutive odd integers whose sum is less than 100 is 47 and 49.

# Chapter 3

# Polynomials: Operations

**1.** $3^4$ means $3 \cdot 3 \cdot 3 \cdot 3$.

**3.** $(1.1)^5$ means $(1.1)(1.1)(1.1)(1.1)(1.1)$.

**5.** $(7p)^2$ means $(7p)(7p)$.

**7.** $8k^3$ means $8 \cdot k \cdot k \cdot k$.

**9.** $a^0 = 1, a \neq 0$

**11.** $b^1 = b$

**13.** $8.38^0 = 1$

**15.** $(ab)^1 = ab$

**17.** $m^3 = 3^3 = 3 \cdot 3 \cdot 3 = 27$

**19.** $p^1 = 19^1 = 19$

**21.** $x^4 = 4^4 = 4 \cdot 4 \cdot 4 \cdot 4 = 256$

**23.** $y^2 - 7 = 10^2 - 7$
$\qquad = 100 - 7 \quad$ Evaluating the power
$\qquad = 93 \qquad\quad$ Subtracting

**25.** $x^1 + 3 = 7^1 + 3$
$\qquad\quad = 7 + 3 \quad (7^1 = 7)$
$\qquad\quad = 10$

$\quad x^0 + 3 = 7^0 + 3$
$\qquad\quad = 1 + 3 \quad (7^0 = 1)$
$\qquad\quad = 4$

**27.** $A = \pi r^2 \approx 3.14 \times (34 \text{ ft})^2$
$\qquad \approx 3.14 \times 1156 \text{ ft}^2 \quad$ Evaluating the power
$\qquad \approx 3629.84 \text{ ft}^2$

**29.** $3^{-2} = \dfrac{1}{3^2} = \dfrac{1}{9}$

**31.** $10^{-3} = \dfrac{1}{10^3} = \dfrac{1}{1000}$

**33.** $7^{-3} = \dfrac{1}{7^3} = \dfrac{1}{343}$

**35.** $a^{-3} = \dfrac{1}{a^3}$

**37.** $\dfrac{1}{y^{-4}} = y^4$

**39.** $\dfrac{1}{z^{-n}} = z^n$

**41.** $\dfrac{1}{4^3} = 4^{-3}$

**43.** $\dfrac{1}{x^3} = x^{-3}$

**45.** $2^4 \cdot 2^3 = 2^{4+3} = 2^7$

**47.** $8^5 \cdot 8^9 = 8^{5+9} = 8^{14}$

**49.** $x^4 \cdot x^3 = x^{4+3} = x^7$

**51.** $9^{17} \cdot 9^{21} = 9^{17+21} = 9^{38}$

**53.** $(3y)^4 (3y)^8 = (3y)^{4+8} = (3y)^{12}$

**55.** $(7y)^1 (7y)^{16} = (7y)^{1+16} = (7y)^{17}$

**57.** $3^{-5} \cdot 3^8 = 3^{-5+8} = 3^3$

**59.** $x^{-2} \cdot x = x^{-2+1} = x^{-1} = \dfrac{1}{x}$

**61.** $x^{14} \cdot x^3 = x^{14+3} = x^{17}$

**63.** $x^{-7} \cdot x^{-6} = x^{-7+(-6)} = x^{-13} = \dfrac{1}{x^{13}}$

**65.** $t^8 \cdot t^{-8} = t^{8+(-8)} = t^0 = 1$

**67.** $\dfrac{7^5}{7^2} = 7^{5-2} = 7^3$

**69.** $\dfrac{8^{12}}{8^6} = 8^{12-6} = 8^6$

**71.** $\dfrac{y^9}{y^5} = y^{9-5} = y^4$

**73.** $\dfrac{16^2}{16^8} = 16^{2-8} = 16^{-6} = \dfrac{1}{16^6}$

**75.** $\dfrac{m^6}{m^{12}} = m^{6-12} = m^{-6} = \dfrac{1}{m^6}$

**77.** $\dfrac{(8x)^6}{(8x)^{10}} = (8x)^{6-10} = (8x)^{-4} = \dfrac{1}{(8x)^4}$

**79.** $\dfrac{(2y)^9}{(2y)^9} = (2y)^{9-9} = (2y)^0 = 1$

**81.** $\dfrac{x}{x^{-1}} = x^{1-(-1)} = x^2$

**83.** $\dfrac{x^7}{x^{-2}} = x^{7-(-2)} = x^9$

**85.** $\dfrac{z^{-6}}{z^{-2}} = z^{-6-(-2)} = z^{-4} = \dfrac{1}{z^4}$

**87.** $\dfrac{x^{-5}}{x^{-8}} = x^{-5-(-8)} = x^3$

**89.** $\dfrac{m^{-9}}{m^{-9}} = m^{-9-(-9)} = m^0 = 1$

**91.** $5^2 = 5 \cdot 5 = 25$

$5^{-2} = \dfrac{1}{5^2} = \dfrac{1}{25}$

$\left(\dfrac{1}{5}\right)^2 = \dfrac{1}{5} \cdot \dfrac{1}{5} = \dfrac{1}{25}$

$\left(\dfrac{1}{5}\right)^{-2} = \dfrac{1}{\left(\dfrac{1}{5}\right)^2} = \dfrac{1}{\dfrac{1}{25}} = 1 \cdot \dfrac{25}{1} = 25$

$-5^2 = -(5)(5) = -25$

$(-5)^2 = (-5)(-5) = 25$

**93.** $64\%t$, or $0.64t$

**95.**
$$\begin{array}{r} 6\,4\phantom{.} \\ 2\,4.3_\wedge \overline{\smash{)}\,1\,5\,5\,5\,5.2_\wedge} \\ \underline{1\,4\,5\,8\phantom{00}} \\ 9\,7\,2\phantom{0} \\ \underline{9\,7\,2\phantom{0}} \\ 0 \end{array}$$

**97.** $3x - 4 + 5x - 10x = x - 8$

$\phantom{3x - 4 + 5x}-2x - 4 = x - 8$     Collecting like terms

$-2x - 4 + 4 = x - 8 + 4$    Adding 4

$\phantom{-2x - 4 + 4}-2x = x - 4$

$\phantom{-2}-2x - x = x - 4 - x$    Subtracting $x$

$\phantom{-2x - x}-3x = -4$

$\phantom{-2x}\dfrac{-3x}{-3} = \dfrac{-4}{-3}$     Dividing by $-3$

$\phantom{-2x - x - x}x = \dfrac{4}{3}$

The solution is $\dfrac{4}{3}$.

**99.** No; $(5y)^0 = 1$, but $5y^0 = 5 \cdot 1 = 5$.

**101.** $a^{5k} \div a^{3k} = a^{5k-3k} = a^{2k}$

**103.** $\dfrac{\left(\dfrac{1}{2}\right)^4}{\left(\dfrac{1}{2}\right)^5} = \left(\dfrac{1}{2}\right)^{4-5} = \left(\dfrac{1}{2}\right)^{-1} = \dfrac{1}{\dfrac{1}{2}} = 1 \cdot \dfrac{2}{1} = 2$

**105.** No; for example, $(2+3)^2 = 5^2 = 25$, but $2^2 + 3^2 = 4 + 9 = 13$.

**107.** Since the bases are the same, the expression with the larger exponent is larger. Thus, $4^2 < 4^3$.

**109.** $4^3 = 64$, $3^4 = 81$, so $4^3 < 3^4$.

**111.** Choose any number except 2. For example, let $x = 1$. Then $\dfrac{x+2}{2} = \dfrac{1+2}{2} = \dfrac{3}{2}$, but $x = 1$.

## Exercise Set 3.2

**1.** $(2^3)^2 = 2^{3 \cdot 2} = 2^6$

**3.** $(5^2)^{-3} = 5^{2(-3)} = 5^{-6} = \dfrac{1}{5^6}$

**5.** $(x^{-3})^{-4} = x^{(-3)(-4)} = x^{12}$

**7.** $(4x^3)^2 = 4^2(x^3)^2$   Raising each factor to the second power

$\phantom{(4x^3)^2} = 16x^6$

**9.** $(x^4 y^5)^{-3} = (x^4)^{-3}(y^5)^{-3} = x^{4(-3)} y^{5(-3)} =$

$x^{-12} y^{-15} = \dfrac{1}{x^{12} y^{15}}$

**11.** $(x^{-6} y^{-2})^{-4} = (x^{-6})^{-4}(y^{-2})^{-4} = x^{(-6)(-4)} y^{(-2)(-4)} =$

$x^{24} y^8$

**13.** $(3x^3 y^{-8} z^{-3})^2 = 3^2 (x^3)^2 (y^{-8})^2 (z^{-3})^2 =$

$9x^6 y^{-16} z^{-6} = \dfrac{9x^6}{y^{16} z^6}$

**15.** $\left(\dfrac{a^2}{b^3}\right)^4 = \dfrac{(a^2)^4}{(b^3)^4} = \dfrac{a^8}{b^{12}}$

**17.** $\left(\dfrac{y^3}{2}\right)^2 = \dfrac{(y^3)^2}{2^2} = \dfrac{y^6}{4}$

**19.** $\left(\dfrac{y^2}{2}\right)^{-3} = \dfrac{(y^2)^{-3}}{2^{-3}} = \dfrac{y^{-6}}{2^{-3}} = \dfrac{\dfrac{1}{y^6}}{\dfrac{1}{2^3}} = \dfrac{1}{y^6} \cdot \dfrac{2^3}{1} = \dfrac{8}{y^6}$

**21.** $\left(\dfrac{x^2 y}{z}\right)^3 = \dfrac{(x^2)^3 y^3}{z^3} = \dfrac{x^6 y^3}{z^3}$

**23.** $\left(\dfrac{a^2 b}{c d^3}\right)^{-2} = \dfrac{(a^2)^{-2} b^{-2}}{c^{-2}(d^3)^{-2}} = \dfrac{a^{-4} b^{-2}}{c^{-2} d^{-6}} = \dfrac{\dfrac{1}{a^4} \cdot \dfrac{1}{b^2}}{\dfrac{1}{c^2} \cdot \dfrac{1}{d^6}} = \dfrac{\dfrac{1}{a^4 b^2}}{\dfrac{1}{c^2 d^6}} =$

$\dfrac{1}{a^4 b^2} \cdot \dfrac{c^2 d^6}{1} = \dfrac{c^2 d^6}{a^4 b^2}$

**25.** $2\underset{\underset{\text{10 places}}{\underbrace{\phantom{.8,000,000,000}}}}{.8,000,000,000.}$

Large number, so the exponent is positive.

$28,000,000,000 = 2.8 \times 10^{10}$

**27.** $9\underset{\underset{\text{17 places}}{\underbrace{\phantom{.07,000,000,000,000,000}}}}{.07,000,000,000,000,000.}$

Large number, so the exponent is positive.

$907,000,000,000,000,000 = 9.07 \times 10^{17}$

**29.** 0.000003.04

└────────┘ 6 places

Small number, so the exponent is negative.

$0.00000304 = 3.04 \times 10^{-6}$

**31.** 0.00000001. 8

└────────┘ 8 places

Small number, so the exponent is negative.

$0.000000018 = 1.8 \times 10^{-8}$

**33.** 1.00,000,000,000.

└────────┘ 11 places

Large number, so the exponent is positive.

$100,000,000,000 = 1.0 \times 10^{11} = 10^{11}$

**35.** $8.74 \times 10^7$

Positive exponent, so the answer is a large number.

8.7400000.

└────────┘ 7 places

$8.74 \times 10^7 = 87,400,000$

**37.** $5.704 \times 10^{-8}$

Negative exponent, so the answer is a small number.

0.00000005.704

└────────┘ 8 places

$5.704 \times 10^{-8} = 0.00000005704$

**39.** $10^7 = 1 \times 10^7$

Positive exponent, so the answer is a large number.

1.0000000.

└────────┘ 7 places

$10^7 = 10,000,000$

**41.** $10^{-5} = 1 \times 10^{-5}$

Negative exponent, so the answer is a small number.

0.00001.

└────────┘ 5 places

$10^{-5} = 0.00001$

**43.** $(3 \times 10^4)(2 \times 10^5) = (3 \cdot 2) \times (10^4 \cdot 10^5)$
$$= 6 \times 10^9$$

**45.** $(5.2 \times 10^5)(6.5 \times 10^{-2}) = (5.2 \cdot 6.5) \times (10^5 \cdot 10^{-2})$
$$= 33.8 \times 10^3$$

The answer at this stage is $33.8 \times 10^3$ but this is not scientific notation since 33.8 is not a number between 1 and 10. We convert 33.8 to scientific notation and simplify.

$33.8 \times 10^3 = (3.38 \times 10^1) \times 10^3 = 3.38 \times (10^1 \times 10^3) = 3.38 \times 10^4$

The answer is $3.38 \times 10^4$.

**47.** $(9.9 \times 10^{-6})(8.23 \times 10^{-8}) = (9.9 \cdot 8.23) \times (10^{-6} \cdot 10^{-8})$
$$= 81.477 \times 10^{-14}$$

The answer at this stage is $81.477 \times 10^{-14}$. We convert 81.477 to scientific notation and simplify.

$81.477 \times 10^{-14} = (8.1477 \times 10^1) \times 10^{-14} =$
$8.1477 \times (10^1 \times 10^{-14}) = 8.1477 \times 10^{-13}.$

The answer is $8.1477 \times 10^{-13}$.

**49.** $\dfrac{8.5 \times 10^8}{3.4 \times 10^{-5}} = \dfrac{8.5}{3.4} \times \dfrac{10^8}{10^{-5}}$
$$= 2.5 \times 10^{8-(-5)}$$
$$= 2.5 \times 10^{13}$$

**51.** $(3.0 \times 10^6) \div (6.0 \times 10^9) = \dfrac{3.0 \times 10^6}{6.0 \times 10^9}$
$$= \dfrac{3.0}{6.0} \times \dfrac{10^6}{10^9}$$
$$= 0.5 \times 10^{6-9}$$
$$= 0.5 \times 10^{-3}$$

The answer at this stage is $0.5 \times 10^{-3}$. We convert 0.5 to scientific notation and simplify.

$0.5 \times 10^{-3} = (5.0 \times 10^{-1}) \times 10^{-3} =$
$5.0 \times (10^{-1} \times 10^{-3}) = 5.0 \times 10^{-4}$

**53.** $\dfrac{7.5 \times 10^{-9}}{2.5 \times 10^{12}} = \dfrac{7.5}{2.5} \times \dfrac{10^{-9}}{10^{12}}$
$$= 3.0 \times 10^{-9-12}$$
$$= 3.0 \times 10^{-21}$$

**55.** First we find the number of seconds in one hour:
$$1 \text{ hr} = 1 \text{ hr} \times \frac{60 \text{ min}}{1 \text{ hr}} \times \frac{60 \text{ sec}}{1 \text{ min}} = 3600 \text{ sec}$$

Then we multiply to find how much water is discharged in one hour:

$(4,200,000)(3600) = (4.2 \times 10^6)(3.6 \times 10^3)$
$$= 15.12 \times 10^9$$
$$= (1.512 \times 10^1) \times 10^9$$
$$= 1.512 \times 10^{10}$$

In one hour, $1.512 \times 10^{10}$ ft$^3$ of water is discharged.

Next we find the number of seconds in one year:

$$1 \text{ yr} = 1 \text{ yr} \times \frac{365 \text{ days}}{1 \text{ yr}} \times \frac{24 \text{ hr}}{1 \text{ day}} \times \frac{3600 \text{ sec}}{1 \text{ hr}} =$$

$31,536,000 \text{ sec}$

Then we multiply to find how much water is discharged in one year:

$(4,200,000)(31,536,000) = (4.2 \times 10^6)(3.1536 \times 10^7)$
$$= 13.24512 \times 10^{13}$$
$$= (1.324512 \times 10^1) \times 10^{13}$$
$$= 1.324512 \times 10^{14}$$

In one year, $1.324512 \times 10^{14}$ ft$^3$ of water is discharged.

**57.**  $(95)(5.98 \times 10^{24}) = (9.5 \times 10^1)(5.98 \times 10^{24})$
$$= 56.81 \times 10^{25}$$
$$= (5.681 \times 10^1) \times 10^{25}$$
$$= 5.681 \times 10^{26}$$

The mass of Saturn is about $5.681 \times 10^{26}$ kg.

**59.** There are 365 days in one year. We multiply to find how much is spent in vending machines each day:

$(6,800,000)(365) = (6.8 \times 10^6)(3.65 \times 10^2)$
$$= 24.82 \times 10^8$$
$$= (2.482 \times 10^1) \times 10^8$$
$$= 2.482 \times 10^9$$

Americans spend $\$2.482 \times 10^9$ in vending machines in one year.

**61.** $9x - 36 = 9 \cdot x - 9 \cdot 4 = 9(x - 4)$

**63.**
$$2x - 4 - 5x + 8 = x - 3$$
$$-3x + 4 = x - 3 \qquad \text{Collecting like terms}$$
$$-3x + 4 - 4 = x - 3 - 4 \qquad \text{Subtracting 4}$$
$$-3x = x - 7$$
$$-3x - x = x - 7 - x \qquad \text{Subtracting } x$$
$$-4x = -7$$
$$\frac{-4x}{-4} = \frac{-7}{-4} \qquad \text{Dividing by } -4$$
$$x = \frac{7}{4}$$

The solution is $\frac{7}{4}$.

**65.**
$$8(2x + 3) - 2(x - 5) = 10$$
$$16x + 24 - 2x + 10 = 10 \qquad \text{Removing parentheses}$$
$$14x + 34 = 10 \qquad \text{Collecting like terms}$$
$$14x + 34 - 34 = 10 - 34 \qquad \text{Subtracting 34}$$
$$14x = -24$$
$$\frac{14x}{14} = \frac{-24}{14} \qquad \text{Dividing by 14}$$
$$x = -\frac{12}{7} \qquad \text{Simplifying}$$

The solution is $-\frac{12}{7}$.

**67.** $\dfrac{(5.2 \times 10^6)(6.1 \times 10^{-11})}{1.28 \times 10^{-3}} = \dfrac{(5.2 \cdot 6.1)}{1.28} \times \dfrac{(10^6 \cdot 10^{-11})}{10^{-3}}$
$$= 24.78125 \times 10^{-2}$$
$$= (2.478125 \times 10^1) \times 10^{-2}$$
$$= 2.478125 \times 10^{-1}$$

**69.** $\dfrac{(5^{12})^2}{5^{25}} = \dfrac{5^{24}}{5^{25}} = 5^{24-25} = 5^{-1} = \dfrac{1}{5}$

**71.** $\dfrac{(3^5)^4}{3^5 \cdot 3^4} = \dfrac{3^{5 \cdot 4}}{3^{5+4}} = \dfrac{3^{20}}{3^9} = 3^{20-9} = 3^{11}$

**73.** $\left(\dfrac{1}{a}\right)^{-n} = \dfrac{1^{-n}}{a^{-n}} = \dfrac{\frac{1}{1^n}}{\frac{1}{a^n}} = \dfrac{1}{1} \cdot \dfrac{a^n}{1} = a^n$

**75.** False; let $x = 2$, $y = 3$, $m = 4$, and $n = 2$:
$$2^4 \cdot 3^2 = 16 \cdot 9 = 144, \text{ but}$$
$$(2 \cdot 3)^{4 \cdot 2} = 6^8 = 1,679,616$$

**77.** False; let $x = 5$, $y = 3$, and $m = 2$:
$$(5 - 3)^2 = 2^2 = 4, \text{ but}$$
$$5^2 - 3^2 = 25 - 9 = 16$$

## Exercise Set 3.3

**1.** $-5x + 2 = -5 \cdot 4 + 2 = -20 + 2 = -18$

**3.** $2x^2 - 5x + 7 = 2 \cdot 4^2 - 5 \cdot 4 + 7 = 2 \cdot 16 - 20 + 7 = 32 - 20 + 7 = 19$

**5.** $x^3 - 5x^2 + x = 4^3 - 5 \cdot 4^2 + 4 = 64 - 5 \cdot 16 + 4 = 64 - 80 + 4 = -12$

**7.** $3x + 5 = 3(-1) + 5 = -3 + 5 = 2$

**9.** $x^2 - 2x + 1 = (-1)^2 - 2(-1) + 1 = 1 + 2 + 1 = 4$

**11.** $-3x^3 + 7x^2 - 3x - 2 =$
$-3(-1)^3 + 7(-1)^2 - 3(-1) - 2 =$
$-3(-1) + 7 \cdot 1 + 3 - 2 = 3 + 7 + 3 - 2 = 11$

**13.** We evaluate the polynomial for $t = 3$:
$$16t^2 = 16(3)^2 = 16 \cdot 9 = 144$$
The building is 144 ft high.

**15.** $0.4a^2 - 40a + 1039 = 0.4(20)^2 - 40(20) + 1039$
$$= 0.4(400) - 40(20) + 1039$$
$$= 160 - 800 + 1039$$
$$= 399$$

The daily number of accidents involving 20-year-old drivers is 399.

**17.** We evaluate the polynomial for $x = 650$:
$5000 + 0.6x^2 = 5000 + 0.6(650)^2$
$$= 5000 + 0.6(422,500)$$
$$= 5000 + 253,500$$
$$= 258,500$$

The total cost of producing 650 stereos is $258,500.

**19.** We evaluate the polynomial for $x = 100$:
$280x - 0.4x^2 = 280(100) - 0.4(100)^2$
$$= 280(100) - 0.4(10,000)$$
$$= 28,000 - 4000$$
$$= 24,000$$

The total revenue from the sale of 100 stereos is $24,000.

**21.** $2 - 3x + x^2 = 2 + (-3x) + x^2$

The terms are $2$, $-3x$, and $x^2$.

**23.** $5x^3 + 6x^2 - 3x^2$

Like terms: $6x^2$ and $-3x^2$   Same variable and exponent

**25.** $2x^4 + 5x - 7x - 3x^4$

Like terms: $2x^4$ and $-3x^4$   Same variable and
Like terms: $5x$ and $-7x$   exponent

**27.** $3x^5 - 7x + 8 + 14x^5 - 2x - 9$

Like terms: $3x^5$ and $14x^5$
Like terms: $-7x$ and $-2x$
Like terms: $8$ and $-9$   Constant terms
are like terms.

**29.** $-3x + 6$

The coefficient of $-3x$, the first term, is $-3$.

The coefficient of $6$, the second term, is $6$.

**31.** $5x^2 + 3x + 3$

The coefficient of $5x^2$, the first term, is $5$.

The coefficient of $3x$, the second term, is $3$.

The coefficient of $3$, the third term, is $3$.

**33.** $-5x^4 + 6x^3 - 3x^2 + 8x - 2$

The coefficient of $-5x^4$, the first term, is $-5$.

The coefficient of $6x^3$, the second term, is $6$.

The coefficient of $-3x^2$, the third term, is $-3$.

The coefficient of $8x$, the fourth term, is $8$.

The coefficient of $-2$, the fifth term, is $-2$.

**35.** $2x - 5x = (2 - 5)x = -3x$

**37.** $x - 9x = 1x - 9x = (1 - 9)x = -8x$

**39.** $5x^3 + 6x^3 + 4 = (5 + 6)x^3 + 4 = 11x^3 + 4$

**41.** $5x^3 + 6x - 4x^3 - 7x = (5 - 4)x^3 + (6 - 7)x =$
$1x^3 + (-1)x = x^3 - x$

**43.** $6b^5 + 3b^2 - 2b^5 - 3b^2 = (6 - 2)b^5 + (3 - 3)b^2 =$
$4b^5 + 0b^2 = 4b^5$

**45.** $\frac{1}{4}x^5 - 5 + \frac{1}{2}x^5 - 2x - 37 =$
$\left(\frac{1}{4} + \frac{1}{2}\right)x^5 - 2x + (-5 - 37) = \frac{3}{4}x^5 - 2x - 42$

**47.** $6x^2 + 2x^4 - 2x^2 - x^4 - 4x^2 =$
$6x^2 + 2x^4 - 2x^2 - 1x^4 - 4x^2 =$
$(6 - 2 - 4)x^2 + (2 - 1)x^4 = 0x^2 + 1x^4 =$
$0 + x^4 = x^4$

**49.** $\frac{1}{4}x^3 - x^2 - \frac{1}{6}x^2 + \frac{3}{8}x^3 + \frac{5}{16}x^3 =$
$\frac{1}{4}x^3 - 1x^2 - \frac{1}{6}x^2 + \frac{3}{8}x^3 + \frac{5}{16}x^3 =$
$\left(\frac{1}{4} + \frac{3}{8} + \frac{5}{16}\right)x^3 + \left(-1 - \frac{1}{6}\right)x^2 =$
$\left(\frac{4}{16} + \frac{6}{16} + \frac{5}{16}\right)x^3 + \left(-\frac{6}{6} - \frac{1}{6}\right)x^2 = \frac{15}{16}x^3 - \frac{7}{6}x^2$

**51.** $x^5 + x + 6x^3 + 1 + 2x^2 = x^5 + 6x^3 + 2x^2 + x + 1$

**53.** $5y^3 + 15y^9 + y - y^2 + 7y^8 =$
$15y^9 + 7y^8 + 5y^3 - y^2 + y$

**55.** $3x^4 - 5x^6 - 2x^4 + 6x^6 = x^4 + x^6 = x^6 + x^4$

**57.** $-2x + 4x^3 - 7x + 9x^3 + 8 = -9x + 13x^3 + 8 =$
$13x^3 - 9x + 8$

**59.** $3x + 3x + 3x - x^2 - 4x^2 = 9x - 5x^2 = -5x^2 + 9x$

**61.** $-x + \frac{3}{4} + 15x^4 - x - \frac{1}{2} - 3x^4 = -2x + \frac{1}{4} + 12x^4 =$
$12x^4 - 2x + \frac{1}{4}$

**63.** $2x - 4 = 2x^1 - 4x^0$

The degree of $2x$ is $1$.

The degree of $-4$ is $0$.

The degree of the polynomial is $1$, the largest exponent.

**65.** $3x^2 - 5x + 2 = 3x^2 - 5x^1 + 2x^0$

The degree of $3x^2$ is $2$.

The degree of $-5x$ is $1$.

The degree of $2$ is $0$.

The degree of the polynomial is $2$, the largest exponent.

**67.** $-7x^3 + 6x^2 + 3x + 7 = -7x^3 + 6x^2 + 3x^1 + 7x^0$

The degree of $-7x^3$ is $3$.

The degree of $6x^2$ is $2$.

The degree of $3x$ is $1$.

The degree of $7$ is $0$.

The degree of the polynomial is $3$, the largest exponent.

**69.** $x^2 - 3x + x^6 - 9x^4 = x^2 - 3x^1 + x^6 - 9x^4$

The degree of $x^2$ is $2$.

The degree of $-3x$ is $1$.

The degree of $x^6$ is $6$.

The degree of $-9x^4$ is $4$.

The degree of the polynomial is $6$, the largest exponent.

**71.** See the answer section in the text.

**73.** In the polynomial $x^3 - 27$, there are no $x^2$ or $x$ terms. The $x^2$ term (or second-degree term) and the $x$ term (or first-degree term) are missing.

**75.** In the polynomial $x^4 - x$, there are no $x^3$, $x^2$, or $x^0$ terms. The $x^3$ term (or third-degree term), the $x^2$ term (or second-degree term), and the $x^0$ term (or zero-degree term) are missing.

**77.** No terms are missing in the polynomial $2x^3 - 5x^2 + x - 3$.

**79.** The polynomial $x^2 - 10x + 25$ is a <u>trinomial</u> because it has just three terms.

**81.** The polynomial $x^3 - 7x^2 + 2x - 4$ is <u>none of these</u> because it has more than three terms.

**83.** The polynomial $4x^2 - 25$ is a <u>binomial</u> because it has just two terms.

**85.** The polynomial $40x$ is a <u>monomial</u> because it has just one term.

**87.** *Familiarize.* Let $a =$ the number of apples the campers had to begin with. Then the first camper ate $\frac{1}{3}a$ apples and $a - \frac{1}{3}a$, or $\frac{2}{3}a$, apples were left. The second camper ate $\frac{1}{3}\left(\frac{2}{3}a\right)$, or $\frac{2}{9}a$, apples, and $\frac{2}{3}a - \frac{2}{9}a$, or $\frac{4}{9}a$, apples were left. The third camper ate $\frac{1}{3}\left(\frac{4}{9}a\right)$, or $\frac{4}{27}a$, apples, and $\frac{4}{9}a - \frac{4}{27}a$, or $\frac{8}{27}a$, apples were left.

*Translate.* We write an equation for the number of apples left after the third camper eats.

Number of apples left is 8.

$$\frac{8}{27}a = 8$$

*Solve.* We solve the equation.

$$\frac{8}{27}a = 8$$
$$a = \frac{27}{8} \cdot 8$$
$$a = 27$$

*Check.* If the campers begin with 27 apples, then the first camper eats $\frac{1}{3} \cdot 27$, or 9, and $27 - 9$, or 18, are left. The second camper then eats $\frac{1}{3} \cdot 18$, or 6 apples and $18 - 6$, or 12, are left. Finally, the third camper eats $\frac{1}{3} \cdot 12$, or 4 apples and $12 - 4$, or 8, are left. The answer checks.

*State.* The campers had 27 apples to begin with.

**89.** $\frac{1}{8} - \frac{5}{6} = \frac{1}{8} + \left(-\frac{5}{6}\right)$, LCM is 24
$$= \frac{1}{8} \cdot \frac{3}{3} + \left(-\frac{5}{6}\right)\left(\frac{4}{4}\right)$$
$$= \frac{3}{24} + \left(-\frac{20}{24}\right)$$
$$= -\frac{17}{24}$$

**91.** $5.6 - 8.2 = 5.6 + (-8.2) = -2.6$

**93.** $\quad (3x^2)^3 + 4x^2 \cdot 4x^4 - x^4(2x)^2 + [(2x)^2]^3 - 100x^2(x^2)^2$
$$= 27x^6 + 4x^2 \cdot 4x^4 - x^4 \cdot 4x^2 + (2x)^6 - 100x^2 \cdot x^4$$
$$= 27x^6 + 16x^6 - 4x^6 + 64x^6 - 100x^6$$
$$= 3x^6$$

**95.** $(5m^5)^2 = 5^2 m^{5 \cdot 2} = 25m^{10}$
The degree is 10.

---

## Exercise Set 3.4

**1.** $(3x + 2) + (-4x + 3) = (3 - 4)x + (2 + 3) = -x + 5$

**3.** $(-6x + 2) + (x^2 + x - 3) =$
$x^2 + (-6 + 1)x + (2 - 3) = x^2 - 5x - 1$

**5.** $(x^2 - 9) + (x^2 + 9) = (1 + 1)x^2 + (-9 + 9) = 2x^2$

**7.** $(3x^2 - 5x + 10) + (2x^2 + 8x - 40) =$
$(3 + 2)x^2 + (-5 + 8)x + (10 - 40) = 5x^2 + 3x - 30$

**9.** $(1.2x^3 + 4.5x^2 - 3.8x) + (-3.4x^3 - 4.7x^2 + 23) =$
$(1.2 - 3.4)x^3 + (4.5 - 4.7)x^2 - 3.8x + 23 =$
$-2.2x^3 - 0.2x^2 - 3.8x + 23$

**11.** $(1 + 4x + 6x^2 + 7x^3) + (5 - 4x + 6x^2 - 7x^3) =$
$(1 + 5) + (4 - 4)x + (6 + 6)x^2 + (7 - 7)x^3 =$
$6 + 0x + 12x^2 + 0x^3 = 6 + 12x^2$, or $12x^2 + 6$

**13.** $\left(\frac{1}{4}x^4 + \frac{2}{3}x^3 + \frac{5}{8}x^2 + 7\right) + \left(-\frac{3}{4}x^4 + \frac{3}{8}x^2 - 7\right) =$
$\left(\frac{1}{4} - \frac{3}{4}\right)x^4 + \frac{2}{3}x^3 + \left(\frac{5}{8} + \frac{3}{8}\right)x^2 + (7 - 7) =$
$-\frac{2}{4}x^4 + \frac{2}{3}x^3 + \frac{8}{8}x^2 + 0 =$
$-\frac{1}{2}x^4 + \frac{2}{3}x^3 + x^2$

**15.** $(0.02x^5 - 0.2x^3 + x + 0.08) + (-0.01x^5 + x^4 - 0.8x - 0.02) =$
$(0.02 - 0.01)x^5 + x^4 - 0.2x^3 + (1 - 0.8)x + (0.08 - 0.02) =$
$0.01x^5 + x^4 - 0.2x^3 + 0.2x + 0.06$

**17.** $9x^8 - 7x^4 + 2x^2 + 5) + (8x^7 + 4x^4 - 2x) +$
$(-3x^4 + 6x^2 + 2x - 1) = 9x^8 + 8x^7 + (-7 + 4 - 3)x^4 +$
$(2 + 6)x^2 + (-2 + 2)x + (5 - 1) =$
$9x^8 + 8x^7 - 6x^4 + 8x^2 + 4$

**19.** Rewrite the problem so the coefficients of like terms have the same number of decimal places.

$$
\begin{array}{r}
0.15x^4 + 0.10x^3 - 0.90x^2 \\
- 0.01x^3 + 0.01x^2 + x \\
1.25x^4 \qquad\qquad + 0.11x^2 \qquad\quad + 0.01 \\
0.27x^3 \qquad\qquad\qquad + 0.99 \\
-0.35x^4 \qquad\qquad + 15.00x^2 \qquad\quad - 0.03 \\
\hline
1.05x^4 + 0.36x^3 + 14.22x^2 + x + 0.97
\end{array}
$$

**21.** Two equivalent expressions for the additive inverse of $-5x$ are

   a) $-(-5x)$ and

   b) $5x$. (Changing the sign)

**23.** Two equivalent expressions for the additive inverse of $-x^2 + 10x - 2$ are

   a) $-(-x^2 + 10x - 2)$ and

   b) $x^2 - 10x + 2$. (Changing the sign of every term)

**25.** Two equivalent expressions for the additive inverse of $12x^4 - 3x^3 + 3$ are

   a) $-(12x^4 - 3x^3 + 3)$ and

   b) $-12x^4 + 3x^3 - 3$. (Changing the sign of every term)

**27.** We change the sign of every term inside parentheses.

   $-(3x - 7) = -3x + 7$

**29.** We change the sign of every term inside parentheses.

   $-(4x^2 - 3x + 2) = -4x^2 + 3x - 2$

**31.** We change the sign of every term inside parentheses.

   $-\left(-4x^4 + 6x^2 + \dfrac{3}{4}x - 8\right) = 4x^4 - 6x^2 - \dfrac{3}{4}x + 8$

**33.** $(3x + 2) - (-4x + 3) = 3x + 2 + 4x - 3$

$$\text{Changing the sign of every term inside parentheses}$$
$$= 7x - 1$$

**35.** $(-6x + 2) - (x^2 + x - 3) = -6x + 2 - x^2 - x + 3$

$$= -x^2 - 7x + 5$$

**37.** $(x^2 - 9) - (x^2 + 9) = x^2 - 9 - x^2 - 9 = -18$

**39.** $(6x^4 + 3x^3 - 1) - (4x^2 - 3x + 3)$

$= 6x^4 + 3x^3 - 1 - 4x^2 + 3x - 3$

$= 6x^4 + 3x^3 - 4x^2 + 3x - 4$

**41.** $(1.2x^3 + 4.5x^2 - 3.8x) - (-3.4x^3 - 4.7x^2 + 23)$

$= 1.2x^3 + 4.5x^2 - 3.8x + 3.4x^3 + 4.7x^2 - 23$

$= 4.6x^3 + 9.2x^2 - 3.8x - 23$

**43.** $\dfrac{5}{8}x^3 - \dfrac{1}{4}x - \dfrac{1}{3} - \left(-\dfrac{1}{8}x^3 + \dfrac{1}{4}x - \dfrac{1}{3}\right)$

$= \dfrac{5}{8}x^3 - \dfrac{1}{4}x - \dfrac{1}{3} + \dfrac{1}{8}x^3 - \dfrac{1}{4}x + \dfrac{1}{3}$

$= \dfrac{6}{8}x^3 - \dfrac{2}{4}x$

$= \dfrac{3}{4}x^3 - \dfrac{1}{2}x$

**45.** $(0.08x^3 - 0.02x^2 + 0.01x) - (0.02x^3 + 0.03x^2 - 1)$

$= 0.08x^3 - 0.02x^2 + 0.01x - 0.02x^3 - 0.03x^2 + 1$

$= 0.06x^3 - 0.05x^2 + 0.01x + 1$

**47.**
$$x^2 + 5x + 6$$
$$\underline{x^2 + 2x\phantom{+6}}$$

$$\begin{array}{ll} x^2 + 5x + 6 & \\ \underline{-x^2 - 2x\phantom{+6}} & \text{Changing signs} \\ \phantom{x^2+}3x + 6 & \text{Adding} \end{array}$$

**49.**
$$5x^4 + 6x^3 - 9x^2$$
$$\underline{-6x^4 - 6x^3 \phantom{-9x^2} + 8x + 9}$$

$$\begin{array}{ll} 5x^4 + 6x^3 - 9x^2 & \\ \underline{6x^4 + 6x^3 \phantom{-9x^2} - 8x - 9} & \text{Changing signs} \\ 11x^4 + 12x^3 - 9x^2 - 8x - 9 & \text{Adding} \end{array}$$

**51.**
$$x^5 \phantom{- x^4 + x^3 - x^2 + x} - 1$$
$$\underline{x^5 - x^4 + x^3 - x^2 + x - 1}$$

$$\begin{array}{ll} x^5 \phantom{- x^4 + x^3 - x^2 + x} - 1 & \\ \underline{-x^5 + x^4 - x^3 + x^2 - x + 1} & \text{Changing signs} \\ \phantom{-x^5 +}x^4 - x^3 + x^2 - x & \text{Adding} \end{array}$$

**53.**

The area of a rectangle is the product of the length and width. The sum of the areas is found as follows:

$$\begin{array}{ccccccc} \text{Area} & & \text{Area} & & \text{Area} & & \text{Area} \\ \text{of } A & + & \text{of } B & + & \text{of } C & + & \text{of } D \\ = 3x \cdot x & + & x \cdot x & + & x \cdot x & + & 4 \cdot x \\ = 3x^2 & + & x^2 & + & x^2 & + & 4x \\ = 5x^2 & + & 4x & & & & \end{array}$$

A polynomial for the sum of the areas is $5x^2 + 4x$.

**55.** We add the lengths of the sides:

$$4a + 7 + a + \dfrac{1}{2}a + 3 + a + 2a + 3a$$

$$= \left(4 + 1 + \dfrac{1}{2} + 1 + 2 + 3\right)a + (7 + 3)$$

$$= 11\dfrac{1}{2}a + 10, \text{ or } \dfrac{23}{2}a + 10$$

**57.** $8x + 3x = 66$

$\phantom{8x +}11x = 66$  Collecting like terms

$\dfrac{11x}{11} = \dfrac{66}{11}$  Dividing by 11

$\phantom{8x + 1}x = 6$

The solution is 6.

**59.**
$$\frac{3}{8}x + \frac{1}{4} - \frac{3}{4}x = \frac{11}{16} + x, \quad \text{LCM is 16}$$

$$16\left(\frac{3}{8}x + \frac{1}{4} - \frac{3}{4}x\right) = 16\left(\frac{11}{16} + x\right) \quad \text{Clearing fractions}$$

$$6x + 4 - 12x = 11 + 16x$$

$$-6x + 4 = 11 + 16x \qquad \text{Collecting like terms}$$

$$-6x + 4 - 4 = 11 + 16x - 4 \quad \text{Subtracting 4}$$

$$-6x = 7 + 16x$$

$$-6x - 16x = 7 + 16x - 16x \quad \text{Subtracting } 16x$$

$$-22x = 7$$

$$\frac{-22x}{-22} = \frac{7}{-22} \qquad \text{Dividing by } -22$$

$$x = -\frac{7}{22}$$

The solution is $-\dfrac{7}{22}$.

**61.**
$$6(y - 3) - 8 = 4(y + 2) + 5$$

$$6y - 18 - 8 = 4y + 8 + 5 \qquad \text{Removing parentheses}$$

$$6y - 26 = 4y + 13 \qquad \text{Collecting like terms}$$

$$6y - 26 + 26 = 4y + 13 + 26 \quad \text{Adding 26}$$

$$6y = 4y + 39$$

$$6y - 4y = 4y + 39 - 4y \quad \text{Subtracting } 4y$$

$$2y = 39$$

$$\frac{2y}{2} = \frac{39}{2} \qquad \text{Dividing by 2}$$

$$y = \frac{39}{2}$$

The solution is $\dfrac{39}{2}$.

**63.**

|        | 4     | $m-4$  |        |
|--------|-------|--------|--------|
| 5      | A     | B      | 5      |
| $m-5$  | C     | D      | $m-5$  |
|        | 4     | $m-4$  |        |

We can add the areas of the four rectangles $A$, $B$, $C$, and $D$.

$5 \cdot 4 + 5(m - 4) + 4(m - 5) + (m - 4)(m - 5)$, or

$20 + 5(m - 4) + 4(m - 5) + (m - 4)(m - 5)$

The length and width of the figure can also be expressed as $4 + (m - 4)$ and $5 + (m - 5)$, or $m$ and $m$. Then the area can be expressed as $m \cdot m$, or $m^2$.

**65.**

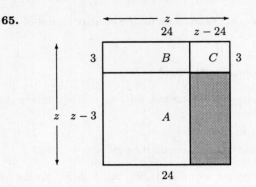

We label the sides $A$, $B$, and $C$ with additional information. The area of the square is $z \cdot z$, or $z^2$. The area of the shaded section is $z^2$ minus the areas of sections $A$, $B$, and $C$.

| Area of shaded section | = | Area of square | − | Area of $A$ | − | Area of $B$ | − | Area of $C$ |
|---|---|---|---|---|---|---|---|---|

$$\begin{aligned}
\text{Area of shaded section} &= z \cdot z - 24(z - 3) - 3 \cdot 24 - 3(z - 24)\\
&= z^2 - 24z + 72 - 72 - 3z + 72\\
&= z^2 - 27z + 72
\end{aligned}$$

A polynomial for the shaded area is $z^2 - 27z + 72$.

**67.**
$$\begin{aligned}
&(3x^2 - 4x + 6) - (-2x^2 + 4) + (-5x - 3)\\
&= 3x^2 - 4x + 6 + 2x^2 - 4 - 5x - 3\\
&= 5x^2 - 9x - 1
\end{aligned}$$

**69.**
$$\begin{aligned}
&(-4 + x^2 + 2x^3) - (-6 - x + 3x^3) - (-x^2 - 5x^3)\\
&= -4 + x^2 + 2x^3 + 6 + x - 3x^3 + x^2 + 5x^3\\
&= 4x^3 + 2x^2 + x + 2
\end{aligned}$$

## Exercise Set 3.5

**1.** $(8x^2)(5) = (8 \cdot 5)x^2 = 40x^2$

**3.** $(-x^2)(-x) = (-1x^2)(-1x) = (-1)(-1)(x^2 \cdot x) = x^3$

**5.** $(8x^5)(4x^3) = (8 \cdot 4)(x^5 \cdot x^3) = 32x^8$

**7.** $(0.1x^6)(0.3x^5) = (0.1)(0.3)(x^6 \cdot x^5) = 0.03x^{11}$

**9.** $\left(-\frac{1}{5}x^3\right)\left(-\frac{1}{3}x\right) = \left(-\frac{1}{5}\right)\left(-\frac{1}{3}\right)(x^3 \cdot x) = \frac{1}{15}x^4$

**11.** $(-4x^2)(0) = 0 \quad$ Any number multiplied by 0 is 0.

**13.** $(3x^2)(-4x^3)(2x^6) = (3)(-4)(2)(x^2 \cdot x^3 \cdot x^6) = -24x^{11}$

**15.** $\begin{aligned}[t] 2x(-x + 5) &= 2x(-x) + 2x(5)\\ &= -2x^2 + 10x \end{aligned}$

**17.** $\begin{aligned}[t] -5x(x - 1) &= -5x(x) - 5x(-1)\\ &= -5x^2 + 5x \end{aligned}$

**19.** $x^2(x^3 + 1) = x^2(x^3) + x^2(1)$
$= x^5 + x^2$

**21.** $3x(2x^2 - 6x + 1) = 3x(2x^2) + 3x(-6x) + 3x(1)$
$= 6x^3 - 18x^2 + 3x$

**23.** $-6x^2(x^2 + x) = -6x^2(x^2) - 6x^2(x)$
$= -6x^4 - 6x^3$

**25.** $3y^2(6y^4 + 8y^3) = 3y^2(6y^4) + 3y^2(8y^3)$
$= 18y^6 + 24y^5$

**27.** $(x + 6)(x + 3) = (x + 6)x + (x + 6)3$
$= x \cdot x + 6 \cdot x + x \cdot 3 + 6 \cdot 3$
$= x^2 + 6x + 3x + 18$
$= x^2 + 9x + 18$

**29.** $(x + 5)(x - 2) = (x + 5)x + (x + 5)(-2)$
$= x \cdot x + 5 \cdot x + x(-2) + 5(-2)$
$= x^2 + 5x - 2x - 10$
$= x^2 + 3x - 10$

**31.** $(x - 4)(x - 3) = (x - 4)x + (x - 4)(-3)$
$= x \cdot x - 4 \cdot x + x(-3) - 4(-3)$
$= x^2 - 4x - 3x + 12$
$= x^2 - 7x + 12$

**33.** $(x + 3)(x - 3) = (x + 3)x + (x + 3)(-3)$
$= x \cdot x + 3 \cdot x + x(-3) + 3(-3)$
$= x^2 + 3x - 3x - 9$
$= x^2 - 9$

**35.** $(5 - x)(5 - 2x) = (5 - x)5 + (5 - x)(-2x)$
$= 5 \cdot 5 - x \cdot 5 + 5(-2x) - x(-2x)$
$= 25 - 5x - 10x + 2x^2$
$= 25 - 15x + 2x^2$

**37.** $(2x + 5)(2x + 5) = (2x + 5)2x + (2x + 5)5$
$= 2x \cdot 2x + 5 \cdot 2x + 2x \cdot 5 + 5 \cdot 5$
$= 4x^2 + 10x + 10x + 25$
$= 4x^2 + 20x + 25$

**39.** $\left(x - \dfrac{5}{2}\right)\left(x + \dfrac{2}{5}\right) = \left(x - \dfrac{5}{2}\right)x + \left(x - \dfrac{5}{2}\right)\dfrac{2}{5}$
$= x \cdot x - \dfrac{5}{2} \cdot x + x \cdot \dfrac{2}{5} - \dfrac{5}{2} \cdot \dfrac{2}{5}$
$= x^2 - \dfrac{5}{2}x + \dfrac{2}{5}x - 1$
$= x^2 - \dfrac{25}{10}x + \dfrac{4}{10}x - 1$
$= x^2 - \dfrac{21}{10}x - 1$

**41.** $(x^2 + x + 1)(x - 1)$
$= (x^2 + x + 1)x + (x^2 + x + 1)(-1)$
$= x^2 \cdot x + x \cdot x + 1 \cdot x + x^2(-1) + x(-1) + 1(-1)$
$= x^3 + x^2 + x - x^2 - x - 1$
$= x^3 - 1$

**43.** $(2x + 1)(2x^2 + 6x + 1)$
$= 2x(2x^2 + 6x + 1) + 1(2x^2 + 6x + 1)$
$= 2x \cdot 2x^2 + 2x \cdot 6x + 2x \cdot 1 + 1 \cdot 2x^2 + 1 \cdot 6x + 1 \cdot 1$
$= 4x^3 + 12x^2 + 2x + 2x^2 + 6x + 1$
$= 4x^3 + 14x^2 + 8x + 1$

**45.** $(y^2 - 3)(3y^2 - 6y + 2)$
$= y^2(3y^2 - 6y + 2) - 3(3y^2 - 6y + 2)$
$= y^2 \cdot 3y^2 + y^2(-6y) + y^2 \cdot 2 - 3 \cdot 3y^2 - 3(-6y) - 3 \cdot 2$
$= 3y^4 - 6y^3 + 2y^2 - 9y^2 + 18y - 6$
$= 3y^4 - 6y^3 - 7y^2 + 18y - 6$

**47.** $(x^3 + x^2)(x^3 + x^2 - x)$
$= x^3(x^3 + x^2 - x) + x^2(x^3 + x^2 - x)$
$= x^3 \cdot x^3 + x^3 \cdot x^2 + x^3(-x) + x^2 \cdot x^3 + x^2 \cdot x^2 + x^2(-x)$
$= x^6 + x^5 - x^4 + x^5 + x^4 - x^3$
$= x^6 + 2x^5 - x^3$

**49.** $(-5x^3 - 7x^2 + 1)(2x^2 - x)$
$= (-5x^3 - 7x^2 + 1)2x^2 + (-5x^3 - 7x^2 + 1)(-x)$
$= -5x^3 \cdot 2x^2 - 7x^2 \cdot 2x^2 + 1 \cdot 2x^2 - 5x^3(-x) - 7x^2(-x) + 1(-x)$
$= -10x^5 - 14x^4 + 2x^2 + 5x^4 + 7x^3 - x$
$= -10x^5 - 9x^4 + 7x^3 + 2x^2 - x$

**51.**

| | |
|---|---|
| $1 + x + x^2$ | Line up like terms |
| $-\,1 - x + x^2$ | in columns |
| $\overline{x^2 + x^3 + x^4}$ | Multiplying the top row by $x^2$ |
| $-\,x - x^2 - x^3$ | Multiplying by $-x$ |
| $\underline{-1 - x - x^2}$ | Multiplying by $-1$ |
| $-1 - 2x - x^2 \qquad + x^4$ | |

**53.**

| | |
|---|---|
| $2t^2 - t - 4$ | |
| $3t^2 + 2t - 1$ | |
| $-\,2t^2 + t + 4$ | Multiplying by $-1$ |
| $4t^3 - 2t^2 - 8t$ | Multiplying by $2t$ |
| $\underline{6t^4 - 3t^3 - 12t^2}$ | Multiplying by $3t^2$ |
| $6t^4 + t^3 - 16t^2 - 7t + 4$ | |

**55.**

| | |
|---|---|
| $x \qquad -x^3 \qquad + x^5$ | |
| $-1 + x^2 \qquad + x^4$ | Rewriting in ascending order |
| $\overline{x^5 - x^7 + x^9}$ | Multiplying by $x^4$ |
| $x^3 - x^5 + x^7$ | Multiplying by $x^2$ |
| $\underline{-x + x^3 - x^5}$ | Multiplying by $-1$ |
| $-x + 2x^3 - x^5 \qquad + x^9$ | |

**57.**

| |
|---|
| $x^3 + x^2 + x + 1$ |
| $x - 1$ |
| $\overline{-x^3 - x^2 - x - 1}$ |
| $\underline{x^4 + x^3 + x^2 + x}$ |
| $x^4 \qquad\qquad - 1$ |

**59.** $-\dfrac{1}{4} - \dfrac{1}{2} = -\dfrac{1}{4} - \dfrac{1}{2} \cdot \dfrac{2}{2} = -\dfrac{1}{4} - \dfrac{2}{4} = -\dfrac{3}{4}$

**61.** $9x - 45y + 15 = 3 \cdot 3x - 3 \cdot 15y + 3 \cdot 5$
$$= 3(3x - 15y + 5)$$

**63.** The shaded area is the area of the large rectangle less the area of the small rectangle:

$$4t(21t + 8) - 2t(3t - 4) = 84t^2 + 32t - 6t^2 + 8t$$
$$= 78t^2 + 40t$$

**65.** Let $b$ = the length of the base.  The $b + 4$ = the height.  Let $A$ represent the area.

$$\text{Area} = \frac{1}{2} \times \text{base} \times \text{height}$$
$$A = \frac{1}{2} \cdot b \cdot (b + 4)$$
$$A = \frac{1}{2}b(b + 4)$$
$$A = \frac{1}{2}b^2 + 2b$$

**67.**     $(x - 2)(x - 7) - (x - 2)(x - 7)$
$$= x(x - 7) - 2(x - 7) - [x(x - 7) - 2(x - 7)]$$
$$= x^2 - 7x - 2x + 14 - (x^2 - 7x - 2x + 14)$$
$$= x^2 - 9x + 14 - (x^2 - 9x + 14)$$
$$= x^2 - 9x + 14 - x^2 + 9x - 14$$
$$= 0$$

## Exercise Set 3.6

**1.**     $(x + 1)(x^2 + 3)$
$$\phantom{=}\ \text{F}\quad\ \text{O}\quad\ \text{I}\quad\ \text{L}$$
$$= x \cdot x^2 + x \cdot 3 + 1 \cdot x^2 + 1 \cdot 3$$
$$= x^3 + 3x + x^2 + 3$$

**3.**     $(x^3 + 2)(x + 1)$
$$\phantom{=}\ \text{F}\quad\ \text{O}\quad\ \text{I}\quad\ \text{L}$$
$$= x^3 \cdot x + x^3 \cdot 1 + 2 \cdot x + 2 \cdot 1$$
$$= x^4 + x^3 + 2x + 2$$

**5.**     $(y + 2)(y - 3)$
$$\phantom{=}\ \text{F}\quad\ \ \text{O}\quad\ \ \text{I}\quad\ \ \text{L}$$
$$= y \cdot y + y \cdot (-3) + 2 \cdot y + 2 \cdot (-3)$$
$$= y^2 - 3y + 2y - 6$$
$$= y^2 - y - 6$$

**7.**     $(3x + 2)(3x + 2)$
$$\phantom{=}\ \text{F}\qquad\ \text{O}\qquad\ \text{I}\qquad\ \text{L}$$
$$= 3x \cdot 3x + 3x \cdot 2 + 2 \cdot 3x + 2 \cdot 2$$
$$= 9x^2 + 6x + 6x + 4$$
$$= 9x^2 + 12x + 4$$

**9.**     $(5x - 6)(x + 2)$
$$\phantom{=}\ \text{F}\qquad\ \ \text{O}\qquad\ \ \text{I}\qquad\ \ \text{L}$$
$$= 5x \cdot x + 5x \cdot 2 + (-6) \cdot x + (-6) \cdot 2$$
$$= 5x^2 + 10x - 6x - 12$$
$$= 5x^2 + 4x - 12$$

**11.**     $(3t - 1)(3t + 1)$
$$\phantom{=}\ \text{F}\qquad\ \ \text{O}\qquad\ \ \text{I}\qquad\ \ \text{L}$$
$$= 3t \cdot 3t + 3t \cdot 1 + (-1) \cdot 3t + (-1) \cdot 1$$
$$= 9t^2 + 3t - 3t - 1$$
$$= 9t^2 - 1$$

**13.**     $(4x - 2)(x - 1)$
$$\phantom{=}\ \text{F}\qquad\ \ \text{O}\qquad\ \ \text{I}\qquad\ \ \text{L}$$
$$= 4x \cdot x + 4x \cdot (-1) + (-2) \cdot x + (-2) \cdot (-1)$$
$$= 4x^2 - 4x - 2x + 2$$
$$= 4x^2 - 6x + 2$$

**15.**     $\left(p - \dfrac{1}{4}\right)\left(p + \dfrac{1}{4}\right)$
$$\phantom{=}\ \text{F}\qquad\ \ \text{O}\qquad\ \ \text{I}\qquad\ \ \text{L}$$
$$= p \cdot p + p \cdot \frac{1}{4} + \left(-\frac{1}{4}\right) \cdot p + \left(-\frac{1}{4}\right) \cdot \frac{1}{4}$$
$$= p^2 + \frac{1}{4}p - \frac{1}{4}p - \frac{1}{16}$$
$$= p^2 - \frac{1}{16}$$

**17.**     $(x - 0.1)(x + 0.1)$
$$\phantom{=}\ \text{F}\qquad\ \ \text{O}\qquad\ \ \text{I}\qquad\ \ \text{L}$$
$$= x \cdot x + x \cdot (0.1) + (-0.1) \cdot x + (-0.1)(0.1)$$
$$= x^2 + 0.1x - 0.1x - 0.01$$
$$= x^2 - 0.01$$

**19.**     $(2x^2 + 6)(x + 1)$
$$\phantom{=}\ \text{F}\quad\ \text{O}\quad\ \text{I}\quad\ \text{L}$$
$$= 2x^3 + 2x^2 + 6x + 6$$

**21.**     $(-2x + 1)(x + 6)$
$$\phantom{=}\ \text{F}\quad\ \text{O}\quad\ \text{I}\quad\ \text{L}$$
$$= -2x^2 - 12x + x + 6$$
$$= -2x^2 - 11x + 6$$

**23.**     $(a + 7)(a + 7)$
$$\phantom{=}\ \text{F}\quad\ \text{O}\quad\ \text{I}\quad\ \text{L}$$
$$= a^2 + 7a + 7a + 49$$
$$= a^2 + 14a + 49$$

**25.**     $(1 + 2x)(1 - 3x)$
$$\phantom{=}\ \text{F}\quad\ \text{O}\quad\ \text{I}\quad\ \text{L}$$
$$= 1 - 3x + 2x - 6x^2$$
$$= 1 - x - 6x^2$$

**27.**     $(x^2 + 3)(x^3 - 1)$
$$\phantom{=}\ \text{F}\quad\ \text{O}\quad\ \text{I}\quad\ \text{L}$$
$$= x^5 - x^2 + 3x^3 - 3$$

**29.**     $(3x^2 - 2)(x^4 - 2)$
$$\phantom{=}\ \text{F}\quad\ \text{O}\quad\ \text{I}\quad\ \text{L}$$
$$= 3x^6 - 6x^2 - 2x^4 + 4$$

**31.**     $(3x^5 + 2)(2x^2 + 6)$
$$\phantom{=}\ \text{F}\quad\ \text{O}\quad\ \text{I}\quad\ \text{L}$$
$$= 6x^7 + 18x^5 + 4x^2 + 12$$

**33.** $(8x^3 + 1)(x^3 + 8)$
$\quad\quad$ F $\quad$ O $\quad$ I $\quad$ L
$= 8x^6 + 64x^3 + x^3 + 8$
$= 8x^6 + 65x^3 + 8$

**35.** $(4x^2 + 3)(x - 3)$
$\quad\quad$ F $\quad$ O $\quad$ I $\quad$ L
$= 4x^3 - 12x^2 + 3x - 9$

**37.** $(4y^4 + y^2)(y^2 + y)$
$\quad\quad$ F $\quad$ O $\quad$ I $\quad$ L
$= 4y^6 + 4y^5 + y^4 + y^3$

**39.** $(x + 4)(x - 4)$ $\quad$ Product of sum and
$\quad\quad\quad\quad\quad\quad\quad\quad$ difference of two terms
$= x^2 - 4^2$
$= x^2 - 16$

**41.** $(2x + 1)(2x - 1)$ $\quad$ Product of sum and
$\quad\quad\quad\quad\quad\quad\quad\quad\quad$ difference of two terms
$= (2x)^2 - 1^2$
$= 4x^2 - 1$

**43.** $(5m - 2)(5m + 2)$ $\quad$ Product of sum and
$\quad\quad\quad\quad\quad\quad\quad\quad\quad$ difference of two terms
$= (5m)^2 - 2^2$
$= 25m^2 - 4$

**45.** $(2x^2 + 3)(2x^2 - 3)$ $\quad$ Product of sum and
$\quad\quad\quad\quad\quad\quad\quad\quad\quad$ difference of two terms
$= (2x^2)^2 - 3^2$
$= 4x^4 - 9$

**47.** $(3x^4 - 4)(3x^4 + 4)$
$= (3x^4)^2 - 4^2$
$= 9x^8 - 16$

**49.** $(x^6 - x^2)(x^6 + x^2)$
$= (x^6)^2 - (x^2)^2$
$= x^{12} - x^4$

**51.** $(x^4 + 3x)(x^4 - 3x)$
$= (x^4)^2 - (3x)^2$
$= x^8 - 9x^2$

**53.** $(x^{12} - 3)(x^{12} + 3)$
$= (x^{12})^2 - 3^2$
$= x^{24} - 9$

**55.** $(2y^8 + 3)(2y^8 - 3)$
$= (2y^8)^2 - 3^2$
$= 4y^{16} - 9$

**57.** $(x + 2)^2 = x^2 + 2 \cdot x \cdot 2 + 2^2$ $\quad$ Square of a binomial
$\quad\quad\quad\quad\quad\quad\quad\quad\quad\quad\quad\quad\quad\quad\quad$ sum
$\quad\quad\quad = x^2 + 4x + 4$

**59.** $(3x^2 + 1)$ $\quad$ Square of a binomial sum
$= (3x^2)^2 + 2 \cdot 3x^2 \cdot 1 + 1^2$
$= 9x^4 + 6x^2 + 1$

**61.** $\left(a - \dfrac{1}{2}\right)^2$ $\quad$ Square of a binomial sum
$= a^2 - 2 \cdot a \cdot \dfrac{1}{2} + \left(\dfrac{1}{2}\right)^2$
$= a^2 - a + \dfrac{1}{4}$

**63.** $(3 + x)^2 = 3^2 + 2 \cdot 3 \cdot x + x^2$
$\quad\quad\quad\quad\quad = 9 + 6x + x^2$

**65.** $(x^2 + 1)^2 = (x^2)^2 + 2 \cdot x^2 \cdot 1 + 1^2$
$\quad\quad\quad\quad\quad\quad = x^4 + 2x^2 + 1$

**67.** $(2 - 3x^4)^2 = 2^2 - 2 \cdot 2 \cdot 3x^4 + (3x^4)^2$
$\quad\quad\quad\quad\quad\quad = 4 - 12x^4 + 9x^8$

**69.** $(5 + 6t^2)^2 = 5^2 + 2 \cdot 5 \cdot 6t^2 + (6t^2)^2$
$\quad\quad\quad\quad\quad\quad = 25 + 60t^2 + 36t^4$

**71.** $(3 - 2x^3)^2 = 3^2 - 2 \cdot 3 \cdot 2x^3 + (2x^3)^2$
$\quad\quad\quad\quad\quad\quad = 9 - 12x^3 + 4x^6$

**73.** $4x(x^2 + 6x - 3)$ $\quad$ Product of a monomial and
$\quad\quad\quad\quad\quad\quad\quad\quad\quad\quad$ a trinomial
$= 4x \cdot x^2 + 4x \cdot 6x + 4x(-3)$
$= 4x^3 + 24x^2 - 12x$

**75.** $\left(2x^2 - \dfrac{1}{2}\right)\left(2x^2 - \dfrac{1}{2}\right)$ $\quad$ Square of a binomial
$\quad\quad\quad\quad\quad\quad\quad\quad\quad\quad\quad\quad$ difference
$= (2x^2)^2 - 2 \cdot 2x^2 \cdot \dfrac{1}{2} + \left(\dfrac{1}{2}\right)^2$
$= 4x^4 - 2x^2 + \dfrac{1}{4}$

**77.** $(-1 + 3p)(1 + 3p)$
$= (3p - 1)(3p + 1)$ $\quad$ Product of the sum and
$\quad\quad\quad\quad\quad\quad\quad\quad$ difference of two terms
$= (3p)^2 - 1^2$
$= 9p^2 - 1$

**79.** $3t^2(5t^3 - t^2 + t)$ $\quad$ Product of a monomial and
$\quad\quad\quad\quad\quad\quad\quad\quad\quad\quad$ a trinomial
$= 3t^2 \cdot 5t^3 + 3t^2(-t^2) + 3t^2 \cdot t$
$= 15t^5 - 3t^4 + 3t^3$

**81.** $(6x^4 + 4)^2$ $\quad\quad\quad\quad$ Square of a binomial sum
$= (6x^4)^2 + 2 \cdot 6x^4 \cdot 4 + 4^2$
$= 36x^8 + 48x^4 + 16$

**83.** $(3x + 2)(4x^2 + 5)$ $\quad$ Product of two binomials;
$\quad\quad\quad\quad\quad\quad\quad\quad\quad\quad$ use FOIL
$= 3x \cdot 4x^2 + 3x \cdot 5 + 2 \cdot 4x^2 + 2 \cdot 5$
$= 12x^3 + 15x + 8x^2 + 10$

**85.**     $(8 - 6x^4)^2$     Square of a binomial difference
$= 8^2 - 2 \cdot 8 \cdot 6x^4 + (6x^4)^2$
$= 64 - 96x^4 + 36x^8$

**87.**
$$
\begin{array}{r}
t^2 + t + 1 \\
t - 1 \\
\hline
-t^2 - t - 1 \\
t^3 + t^2 + t \\
\hline
t^3 \qquad -1
\end{array}
$$

**89.** $3^2 + 4^2 = 9 + 16 = 25$
$(3 + 4)^2 = 7^2 = 49$

**91.** $9^2 - 5^2 = 81 - 25 = 56$
$(9 - 5)^2 = 4^2 = 16$

**93.** *Familiarize*. Let $t =$ the number of watts used by the television set. Then $10t =$ the number of watts used by the lamps, and $40t =$ the number of watts used by the air conditioner.

*Translate*.

| Lamp watts | + | Air conditioner watts | + | Television watts | = | Total watts |
|---|---|---|---|---|---|---|
| ↓ | ↓ | ↓ | ↓ | ↓ | ↓ | ↓ |
| $10t$ | $+$ | $40t$ | $+$ | $t$ | $=$ | $2550$ |

*Solve*. We solve the equation.
$$10t + 40t + t = 2550$$
$$51t = 2550$$
$$t = 50$$

The possible solution is:

Television, $t$: 50 watts

Lamps, $10t$: $10 \cdot 50$, or 500 watts

Air conditioner, $40t$: $40 \cdot 50$, or 2000 watts

*Check*. The number of watts used by the lamps, 500, is 10 times 50, the number used by the television. The number of watts used by the air conditioner, 2000, is 40 times 50, the number used by the television. Also, $50 + 500 + 2000 = 2550$, the total wattage used.

*State*. The television uses 50 watts, the lamps use 500 watts, and the air conditioner uses 2000 watts.

**95.**     $3(x - 2) = 5(2x + 7)$
$3x - 6 = 10x + 35$     Removing parentheses
$3x - 6 + 6 = 10x + 35 + 6$     Adding 6
$3x = 10x + 41$
$3x - 10x = 10x + 41 - 10x$     Subtracting $10x$
$-7x = 41$
$\dfrac{-7x}{-7} = \dfrac{41}{-7}$     Dividing by $-7$
$x = -\dfrac{41}{7}$

The solution is $-\dfrac{41}{7}$.

**97.**     $4y(y + 5)(2y + 8)$
$= 4y(2y^2 + 8y + 10y + 40)$
$= 4y(2y^2 + 18y + 40)$
$= 8y^3 + 72y^2 + 160y$

**99.**     $[(x + 1) - x^2][(x - 2) + 2x^2]$
$\qquad\quad$ F $\qquad\quad$ O $\qquad\quad$ I $\qquad\quad$ L
$= (x+1)(x-2) + (x+1)(2x^2) + (-x^2)(x-2) + (-x^2)(2x^2)$
$= x^2 - 2x + x - 2 + 2x^3 + 2x^2 - x^3 + 2x^2 - 2x^4$
$= -2x^4 + x^3 + 5x^2 - x - 2$

**101.**     $(x + 2)(x - 5) = (x + 1)(x - 3)$
$x^2 - 5x + 2x - 10 = x^2 - 3x + x - 3$
$x^2 - 3x - 10 = x^2 - 2x - 3$
$-3x - 10 = -2x - 3$     Adding $-x^2$
$-3x + 2x = 10 - 3$     Adding $2x$ and 10
$-x = 7$
$x = -7$

The solution is $-7$.

**103.** If $w =$ the width, then $w + 1 =$ the length, and $(w + 1) + 1$, or $w + 2 =$ the height.

Volume $=$ length $\times$ width $\times$ height
$= (w + 1) \cdot w \cdot (w + 2)$
$= (w^2 + w)(w + 2)$
$= w^3 + 2w^2 + w^2 + 2w$
$= w^3 + 3w^2 + 2w$

**105.** If $h =$ the height, then $h - 1 =$ the length, and $(h - 1) - 1$, or $h - 2 =$ the width.

Volume $=$ length $\times$ width $\times$ height
$= (h - 1) \cdot (h - 2) \cdot h$
$= (h^2 - 2h - h + 2)h$
$= (h^2 - 3h + 2)h$
$= h^3 - 3h^2 + 2h$

**107.**     $93 \times 107$
$= (100 - 7)(100 + 7)$
$= 100^2 - 7^2$
$= 10,000 - 49$
$= 9951$

**109.**     $[3a - (2a - 3)][3a + (2a - 3)]$
$= (3a)^2 - (2a - 3)^2$
$= 9a^2 - (4a^2 - 12a + 9)$
$= 9a^2 - 4a^2 + 12a - 9$
$= 5a^2 + 12a - 9$

**111.** a)

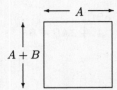

The area of the entire rectangle is $A(A+B)$, or $A^2 + AB$.

b)

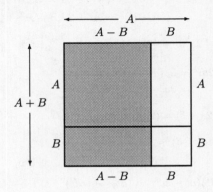

The sum of the areas of the two unshaded rectangles is $A \cdot B + B \cdot B$, or $AB + B^2$.

c)    Area in part (a) - area in part (b)

$= A^2 + AB - (AB + B^2)$

$= A^2 + AB - AB - B^2$

$= A^2 - B^2$

d) The area of the shaded region is $(A+B)(A-B) = A^2 - B^2$. This is the same as the polynomial found in part (c).

---

## Exercise Set 3.7

---

**1.** We replace $x$ by 3 and $y$ by $-2$.

$x^2 - y^2 + xy = 3^2 - (-2)^2 + 3(-2) = 9 - 4 - 6 = -1$

**3.** We replace $x$ by 2, $y$ by $-3$, and $z$ by $-1$.

$xyz^2 + z = 2(-3)(-1)^2 + (-1) = -6 - 1 = -7$

**5.** Evaluate the polynomial for $P = 10,000$ and $i = 0.06$.

$P(1+i)^2 = 10,000(1 + 0.06)^2$

$\qquad = 10,000(1.06)^2$

$\qquad = 10,000(1.1236)$

$\qquad = 11,236$

At 6% interest for 2 years, $10,000 will grow to $11,236.

**7.** Evaluate the polynomial for $P = 10,000$ and $i = 0.06$.

$P(1+i)^3 = 10,000(1 + 0.06)^3$

$\qquad = 10,000(1.06)^3$

$\qquad = 10,000(1.191016)$

$\qquad = 11,910.16$

At 6% interest for 3 years, $10,000 will grow to $11,910.16.

**9.** Replace $h$ by 6.3, $r$ by 1.2, and $\pi$ by 3.14.

$2\pi rh + 2\pi r^2 \approx 2(3.14)(1.2)(6.3) + 2(3.14)(1.2)^2$

$\qquad \approx 2(3.14)(1.2)(6.3) + 2(3.14)(1.44)$

$\qquad \approx 47.4768 + 9.0432$

$\qquad \approx 56.52$

The surface area of the can is about $56.52$ in$^2$.

**11.** $x^3y - 2xy + 3x^2 - 5$

| Term | Coefficient | Degree | |
|------|-------------|--------|--|
| $x^3y$ | 1 | 4 | (Think: $x^3y = x^3y^1$) |
| $-2xy$ | $-2$ | 2 | (Think: $-2xy = -2x^1y^1$) |
| $3x^2$ | 3 | 2 | |
| $-5$ | $-5$ | 0 | (Think: $-5 = -5x^0$) |

The degree of the polynomial is the degree of the term of highest degree.

The term of highest degree is $x^3y$. Its degree is 4. The degree of the polynomial is 4.

**13.** $17x^2y^3 - 3x^3yz - 7$

| Term | Coefficient | Degree | |
|------|-------------|--------|--|
| $17x^2y^3$ | 17 | 5 | |
| $-3x^3yz$ | $-3$ | 5 | (Think: $-3x^3yz = -3x^3y^1z^1$) |
| $-7$ | $-7$ | 0 | (Think: $-7 = -7x^0$) |

The terms of highest degree are $17x^2y^3$ and $-3x^3yz$. Each has degree 5. The degree of the polynomial is 5.

**15.** $a + b - 2a - 3b = (1-2)a + (1-3)b = -a - 2b$

**17.** $3x^2y - 2xy^2 + x^2$

There are <u>no</u> like terms, so none of the terms can be collected.

**19.**  $2u^2v - 3uv^2 + 6u^2v - 2uv^2$

$= (2+6)u^2v + (-3-2)uv^2$

$= 8u^2v - 5uv^2$

**21.**  $6au + 3av + 14au + 7av$

$= (6+14)au + (3+7)av$

$= 20au + 10av$

**23.**  $(2x^2 - xy + y^2) + (-x^2 - 3xy + 2y^2)$

$= (2-1)x^2 + (-1-3)xy + (1+2)y^2$

$= x^2 - 4xy + 3y^2$

**25.**  $(r - 2s + 3) + (2r + s) + (s + 4)$

$= (1+2)r + (-2+1+1)s + (3+4)$

$= 3r + 0s + 7$

$= 3r + 7$

**27.**  $(2x^2 - 3xy + y^2) + (-4x^2 - 6xy - y^2) + (x^2 + xy - y^2)$

$= (2-4+1)x^2 + (-3-6+1)xy + (1-1-1)y^2$

$= -x^2 - 8xy - y^2$

**29.**  $(xy - ab - 8) - (xy - 3ab - 6)$

$= xy - ab - 8 - xy + 3ab + 6$

$= (1-1)xy + (-1+3)ab + (-8+6)$

$= 2ab - 2$

**31.**    $(-2a + 7b - c) - (-3b + 4c - 8d)$
$= -2a + 7b - c + 3b - 4c + 8d$
$= -2a + (7 + 3)b + (-1 - 4)c + 8d$
$= -2a + 10b - 5c + 8d$

$\qquad\qquad\qquad$ F$\qquad$O$\qquad$I$\qquad$L
**33.**  $(3z - u)(2z + 3u) = 6z^2 + 9zu - 2uz - 3u^2$
$\qquad\qquad\qquad\qquad = 6z^2 + 7zu - 3u^2$

$\qquad\qquad\qquad$ F$\qquad$O$\qquad$I$\qquad$L
**35.**  $(a^2b - 2)(a^2b - 5) = a^4b^2 - 5a^2b - 2a^2b + 10$
$\qquad\qquad\qquad\qquad\quad = a^4b^2 - 7a^2b + \ 10$

**37.**    $(a + a^2 - 1)(a^2 + 1 - y)$
$= (a^2 + a - 1)(a^2 - y + 1)$

$\qquad\qquad\qquad\qquad a^2 + a - 1$
$\qquad\qquad\qquad\qquad a^2 - y + 1$
$\qquad\qquad\qquad\overline{\qquad a^2 + a - 1}$
$\qquad\qquad -a^2y - ay + y$
$\underline{a^4 + a^3 \qquad\qquad -a^2}$
$a^4 + a^3 - a^2y - ay + y \ \ +a - 1$

**39.**  $(a^3 + bc)(a^3 - bc) = (a^3)^2 - (bc)^2$
$\qquad\qquad [(A + B)(A - B) = A^2 - B^2]$
$\qquad\qquad\qquad\qquad = a^6 - b^2c^2$

**41.**
$\qquad\qquad\qquad y^4x + y^2 \ + 1$
$\qquad\qquad\qquad\qquad\quad y^2 \ + 1$
$\qquad\qquad\overline{\qquad y^4x + y^2 \ + 1}$
$\underline{y^6x + y^4 \qquad\ + y^2}$
$y^6x + y^4 + y^4x + 2y^2 + 1$

**43.**    $(3xy - 1)(4xy + 2)$
$\qquad\quad$ F$\qquad$O$\qquad$I$\qquad$L
$= 12x^2y^2 + 6xy - 4xy - 2$
$= 12x^2y^2 + 2xy - 2$

**45.**  $(3 - c^2d^2)(4 + c^2d^2)$
$\qquad\quad$ F$\qquad$O$\qquad$I$\qquad$L
$= 12 + 3c^2d^2 - 4c^2d^2 - c^4d^4$
$= 12 - c^2d^2 - c^4d^4$

**47.**  $(m^2 - n^2)(m + n)$
$\qquad\quad$ F$\qquad$O$\qquad$I$\qquad$L
$= m^3 + m^2n - mn^2 - n^3$

**49.**  $(xy + x^5y^5)(x^4y^4 - xy)$
$\qquad\quad$ F$\qquad$O$\qquad$I$\qquad$L
$= x^5y^5 - x^2y^2 + x^9y^9 - x^6y^6$
$= x^9y^9 - x^6y^6 + x^5y^5 - x^2y^2$

**51.**    $(x + h)^2$
$= x^2 + 2xh + h^2 \quad [(A + B)^2 = A^2 + 2AB + B^2]$

**53.**    $(r^3t^2 - 4)^2$
$= (r^3t^2)^2 - 2 \cdot r^3t^2 \cdot 4 + 4^2$
$\qquad\qquad [(A - B)^2 = A^2 - 2AB + B^2]$
$= r^6t^4 - 8r^3t^2 + 16$

**55.**    $(p^4 + m^2n^2)^2$
$= (p^4)^2 + 2 \cdot p^4 \cdot m^2n^2 + (m^2n^2)^2$
$\qquad\qquad [(A + B)^2 = A^2 + 2AB + B^2]$
$= p^8 + 2p^4m^2n^2 + m^4n^4$

**57.**    $\left(2a^3 - \dfrac{1}{2}b^3\right)^2$
$= (2a^3)^2 - 2 \cdot 2a^3 \cdot \dfrac{1}{2}b^3 + \left(\dfrac{1}{2}b^3\right)^2$
$\qquad\qquad [(A - B)^2 = A^2 - 2AB + B^2]$
$= 4a^6 - 2a^3b^3 + \dfrac{1}{4}b^6$

**59.**  $3a(a - 2b)^2 = 3a(a^2 - 4ab + 4b^2)$
$\qquad\qquad\qquad = 3a^3 - 12a^2b + 12ab^2$

**61.**  $(2a - b)(2a + b) = (2a)^2 - b^2 = 4a^2 - b^2$

**63.**  $(c^2 - d)(c^2 + d) = (c^2)^2 - d^2$
$\qquad\qquad\qquad\qquad = c^4 - d^2$

**65.**  $(ab + cd^2)(ab - cd^2) = (ab)^2 - (cd^2)^2$
$\qquad\qquad\qquad\qquad\qquad = a^2b^2 - c^2d^4$

**67.**    $(x + y - 3)(x + y + 3)$
$= [(x + y) - 3][(x + y) + 3]$
$= (x + y)^2 - 3^2$
$= x^2 + 2xy + y^2 - 9$

**69.**    $[x + y + z][x - (y + z)]$
$= [x + (y + z)][x - (y + z)]$
$= x^2 - (y + z)^2$
$= x^2 - (y^2 + 2yz + z^2)$
$= x^2 - y^2 - 2yz - z^2$

**71.**    $(a + b + c)(a - b - c)$
$= [a + (b + c)][a - (b + c)]$
$= a^2 - (b + c)^2$
$= a^2 - (b^2 + 2bc + c^2)$
$= a^2 - b^2 - 2bc - c^2$

**73.** It is helpful to add additional labels to the figure.

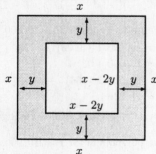

The area of the large square is $x \cdot x$, or $x^2$. The area of the small square is $(x - 2y)(x - 2y)$, or $(x - 2y)^2$.

$$\begin{array}{ccccc}
\text{Area of shaded} & = & \text{Area of large} & - & \text{Area of small}\\
\text{region} & & \text{square} & & \text{square}\\
\text{Area of shaded} & = & x^2 & - & (x-2y)^2\\
\text{region}
\end{array}$$

$$= x^2 - (x^2 - 4xy + 4y^2)$$
$$= x^2 - x^2 + 4xy - 4y^2$$
$$= 4xy - 4y^2$$

**75.** It is helpful to add additional labels to the figure.

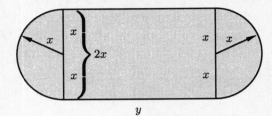

The two semicircles make a circle with radius $x$. The area of that circle is $\pi x^2$. The area of the rectangle is $2x \cdot y$. The sum of the two regions, $\pi x^2 + 2xy$, is the area of the shaded region.

**77.** Evaluate the polynomial for $G = 162$, $W_1 = 77$, and $L_2 = 53$.

$$G - W_1 - L_2 + 1 = 162 - 77 - 53 + 1 = 33$$

The magic number for the Philadelphia Phillies is 33.

**79.** $(A + B)^3 = (A + B)(A + B)^2$
$(A + B)^3 = (A + B)(A^2 + 2AB + B^2)$
$(A + B)^3 = (A + B)A^2 + (A + B)2AB + (A + B)^2$
$(A + B)^3 = A^3 + A^2B + 2A^2B + 2AB^2 + AB^2 + B^3$
$(A + B)^3 = A^3 + 3A^2B + 3AB^2 + B^3$

## Exercise Set 3.8

**1.** $\dfrac{24x^4 - 4x^3 + x^2 - 16}{8}$

$$= \frac{24x^4}{8} - \frac{4x^3}{8} + \frac{x^2}{8} - \frac{16}{8}$$
$$= 3x^4 - \frac{1}{2}x^3 + \frac{1}{8}x^2 - 2$$

Check: We multiply.

$$\begin{array}{r}
3x^4 - \frac{1}{2}x^3 + \frac{1}{8}x^2 - 2\\
8\\
\hline
24x^4 - 4x^3 + x^2 - 16
\end{array}$$

**3.** $\dfrac{u - 2u^2 - u^5}{u}$

$$= \frac{u}{u} - \frac{2u^2}{u} - \frac{u^5}{u}$$
$$= 1 - 2u - u^4$$

Check: We multiply.

$$\begin{array}{r}
1 - 2u - u^4\\
u\\
\hline
u - 2u^2 - u^5
\end{array}$$

**5.** $(15t^3 + 24t^2 - 6t) \div (3t)$

$$= \frac{15t^3 + 24t^2 - 6t}{3t}$$
$$= \frac{15t^3}{3t} + \frac{24t^2}{3t} - \frac{6t}{3t}$$
$$= 5t^2 + 8t - 2$$

Check: We multiply.

$$\begin{array}{r}
5t^2 + 8t - 2\\
3t\\
\hline
15t^3 + 24t^2 - 6t
\end{array}$$

**7.** $(20x^6 - 20x^4 - 5x^2) \div (-5x^2)$

$$= \frac{20x^6 - 20x^4 - 5x^2}{-5x^2}$$
$$= \frac{20x^6}{-5x^2} - \frac{20x^4}{-5x^2} - \frac{5x^2}{-5x^2}$$
$$= -4x^4 - (-4x^2) - (-1)$$
$$= -4x^4 + 4x^2 + 1$$

Check: We multiply.

$$\begin{array}{r}
-4x^4 + 4x^2 + 1\\
-5x^2\\
\hline
20x^6 - 20x^4 - 5x^2
\end{array}$$

**9.** $(24x^5 - 40x^4 + 6x^3) \div (4x^3)$

$$= \frac{24x^5 - 40x^4 + 6x^3}{4x^3}$$
$$= \frac{24x^5}{4x^3} - \frac{40x^4}{4x^3} + \frac{6x^3}{4x^3}$$
$$= 6x^2 - 10x + \frac{3}{2}$$

Check: We multiply.

$$\begin{array}{r}
6x^2 - 10x + \frac{3}{2}\\
4x^3\\
\hline
24x^5 - 40x^4 + 6x^3
\end{array}$$

**11.** $\dfrac{18x^2 - 5x + 2}{2}$

$$= \frac{18x^2}{2} - \frac{5x}{2} + \frac{2}{2}$$
$$= 9x^2 - \frac{5}{2}x + 1$$

Check: We multiply.

$$\begin{array}{r}
9x^2 - \frac{5}{2}x + 1\\
2\\
\hline
18x^2 - 5x + 2
\end{array}$$

**13.** $\dfrac{12x^3 + 26x^2 + 8x}{2x}$

$$= \frac{12x^3}{2x} + \frac{26x^2}{2x} + \frac{8x}{2x}$$
$$= 6x^2 + 13x + 4$$

Check: We multiply.

$$6x^2 \;+\; 13x \;+\; 4$$
$$\underline{\hspace{4em} 2x}$$
$$12x^3 + 26x^2 + 8x$$

**15.**
$$\frac{9r^2s^2 + 3r^2s - 6rs^2}{3rs}$$
$$= \frac{9r^2s^2}{3rs} + \frac{3r^2s}{3rs} - \frac{6rs^2}{3rs}$$
$$= 3rs + r - 2s$$

Check: We multiply.

$$3rs \;+\; r \;-\; 2s$$
$$\underline{\hspace{8em} 3rs}$$
$$9r^2s^2 + 3r^2s - 6rs^2$$

**17.**
$$\begin{array}{r} x + 2 \\ x+2 \,\overline{)\, x^2+4x+4} \\ \underline{x^2+2x} \\ 2x+4 \;\leftarrow (x^2+4x)-(x^2+2x) \\ \underline{2x+4} \\ 0 \;\leftarrow (2x+4)-(2x+4) \end{array}$$

The answer is $x + 2$.

**19.**
$$\begin{array}{r} x - 5 \\ x-5 \,\overline{)\, x^2-10x-25} \\ \underline{x^2-5x} \\ -5x-25 \;\leftarrow (x^2-10x)-(x^2-5x) \\ \underline{-5x+25} \\ -50 \;\leftarrow (-5x-25)-(-5x+25) \end{array}$$

The answer is $x - 5 + \dfrac{-50}{x-5}$.

**21.**
$$\begin{array}{r} x - 2 \\ x+6 \,\overline{)\, x^2+4x-14} \\ \underline{x^2+6x} \\ -2x-14 \;\leftarrow (x^2+4x)-(x^2+6x) \\ \underline{-2x-12} \\ -2 \;\leftarrow (-2x-14)-(-2x-12) \end{array}$$

The answer is $x - 2 + \dfrac{-2}{x+6}$.

**23.**
$$\begin{array}{r} x - 3 \\ x+3 \,\overline{)\, x^2+0x-9} \leftarrow \text{Filling in the missing term} \\ \underline{x^2+3x} \\ -3x-9 \;\leftarrow x^2-(x^2+3x) \\ \underline{-3x-9} \\ 0 \;\leftarrow (-3x-9)-(-3x-9) \end{array}$$

The answer is $x - 3$.

**25.**
$$\begin{array}{r} x^4- x^3+ x^2 - x +1 \\ x+1 \,\overline{)\, x^5+0x^4+0x^3+0x^2+0x+1} \leftarrow \text{Filling in missing terms} \\ \underline{x^5+ x^4} \\ -x^4 \;\leftarrow x^5-(x^5+x^4) \\ \underline{-x^4 - x^3} \\ x^3 \;\leftarrow -x^4-(-x^4-x^3) \\ \underline{x^3 + x^2} \\ -x^2 \;\leftarrow x^3-(x^3+x^2) \\ \underline{-x^2 - x} \\ x+1 \;\leftarrow -x^2-(-x^2-x) \\ \underline{x+1} \\ 0 \;\leftarrow (x+1)-(x+1) \end{array}$$

The answer is $x^4 - x^3 + x^2 - x + 1$.

**27.**
$$\begin{array}{r} 2x^2- 7x + 4 \\ 4x+3 \,\overline{)\, 8x^3-22x^2- 5x +12} \\ \underline{8x^3+6x^2} \\ -28x^2- 5x \;\leftarrow (8x^3-22x^2)-(8x^3+6x^2) \\ \underline{-28x^2-21x} \\ 16x+12 \;\leftarrow (-28x^2-5x)-(-28x^2-21x) \\ \underline{16x+12} \\ 0 \;\leftarrow (16x+12)-(16x+12) \end{array}$$

The answer is $2x^2 - 7x + 4$.

**29.**
$$\begin{array}{r} x^3- 6 \\ x^3-7 \,\overline{)\, x^6-13x^3+42} \\ \underline{x^6- 7x^3} \\ -6x^3+42 \;\leftarrow (x^6-13x^3)-(x^6-7x^3) \\ \underline{-6x^3+42} \\ 0 \;\leftarrow (-6x^3+42)-(-6x^3+42) \end{array}$$

The answer is $x^3 - 6$.

**31.**
$$\begin{array}{r} x^3+2x^2+ 4x + 8 \\ x-2 \,\overline{)\, x^4+0x^3+0x^2+0x-16} \\ \underline{x^4-2x^3} \\ 2x^3 \;\leftarrow x^4-(x^4-2x^3) \\ \underline{2x^3-4x^2} \\ 4x^2 \;\leftarrow 2x^3-(2x^3-4x^2) \\ \underline{4x^2-8x} \\ 8x-16 \;\leftarrow 4x^2-(4x^2-8x) \\ \underline{8x-16} \\ 0 \;\leftarrow (8x-16)-(8x-16) \end{array}$$

The answer is $x^3 + 2x^2 + 4x + 8$.

**33.**
$$\begin{array}{r} t^2+1 \\ t-1 \,\overline{)\, t^3-t^2+t-1} \\ \underline{t^3-t^2} \;\leftarrow (t^3-t^2)-(t^3-t^2) \\ 0+t-1 \\ \underline{t-1} \;\leftarrow (t-1)-(t-1) \\ 0 \end{array}$$

The answer is $t^2 + 1$.

**35.** $-2.3 - (-9.1) = -2.3 + 9.1 = 6.8$

**37.** *Familiarize*. Let $w$ = the width. Then $w + 15$ = the length. We draw a picture.

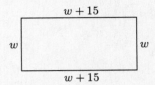

We will use the fact that the perimeter is 640 ft to find $w$ (the width). Then we can find $w + 15$ (the length) and multiply the length and the width to find the area.

*Translate*.

Width+Width+ Length + Length =Perimeter
$\ \ w\ \ +\ \ w\ \ +(w+15)+(w+15)=\ \ \ \ 640$

*Solve*.

$$w + w + (w + 15) + (w + 15) = 640$$
$$4w + 30 = 640$$
$$4w = 610$$
$$w = 152.5$$

If the width is 152.5, then the length is $152.5+15$, or 167.5. The area is $(167.5)(152.5)$, or $25,543.75$ ft$^2$.

*Check*. The length, 167.5 ft, is 15 ft greater than the width, 152.5 ft. The perimeter is $152.5 + 152.5 + 167.5 + 167.5$, or 640 ft. We should also recheck the computation we used to find the area. The answer checks.

*State*. The area is $25,543.75$ ft$^2$.

**39.**
$$-10(x - 4) = 5(2x + 5) - 7$$
$$-10x + 40 = 10x + 25 - 7$$
$$-10x + 40 = 10x + 18 \qquad \text{Collecting like terms}$$
$$-10x + 40 - 40 = 10x + 18 - 40 \qquad \text{Subtracting 40}$$
$$-10x = 10x - 22$$
$$-10x - 10x = 10x - 22 - 10x \qquad \text{Subtracting } 10x$$
$$-20x = -22$$
$$\frac{-20x}{-20} = \frac{-22}{-20} \qquad \text{Dividing by } -20$$
$$x = \frac{11}{10}$$

The solution is $\dfrac{11}{10}$.

**41.**
$$
\begin{array}{r}
x^2+\ 5 \\
x^2+4\ \overline{\smash{\big)}\ x^4+9x^2+20} \\
\underline{x^4+4x^2\phantom{+20}} \\
5x^2+20 \\
\underline{5x^2+20} \\
0
\end{array}
$$

The answer is $x^2 + 5$.

**43.**
$$
\begin{array}{r}
a+\ 3 \\
5a^2 - 7a - 2\ \overline{\smash{\big)}\ 5a^3+ 8a^2 -23a-1} \\
\underline{5a^3- 7a^2 -\ 2a} \\
15a^2-21a-1 \\
\underline{15a^2-21a-6} \\
5
\end{array}
$$

The answer is $a + 3 + \dfrac{5}{5a^2 - 7a - 2}$.

**45.** We rewrite the dividend in descending order.
$$
\begin{array}{r}
2x^2+\ x\ -\ 3 \\
3x^3 - 2x - 1\ \overline{\smash{\big)}\ 6x^5+3x^4 -13x^3 -4x^2+5x+3} \\
\underline{6x^5\phantom{+3x^4}\ -\ 4x^3\ -2x^2\phantom{+5x}} \\
3x^4-\ 9x^3\ -2x^2+5x \\
\underline{3x^4\phantom{-9x^3}\ -2x^2-\ x} \\
-9x^3\ \ \ \ +6x+3 \\
\underline{-9x^3\ \ \ \ +6x+3} \\
0
\end{array}
$$

The answer is $2x^2 + x - 3$.

**47.**
$$
\begin{array}{r}
a^5+ a^4b + a^3b^2 + a^2b^3 +\ ab^4\ +\ b^5 \\
a - b\ \overline{\smash{\big)}\ a^6+0a^5b+0a^4b^2+0a^3b^3+0a^2b^4+0ab^5-b^6} \\
\underline{a^6- a^5b} \\
a^5b \\
\underline{a^5b - a^4b^2} \\
a^4b^2 \\
\underline{a^4b^2 - a^3b^3} \\
a^3b^3 \\
\underline{a^3b^3 - a^2b^4} \\
a^2b^4 \\
\underline{a^2b^4 - ab^5} \\
ab^5\ -b^6 \\
\underline{ab^5\ -b^6} \\
0
\end{array}
$$

The answer is $a^5 + a^4b + a^3b^2 + a^2b^3 + ab^4 + b^5$.

**49.**
$$
\begin{array}{r}
x+5 \\
x - 1\ \overline{\smash{\big)}\ x^2+4x+c} \\
\underline{x^2- x} \\
5x+c \\
\underline{5x-5} \\
c+5
\end{array}
$$

We set the remainder equal to 0.

$$c + 5 = 0$$
$$c = -5$$

Thus, $c$ must be $-5$.

**51.**
$$
\begin{array}{r}
c^2x +(-2c+c^2) \\
x - 1\ \overline{\smash{\big)}\ c^2x^2 -\ 2cx+1} \\
\underline{c^2x^2 -\ c^2x} \\
(-2c+ c^2)x+1 \\
\underline{(-2c+ c^2)x-(-2c+c^2)} \\
1+(-2c+c^2)
\end{array}
$$

We set the remainder equal to 0.

$$c^2 - 2c + 1 = 0$$
$$(c - 1)^2 = 0$$
$$c = 1$$

Thus, $c$ must be 1.

# Chapter 4

# Polynomials: Factoring

## Exercise Set 4.1

**1.** Answers may vary. $8x^3 = (4x^2)(2x) = (-8)(-x^3) = (2x^2)(4x)$

**3.** Answers may vary. $-10a^6 = (-5a^5)(2a) = (10a^3)(-a^3) = (-2a^2)(5a^4)$

**5.** Answers may vary. $24x^4 = (6x)(4x^3) = (-3x^2)(-8x^2) = (2x^3)(12x)$

**7.** $x^2 - 6x = x \cdot x - x \cdot 6$    Factoring each term

$= x(x - 6)$    Factoring out the common factor $x$

**9.** $2x^2 + 6x = 2x \cdot x + 2x \cdot 3$    Factoring each term

$= 2x(x + 3)$    Factoring out the common factor $2x$

**11.** $x^3 + 6x^2 = x^2 \cdot x + x^2 \cdot 6$    Factoring each term

$= x^2(x + 6)$    Factoring out $x^2$

**13.** $8x^4 - 24x^2 = 8x^2 \cdot x^2 - 8x^2 \cdot 3$

$= 8x^2(x^2 - 3)$    Factoring out $8x^2$

**15.** $2x^2 + 2x - 8 = 2 \cdot x^2 + 2 \cdot x - 2 \cdot 4$

$= 2(x^2 + x - 4)$    Factoring out 2

**17.** $17x^5y^3 + 34x^3y^2 + 51xy$

$= 17xy \cdot x^4y^2 + 17xy \cdot 2x^2y + 17xy \cdot 3$

$= 17xy(x^4y^2 + 2x^2y + 3)$

**19.** $6x^4 - 10x^3 + 3x^2 = x^2 \cdot 6x^2 - x^2 \cdot 10x + x^2 \cdot 3$

$= x^2(6x^2 - 10x + 3)$

**21.** $x^5y^5 + x^4y^3 + x^3y^3 - x^2y^2$

$= x^2y^2 \cdot x^3y^3 + x^2y^2 \cdot x^2y + x^2y^2 \cdot xy + x^2y^2(-1)$

$= x^2y^2(x^3y^3 + x^2y + xy - 1)$

**23.** $2x^7 - 2x^6 - 64x^5 + 4x^3$

$= 2x^3 \cdot x^4 - 2x^3 \cdot x^3 - 2x^3 \cdot 32x^2 + 2x^3 \cdot 2$

$= 2x^3(x^4 - x^3 - 32x^2 + 2)$

**25.** $1.6x^4 - 2.4x^3 + 3.2x^2 + 6.4x$

$= 0.8x(2x^3) - 0.8x(3x^2) + 0.8x(4x) + 0.8x(8)$

$= 0.8x(2x^3 - 3x^2 + 4x + 8)$

**27.** $\dfrac{5}{3}x^6 + \dfrac{4}{3}x^5 + \dfrac{1}{3}x^4 + \dfrac{1}{3}x^3$

$= \dfrac{1}{3}x^3(5x^3) + \dfrac{1}{3}x^3(4x^2) + \dfrac{1}{3}x^3(x) + \dfrac{1}{3}x^3(1)$

$= \dfrac{1}{3}x^3(5x^3 + 4x^2 + x + 1)$

**29.** Factor: $x^2(x + 3) + 2(x + 3)$

The binomial $x + 3$ is common to both terms:

$x^2(x + 3) + 2(x + 3) = (x^2 + 2)(x + 3)$

**31.** $x^3 + 3x^2 + 2x + 6$

$= (x^3 + 3x^2) + (2x + 6)$

$= x^2(x + 3) + 2(x + 3)$    Factoring each binomial

$= (x^2 + 2)(x + 3)$    Factoring out the common factor $x + 3$

**33.** $2x^3 + 6x^2 + x + 3$

$= (2x^3 + 6x^2) + (x + 3)$

$= 2x^2(x + 3) + 1(x + 3)$    Factoring each binomial

$= (2x^2 + 1)(x + 3)$

**35.** $8x^3 - 12x^2 + 6x - 9 = 4x^2(2x - 3) + 3(2x - 3)$

$= (4x^2 + 3)(2x - 3)$

**37.** $12x^3 - 16x^2 + 3x - 4$

$= 4x^2(3x - 4) + 1(3x - 4)$    Factoring 1 out of the second binomial

$= (4x^2 + 1)(3x - 4)$

**39.** $5x^3 - 5x^2 - x + 1$

$= (5x^3 - 5x^2) + (-x + 1)$

$= 5x^2(x - 1) - 1(x - 1)$    Check: $-1(x-1) = -x + 1$

$= (5x^2 - 1)(x - 1)$

**41.** $x^3 + 8x^2 - 3x - 24 = x^2(x + 8) - 3(x + 8)$

$= (x^2 - 3)(x + 8)$

**43.** $2x^3 - 8x^2 - 9x + 36 = 2x^2(x - 4) - 9(x - 4)$

$= (2x^2 - 9)(x - 4)$

**45.** $-2x < 48$

$x > -24$    Dividing by $-2$ and reversing the inequality symbol

The solution set is $\{x | x > -24\}$.

**47.** $\dfrac{-108}{-4} = 27$    (The quotient of two negative numbers is positive.)

**49.** $(y+5)(y+7) = y^2 + 7y + 5y + 35$   Using FOIL
$$= y^2 + 12y + 35$$

**51.** $(y+7)(y-7) = y^2 - 7^2 = y^2 - 49$
$$[(A+B))(A-B) = A^2 - B^2]$$

**53.** $4x^5 + 6x^3 + 6x^2 + 9 = 2x^3(2x^2 + 3) + 3(2x^2 + 3)$
$$= (2x^3 + 3)(2x^2 + 3)$$

**55.** $x^{12} + x^7 + x^5 + 1 = x^7(x^5 + 1) + (x^5 + 1)$
$$= (x^7 + 1)(x^5 + 1)$$

**57.** $p^3 + p^2 - 3p + 10 = p^2(p+1) - (3p - 10)$

This polynomial is not factorable using factoring by grouping.

## Exercise Set 4.2

**1.** $x^2 + 8x + 15$

Since the constant term and coefficient of the middle term are both positive, we look for a factorization of 15 in which both factors are positive. Their sum must be 8.

| Pairs of factors | Sums of factors |
|---|---|
| 1, 15 | 16 |
| 3, 5 | 8 |

The numbers we want are 3 and 5.
$x^2 + 8x + 15 = (x+3)(x+5)$.

**3.** $x^2 + 7x + 12$

Since the constant term is positive and the coefficient of the middle term is positive, we look for a factorization of 12 in which both factors are positive. Their sum must be 7.

| Pairs of factors | Sums of factors |
|---|---|
| 1, 12 | 13 |
| 2, 6 | 8 |
| 3, 4 | 7 |

The numbers we want are 3 and 4.
$x^2 + 7x + 12 = (x+3)(x+4)$.

**5.** $x^2 - 6x + 9$

Since the constant term is positive and the coefficient of the middle term is negative, we look for a factorization of 9 in which both factors are negative. Their sum must be $-6$.

| Pairs of factors | Sums of factors |
|---|---|
| $-1, -9$ | $-10$ |
| $-3, -3$ | $-6$ |

The numbers we want are $-3$ and $-3$.
$x^2 - 6x + 9 = (x-3)(x-3)$, or $(x-3)^2$.

**7.** $x^2 + 9x + 14$

Since the constant term is positive and the coefficient of the middle term is positive, we look for a factorization of 14 in which both factors are positive. Their sum must be 9.

| Pairs of factors | Sums of factors |
|---|---|
| 1, 14 | 15 |
| 2, 7 | 9 |

The numbers we want are 2 and 7.
$x^2 + 9x + 14 = (x+2)(x+7)$.

**9.** $b^2 + 5b + 4$

Since the constant term is positive and the coefficient of the middle term is positive, we look for a factorization of 4 in which both factors are positive. Their sum must be 5.

| Pairs of factors | Sums of factors |
|---|---|
| 1, 4 | 5 |
| 2, 2 | 4 |

The numbers we want are 1 and 4.
$b^2 + 5b + 4 = (b+1)(b+4)$.

**11.** $x^2 + \dfrac{2}{3}x + \dfrac{1}{9}$

Since the constant term is positive and the coefficient of the middle term is positive, we look for a factorization of $\frac{1}{9}$ in which both factors are positive. Their sum must be $\frac{2}{3}$.

| Pairs of factors | Sums of factors |
|---|---|
| $1, \dfrac{1}{9}$ | $\dfrac{10}{9}$ |
| $\dfrac{1}{3}, \dfrac{1}{3}$ | $\dfrac{2}{3}$ |

The numbers we want are $\frac{1}{3}$ and $\frac{1}{3}$.
$x^2 + \dfrac{2}{3}x + \dfrac{1}{9} = \left(x + \dfrac{1}{3}\right)\left(x + \dfrac{1}{3}\right)$, or $\left(x + \dfrac{1}{3}\right)^2$.

**13.** $d^2 - 7d + 10$

Since the constant term is positive and the coefficient of the middle term is negative, we look for a factorization of 10 in which both factors are negative. Their sum must be $-7$.

| Pairs of factors | Sums of factors |
|---|---|
| $-1, -10$ | $-11$ |
| $-2, -5$ | $-7$ |

The numbers we want are $-2$ and $-5$.
$d^2 - 7d + 10 = (d-2)(d-5)$.

**15.** $y^2 - 11y + 10$

Since the constant term is positive and the coefficient of the middle term is negative, we look for a factorization of 10 in which both factors are negative. Their sum must be $-11$.

| Pairs of factors | Sums of factors |
|:---:|:---:|
| $-1, \; -10$ | $-11$ |
| $-2, \; -5$ | $-7$ |

The numbers we want are $-1$ and $-10$.

$y^2 - 11y + 10 = (y - 1)(y - 10)$.

**17.** $x^2 + x - 42$

Since the constant term is negative, we look for a factorization of $-42$ in which one factor is positive and one factor is negative. Their sum must be 1, the coefficient of the middle term.

| Pairs of factors | Sums of factors |
|:---:|:---:|
| $-1, \; 42$ | $41$ |
| $1, \; -42$ | $-41$ |
| $-2, \; 21$ | $19$ |
| $2, \; -21$ | $-19$ |
| $-3, \; 14$ | $11$ |
| $3, \; -14$ | $-11$ |
| $-6, \; 7$ | $1$ |
| $6, \; -7$ | $-1$ |

The numbers we want are $-6$ and $7$.

$x^2 + x - 42 = (x - 6)(x + 7)$.

**19.** $x^2 - 7x - 18$

Since the constant term is negative, we look for a factorization of $-18$ in which one factor is positive and one factor is negative. Their sum must be $-7$, the coefficient of the middle term.

| Pairs of factors | Sums of factors |
|:---:|:---:|
| $-1, \; 18$ | $17$ |
| $1, \; -18$ | $-17$ |
| $-2, \; 9$ | $7$ |
| $2, \; -9$ | $-7$ |
| $-3, \; 6$ | $3$ |
| $3, \; -6$ | $-3$ |

The numbers we want are $2$ and $-9$.

$x^2 - 7x - 18 = (x + 2)(x - 9)$.

**21.** $x^3 - 6x^2 - 16x = x(x^2 - 6x - 16)$

After factoring out the common factor, $x$, we consider $x^2 - 6x - 16$. Since the constant term is negative, we look for a factorization of $-16$ in which one factor is positive and one factor is negative. Their sum must be $-6$, the coefficient of the middle term.

| Pairs of factors | Sums of factors |
|:---:|:---:|
| $-1, \; 16$ | $15$ |
| $1, \; -16$ | $-15$ |
| $-2, \; 8$ | $6$ |
| $2, \; -8$ | $-6$ |
| $-4, \; 4$ | $0$ |

The numbers we want are $2$ and $-8$.

Then $x^2 - 6x - 16 = (x + 2)(x - 8)$, so $x^3 - 6x^2 - 16x = x(x + 2)(x - 8)$.

**23.** $y^2 - 4y - 45$

Since the constant term is negative, we look for a factorization of $-45$ in which one factor is positive and one factor is negative. Their sum must be $-4$, the coefficient of the middle term.

| Pairs of factors | Sums of factors |
|:---:|:---:|
| $-1, \; 45$ | $44$ |
| $1, \; -45$ | $-44$ |
| $-3, \; 15$ | $12$ |
| $3, \; -15$ | $-12$ |
| $-5, \; 9$ | $4$ |
| $5, \; -9$ | $-4$ |

The numbers we want are $5$ and $-9$.

$y^2 - 4y - 45 = (y + 5)(y - 9)$.

**25.** $-2x - 99 + x^2 = x^2 - 2x - 99$

Since the constant term is negative, we look for a factorization of $-99$ in which one factor is positive and one factor is negative. Their sum must be $-2$, the coefficient of the middle term.

| Pairs of factors | Sums of factors |
|:---:|:---:|
| $-1, \; 99$ | $98$ |
| $1, \; -99$ | $-98$ |
| $-3, \; 33$ | $30$ |
| $3, \; -33$ | $-30$ |
| $-9, \; 11$ | $2$ |
| $9, \; -11$ | $-2$ |

The numbers we want are $9$ and $-11$.

$-2x - 99 + x^2 = (x + 9)(x - 11)$.

**27.** $c^4 + c^2 - 56$

Consider this trinomial as $(c^2)^2 + c^2 - 56$. We look for numbers $p$ and $q$ such that $c^4 + c^2 - 56 = (c^2 + p)(c^2 + q)$. Since the constant term is negative, we look for a factorization of $-56$ in which one factor is positive and one factor is negative. Their sum must be 1.

| Pairs of factors | Sums of factors |
|:---:|:---:|
| −1,   56 | 55 |
| 1,  −56 | −55 |
| −2,   28 | 26 |
| 2,  −28 | −26 |
| −4,   14 | 12 |
| 4,  −14 | −12 |
| −7,    8 | 1 |
| 7,   −8 | −1 |

The numbers we want are −7 and 8.

$c^4 + c^2 - 56 = (c^2 - 7)(c^2 + 8)$.

**29.** $a^4 + 2a^2 - 35$

Consider this trinomial as $(a^2)^2 + 2a^2 - 35$. We look for numbers $p$ and $q$ such that $a^4 + 2a^2 - 35 = (a^2 + p)(a^2 + q)$. Since the constant term is negative, we look for a factorization of −35 in which one factor is positive and one factor is negative. Their sum must be 2.

| Pairs of factors | Sums of factors |
|:---:|:---:|
| −1,   35 | 34 |
| 1,  −35 | −34 |
| −5,    7 | 2 |
| 5,   −7 | −2 |

The numbers we want are −5 and 7.

$a^4 + 2a^2 - 35 = (a^2 - 5)(a^2 + 7)$.

**31.** $x^2 + x + 1$

Since the constant term and the coefficient of the middle term are both positive, we look for a factorization of 1 in which both factors are positive. The sum must be 1. The only possible pair of factors is 1 and 1, but their sum is not 1. Thus, this polynomial is not factorable into binomials.

**33.** $7 - 2p + p^2 = p^2 - 2p + 7$

Since the constant term is positive and the coefficient of the middle term is negative, we look for a factorization of 7 in which both factors are negative. The sum must be −2. The only possible pair of factors is −1 and −7, but their sum is not −2. Thus, this polynomial is not factorable into binomials.

**35.** $x^2 + 20x + 100$

We look for two factors, both positive, whose product is 100 and whose sum is 20.

They are 10 and 10. $10 \cdot 10 = 100$ and $10 + 10 = 20$.

$x^2 + 20x + 100 = (x + 10)(x + 10)$, or $(x + 10)^2$.

**37.** $x^2 - 21x - 100$

We look for two factors, one positive and one negative, whose product is −100 and whose sum is −21.

They are 4 and −25. $4 \cdot (-25) = -100$ and $4 + (-25) = -21$.

$x^2 - 21x - 100 = (x + 4)(x - 25)$.

**39.** $x^2 - 21x - 72$

We look for two factors, one positive and one negative, whose product is −72 and whose sum is −21. They are 3 and −24.

$x^2 - 21x - 72 = (x + 3)(x - 24)$.

**41.** $x^2 - 25x + 144$

We look for two factors, both negative, whose product is 144 and whose sum is −25. They are −9 and −16.

$x^2 - 25x + 144 = (x - 9)(x - 16)$.

**43.** $a^2 + a - 132$

We look for two factors, one positive and one negative, whose product is −132 and whose sum is 1. They are −11 and 12.

$a^2 + a - 132 = (a - 11)(a + 12)$.

**45.** $120 - 23x + x^2 = x^2 - 23x + 120$

We look for two factors, both negative, whose product is 120 and whose sum is −23. They are −8 and −15.

$x^2 - 23x + 120 = (x - 8)(x - 15)$.

**47.** First write the polynomial in descending order and factor out −1.

$108 - 3x - x^2 = -x^2 - 3x + 108 = -1(x^2 + 3x - 108)$

Now we factor the polynomial $x^2 + 3x - 108$. We look for two factors, one positive and one negative, whose product is −108 and whose sum is 3. They are −9 and 12.

$x^2 + 3x - 108 = (x - 9)(x + 12)$

The final answer must include −1 which was factored out above.

$-x^2 - 3x + 108 = -1(x - 9)(x + 12)$, or $-(x - 9)(x + 12)$

Using the distributive law to find $-1(x - 9)$, we see that $-1(x-9)(x+12)$ can also be expressed as $(-x+9)(x+12)$, or $(9 - x)(12 + x)$.

**49.** $y^2 - 0.2y - 0.08$

We look for two factors, one positive and one negative, whose product is −0.08 and whose sum is −0.2. They are −0.4 and 0.2.

$y^2 - 0.2y - 0.08 = (y - 0.4)(y + 0.2)$.

**51.** $p^2 + 3pq - 10q^2 = p^2 + 3pq - 10q^2$

Think of $3q$ as a "coefficient" of $p$. Then we look for factors of $-10q^2$ whose sum is $3q$. They are $5q$ and $-2q$.

$p^2 + 3pq - 10q^2 = (p + 5q)(p - 2q)$.

**53.** $m^2 + 5mn + 4n^2 = m^2 + 5nm + 4n^2$

We look for factors of $4n^2$ whose sum is $5n$. They are $4n$ and $n$.

$m^2 + 5mn + 4n^2 = (m + 4n)(m + 4)$

**55.** $s^2 - 2st - 15t^2 = s^2 - 2ts - 15t^2$

We look for factors of $-15t^2$ whose sum is $-2t$. They are $-5t$ and $3t$.

$s^2 - 2st - 15t^2 = (s - 5t)(s + 3t)$

**57.** $8x(2x^2 - 6x + 1) = 8x \cdot 2x^2 - 8x \cdot 6x + 8x \cdot 1 =$
$16x^3 - 48x^2 + 8x$

**59.** $(7w + 6)^2 = (7w)^2 + 2 \cdot 7w \cdot 6 + 6^2 = 49w^2 + 84w + 36$

**61.** $(4w - 11)(4w + 11) = (4w)^2 - (11)^2 = 16w^2 - 121$

**63.** $y^2 + my + 50$

We look for pairs of factors whose product is 50. The sum of each pair is represented by $m$.

| Pairs of factors whose product is −50 | Sums of factors |
|---|---|
| 1,  50 | 51 |
| −1, −50 | −51 |
| 2,  25 | 27 |
| −2, −25 | −27 |
| 5,  10 | 15 |
| −5, −10 | −15 |

The polynomial $y^2 + my + 50$ can be factored if $m$ is 51, −51, 27, −27, 15, or −15.

**65.** $x^2 - \dfrac{1}{2}x - \dfrac{3}{16}$

We look for two factors, one positive and one negative, whose product is $-\dfrac{3}{16}$ and whose sum is $-\dfrac{1}{2}$.

They are $-\dfrac{3}{4}$ and $\dfrac{1}{4}$.

$-\dfrac{3}{4} \cdot \dfrac{1}{4} = -\dfrac{3}{16}$ and $-\dfrac{3}{4} + \dfrac{1}{4} = -\dfrac{2}{4} = -\dfrac{1}{2}$.

$x^2 - \dfrac{1}{2}x - \dfrac{3}{16} = \left(x - \dfrac{3}{4}\right)\left(x + \dfrac{1}{4}\right)$

**67.** $x^2 + \dfrac{30}{7}x - \dfrac{25}{7}$

We look for two factors, one positive and one negative, whose product is $-\dfrac{25}{7}$ and whose sum is $\dfrac{30}{7}$.

They are 5 and $-\dfrac{5}{7}$.

$5 \cdot \left(-\dfrac{5}{7}\right) = -\dfrac{25}{7}$ and $5 + \left(-\dfrac{5}{7}\right) = \dfrac{35}{7} + \left(-\dfrac{5}{7}\right) = \dfrac{30}{7}$.

$x^2 + \dfrac{30}{7}x - \dfrac{25}{7} = (x + 5)\left(x - \dfrac{5}{7}\right)$

**69.** $b^{2n} + 7b^n + 10$

Consider this trinomial as $(b^n)^2 + 7b^n + 10$. We look for numbers $p$ and $q$ such that $b^{2n} + 7b^n + 10 = (b^n + p)(b^n + q)$. We find two factors, both positive, whose product is 10 and whose sum is 7. They are 5 and 2.

$b^{2n} + 7b^n + 10 = (b^n + 5)(b^n + 2)$

**71.** We first label the drawing with additional information.

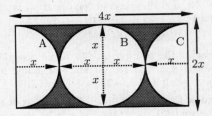

$4x$ represents the length of the rectangle and $2x$ the width. The area of the rectangle is $4x \cdot 2x$, or $8x^2$.

The area of semicircle $A$ is $\dfrac{1}{2}\pi x^2$.

The area of circle $B$ is $\pi x^2$.

The area of semicircle $C$ is $\dfrac{1}{2}\pi x^2$.

| Area of shaded region | = | Area of rectangle | − | Area of A | − | Area of B | − | Area of C |
|---|---|---|---|---|---|---|---|---|
| Area of shaded region | = | $8x^2$ | − | $\dfrac{1}{2}\pi x^2$ | − | $\pi x^2$ | − | $\dfrac{1}{2}\pi x^2$ |

$$= 8x^2 - 2\pi x^2$$
$$= 2x^2(4 - \pi)$$

The shaded area can be represented by $2x^2(4 - \pi)$.

## Exercise Set 4.3

**1.** $2x^2 - 7x - 4$

(1) Look for a common factor. There is none (other than 1 or −1).

(2) Factor the first term, $2x^2$. The only possibility is $2x$, $x$. The desired factorization is of the form:

$$(2x + \quad)(x + \quad)$$

(3) Factor the last term, −4, which is negative. The possibilities are −4, 1 and 4, −1 and 2, −2.

(4) Look for combinations of factors from steps (2) and (3) such that the sum of their products is the middle term, −7x. We try some possibilities:

$$(2x - 4)(x + 1) = 2x^2 - 2x - 4$$
$$(2x + 4)(x - 1) = 2x^2 + 2x - 4$$
$$(2x + 2)(x - 2) = 2x^2 - 2x - 4$$
$$(2x + 1)(x - 4) = 2x^2 - 7x - 4$$

The factorization is $(2x + 1)(x - 4)$.

**3.** $5x^2 - x - 18$

(1) There is no common factor (other than 1 or −1).

(2) Factor the first term, $5x^2$. The only possibility is $5x$, $x$. The desired factorization is of the form:

$$(5x+\ \ )(x+\ \ )$$

(3) Factor the last term, $-18$. The possibilities are $-18$, 1 and 18, $-1$ and $-9$, 2 and 9, $-2$ and $-6$, 3 and 6, $-3$.

(4) Look for combinations of factors from steps (2) and (3) such that the sum of their products is the middle term, $x$. We try some possibilities:

$$(5x-18)(x+1) = 5x^2 - 13x - 18$$
$$(5x+18)(x-1) = 5x^2 + 13x - 18$$
$$(5x+9)(x-2) = 5x^2 - x - 18$$

The factorization is $(5x+9)(x-2)$.

**5.** $6x^2 + 23x + 7$

(1) There is no common factor (other than 1 or $-1$).

(2) Factor the first term, $6x^2$. The possibilities are $6x$, $x$ and $3x$, $2x$. We have these as possibilities for factorizations:

$$(6x+\ \ )(x+\ \ ) \text{ and } (3x+\ \ )(2x+\ \ )$$

(3) Factor the last term, 7. The possibilities are 7, 1 and $-7$, $-1$.

(4) Look for combinations of factors from steps (2) and (3) such that the sum of their products is the middle term, $23x$. Since all signs are positive, we need consider only plus signs. We try some possibilities:

$$(6x+7)(x+1) = 6x^2 + 13x + 7$$
$$(3x+7)(2x+1) = 6x^2 + 17x + 7$$
$$(6x+1)(x+7) = 6x^2 + 43x + 7$$
$$(3x+1)(2x+7) = 6x^2 + 23x + 7$$

The factorization is $(3x+1)(2x+7)$.

**7.** $3x^2 + 4x + 1$

(1) There is no common factor (other than 1 or $-1$).

(2) Factor the first term, $3x^2$. The only possibility is $3x$, $x$. The desired factorization is of the form:

$$(3x+\ \ )(x+\ \ )$$

(3) Factor the last term, 1. The possibilities are 1, 1 and $-1$, $-1$.

(4) Look for combinations of factors from steps (2) and (3) such that the sum of their products is the middle term, $4x$. Since all signs are positive, we need consider only plus signs. There is only one such possibility:

$$(3x+1)(x+1) = 3x^2 + 4x + 1$$

The factorization is $(3x+1)(x+1)$.

**9.** $4x^2 + 4x - 15$

(1) There is no common factor (other than 1 or $-1$).

(2) Factor the first term, $4x^2$. The possibilities are $4x$, $x$ and $2x$, $2x$. We have these as possibilities for factorizations:

$$(4x+\ \ )(x+\ \ ) \text{ and } (2x+\ \ )(2x+\ \ )$$

(3) Factor the last term, $-15$. The possibilities are 15, $-1$ and $-15$, 1 and 5, $-3$ and $-5$, 3.

(4) We try some possibilities:

$$(4x+15)(x-1) = 4x^2 + 11x - 15$$
$$(2x+15)(2x-1) = 4x^2 + 28x - 15$$
$$(4x-15)(x+1) = 4x^2 - 11x - 15$$
$$(2x-15)(2x+1) = 4x^2 - 28x - 15$$
$$(4x+5)(x-3) = 4x^2 - 7x - 15$$
$$(2x+5)(2x-3) = 4x^2 + 4x - 15$$

The factorization is $(2x+5)(2x-3)$.

**11.** $2x^2 - x - 1$

(1) There is no common factor (other than 1 or $-1$).

(2) Factor the first term, $2x^2$. The only possibility is $2x$, $x$. The desired factorization is of the form:

$$(2x+\ \ )(x+\ \ )$$

(3) Factor the last term, $-1$. The only possibility is $-1$, 1.

(4) We try the possibilities:

$$(2x-1)(x+1) = 2x^2 + x - 1$$
$$(2x+1)(x-1) = 2x^2 - x - 1$$

The factorization is $(2x+1)(x-1)$.

**13.** $9x^2 + 18x - 16$

(1) There is no common factor (other than 1 or $-1$).

(2) Factor the first term, $9x^2$. The possibilities are $9x$, $x$ and $3x$, $3x$. We have these as possibilities for factorizations:

$$(9x+\ \ )(x+\ \ ) \text{ and } (3x+\ \ )(3x+\ \ )$$

(3) Factor the last term, $-16$. The possibilities are 16, $-1$ and $-16$, 1 and 8, $-2$ and $-8$, 2 and 4, $-4$.

(4) We try some possibilities:

$$(9x+16)(x-1) = 9x^2 + 7x - 16$$
$$(3x+16)(3x-1) = 9x^2 + 45x - 16$$
$$(9x-16)(x+1) = 9x^2 - 7x - 16$$
$$(3x-16)(3x+1) = 9x^2 - 45x - 16$$

$$(9x + 8)(x - 2) = 9x^2 - 10x - 16$$
$$(3x + 8)(3x - 2) = 9x^2 + 18x - 16$$

The factorization is $(3x + 8)(3x - 2)$.

**15.** $3x^2 - 5x - 2$

(1) There is no common factor (other than 1 or $-1$).

(2) Factor the first term, $3x^2$. The only possibility is $3x$, $x$. The desired factorization is of the form:

$$(3x + \quad)(x + \quad)$$

(3) Factor the last term, $-2$. The possibilities are 2, $-1$ and $-2$ and 1.

(4) We try some possibilities:

$$(3x + 2)(x - 1) = 3x^2 - x - 2$$
$$(3x - 2)(x + 1) = 3x^2 + x - 2$$
$$(3x - 1)(x + 2) = 3x^2 + 5x - 2$$
$$(3x + 1)(x - 2) = 3x^2 - 5x - 2$$

The factorization is $(3x + 1)(x - 2)$.

**17.** $12x^2 + 31x + 20$

(1) There is no common factor (other than 1 or $-1$).

(2) Factor the first term, $12x^2$. The possibilities are $12x$, $x$ and $6x$, $2x$ and $4x$, $3x$. We have these as possibilities for factorizations:

$$(12x + \quad)(x + \quad) \text{ and } (6x + \quad)(2x + \quad) \text{ and}$$
$$(4x + \quad)(3x + \quad)$$

(3) Factor the last term, 20. Since all signs are positive, we need consider only positive pairs of factors. Those factor pairs are 20, 1 and 10, 2 and 5, 4.

(4) We can immediately reject all possibilities in which either factor has a common factor, such as $(12x + 20)$ or $(6x + 4)$, because we determined at the outset that there are no common factors. We try some of the remaining possibilities:

$$(12x + 1)(x + 20) = 12x^2 + 241x + 20$$
$$(12x + 5)(x + 4) = 12x^2 + 53x + 20$$
$$(6x + 1)(2x + 20) = 12x^2 + 122x + 20$$
$$(4x + 5)(3x + 4) = 12x^2 + 31x + 20$$

The factorization is $(4x + 5)(3x + 4)$.

**19.** $14x^2 + 19x - 3$

(1) There is no common factor (other than 1 or $-1$).

(2) Factor the first term, $14x^2$. The possibilities are $14x$, $x$ and $7x$, $2x$. We have these as possibilities for factorizations:

$$(14x + \quad)(x + \quad) \text{ and } (7x + \quad)(2x + \quad)$$

(3) Factor the last term, $-3$. The possibilities are $-1$, 3 and 3, $-1$.

(4) We try some possibilities:

$$(14x - 1)(x + 3) = 14x^2 + 41x - 3$$
$$(7x - 1)(2x + 3) = 7x^2 + 19x - 3$$

The factorization is $(7x - 1)(2x + 3)$.

**21.** $9x^2 + 18x + 8$

(1) There is no common factor (other than 1 or $-1$).

(2) Factor the first term, $9x^2$. The possibilities are $9x$, $x$ and $3x$, $3x$. We have these as possibilities for factorizations:

$$(9x + \quad)(x + \quad) \text{ and } (3x + \quad)(3x + \quad)$$

(3) Factor the last term, 8. Since all signs are positive, we need consider only positive pairs of factors. Those factor pairs are 8, 1 and 4, 2.

(4) We try some possibilities:

$$(9x + 8)(x + 1) = 9x^2 + 17x + 8$$
$$(3x + 8)(3x + 1) = 9x^2 + 27x + 8$$
$$(9x + 4)(x + 2) = 9x^2 + 22x + 8$$
$$(3x + 4)(3x + 2) = 9x^2 + 18x + 8$$

The factorization is $(3x + 4)(3x + 2)$.

**23.** $49 - 42x + 9x^2 = 9x^2 - 42x + 49$

(1) There is no common factor (other than 1 or $-1$).

(2) Factor the first term, $9x^2$. The possibilities are $9x$, $x$ and $3x$, $3x$. We have these as possibilities for factorizations:

$$(9x + \quad)(x + \quad) \text{ and } (3x + \quad)(3x + \quad)$$

(3) Factor 49. Since 49 is positive and the middle term is negative, we need consider only negative pairs of factors. Those factor pairs are $-49$, $-1$ and $-7$, $-7$.

(4) We try some possibilities:

$$(9x - 49)(x - 1) = 9x^2 - 58x + 49$$
$$(3x - 49)(3x - 1) = 9x^2 - 150x + 49$$
$$(9x - 7)(x - 7) = 9x^2 - 70x + 49$$
$$(3x - 7)(3x - 7) = 9x^2 - 42x + 49$$

The factorization is $(3x - 7)(3x - 7)$, or $(3x - 7)^2$. This can also be expressed as follows:
$$(3x - 7)^2 = (-1)^2(3x - 7)^2 = [-1 \cdot (3x - 7)]^2 =$$
$$(-3x + 7)^2, \text{ or } (7 - 3x)^2$$

**25.** $24x^2 + 47x - 2$

    (1) There is no common factor (other than 1 or $-1$).

    (2) Factor the first term, $24x^2$. The possibilities are $24x$, $x$ and $12x$, $2x$ and $6x$, $4x$ and $3x$, $8x$. We have these as possibilities for factorizations:

        $(24x+\quad)(x+\quad)$ and $(12x+\quad)(2x+\quad)$ and $(6x+\quad)(4x+\quad)$ and $(3x+\quad)(8x+\quad)$

    (3) Factor the last term, $-2$. The possibilities are $2$, $-1$ and $-2$, $1$.

    (4) We can immediately reject all possibilities in which either factor has a common factor, such as $(24x + 2)$ or $(12x - 2)$, because we determined at the outset that there are no common factors. We try some of the remaining possibilities:

        $(24x - 1)(x + 2) = 24x^2 + 47x - 2$

The factorization is $(24x - 1)(x + 2)$.

**27.** $35x^2 - 57x - 44$

    (1) There is no common factor (other than 1 or $-1$).

    (2) Factor the first term, $35x^2$. The possibilities are $35x$, $x$ and $7x$, $5x$. We have these as possibilities for factorizations:

        $(35x+\quad)(x+\quad)$ and $(7x+\quad)(5x+\quad)$

    (3) Factor the last term, $-44$. The possibilities are $1$, $-44$ and $-1$, $44$ and $2$, $-22$ and $-2$, $22$ and $4$, $-11$, and $-4$, $11$.

    (4) We try some possibilities:

        $(35x + 1)(x - 44) = 35x^2 - 1539x - 44$

        $(7x + 1)(5x - 44) = 35x^2 - 303x - 44$

        $(35x + 2)(x - 22) = 35x^2 - 768x - 44$

        $(7x + 2)(5x - 22) = 35x^2 - 144x - 44$

        $(35x + 4)(x - 11) = 35x^2 - 381x - 44$

        $(7x + 4)(5x - 11) = 35x^2 - 57x - 44$

The factorization is $(7x + 4)(5x - 11)$.

**29.** $20 + 6x - 2x^2$

    (1) We factor out the common factor, 2:
        $2(10 + 3x - x^2)$

    Then we factor the trinomial $10 + 3x - x^2$.

    (2) Factor 10. The possibilities are 10, 1 and 5, 2. We have these as possibilities for factorizations:

        $(10+\quad)(1+\quad)$ and $(5+\quad)(2+\quad)$

    Note that the second term of each factor is an $x$-term.

    (3) Factor $-x^2$. The only possibility is $x$, $-x$.

    (4) We try some possibilities:

        $(10 + x)(1 - x) = 10 - 9x - x^2$

        $(5 + x)(2 - x) = 10 - 3x - x^2$

        $(5 - x)(2 + x) = 10 + 3x - x^2$

The factorization of $10 + 3x - x^2$ is $(5 - x)(2 + x)$. We must include the common factor in order to get a factorization of the original trinomial.

$20 + 6x - 2x^2 = 2(5 - x)(2 + x)$

**31.** $12x^2 + 28x - 24$

    (1) We factor out the common factor, 4:
        $4(3x^2 + 7x - 6)$

    Then we factor the trinomial $3x^2 + 7x - 6$.

    (2) Factor $3x^2$. The only possibility is $3x$, $x$. The desired factorization is of the form:

        $(3x+\quad)(x+\quad)$

    (3) Factor $-6$. The possibilities are $6$, $-1$ and $-6$, $1$ and $3$, $-2$ and $-3$, $2$.

    (4) We can immediately reject all possibilities in which either factor has a common factor, such as $(3x + 6)$ or $(3x - 3)$, because we factored out the largest common factor at the outset. We try some of the remaining possibilities:

        $(3x - 1)(x + 6) = 3x^2 + 17x - 6$

        $(3x - 2)(x + 3) = 3x^2 + 7x - 6$

The factorization of $3x^2 + 7x - 6$ is $(3x - 2)(x + 3)$. We must include the common factor in order to get a factorization of the original trinomial.

$12x^2 + 28x - 24 = 4(3x - 2)(x + 3)$

**33.** $30x^2 - 24x - 54$

    (1) We factor out the common factor, 6:
        $6(5x^2 - 4x - 9)$

    Then we factor the trinomial $5x^2 - 4x - 9$.

    (2) Factor $5x^2$. The only possibility is $5x$, $x$. The desired factorization is of the form:

        $(5x+\quad)(x+\quad)$

    (3) Factor $-9$. The possibilities are $9$, $-1$ and $-9$, $1$ and $3$, $-3$.

    (4) We try some possibilities:

        $(5x + 9)(x - 1) = 5x^2 + 4x - 9$

        $(5x - 9)(x + 1) = 5x^2 - 4x - 9$

The factorization of $5x^2 - 4x - 9$ is $(5x - 9)(x + 1)$. We must include the common factor in order to get a factorization of the original trinomial.

$30x^2 - 24x - 54 = 6(5x - 9)(x + 1)$

**35.** $4x + 6x^2 - 10 = 6x^2 + 4x - 10$

(1) We factor out the common factor, 2:

$2(3x^2 + 2x - 5)$

Then we factor the trinomial $3x^2 + 2x - 5$.

(2) Factor $3x^2$. The only possibility is $3x$, $x$. The desired factorization is of the form:

$(3x+\quad)(x+\quad)$

(3) Factor $-5$. The possibilities are 5, $-1$ and $-5$, 1.

(4) We try some possibilities:

$(3x + 5)(x - 1) = 3x^2 + 2x - 5$

Then $3x^2 + 2x - 5 = (3x + 5)(x - 1)$, so $6x^2 + 4x - 10 = 2(3x + 5)(x - 1)$.

**37.** $3x^2 - 4x + 1$

(1) There is no common factor (other than 1 or $-1$).

(2) Factor the first term, $3x^2$. The only possibility is $3x$, $x$. The desired factorization is of the form:

$(3x+\quad)(x+\quad)$

(3) Factor the last term, 1. Since 1 is positive and the middle term is negative, we need consider only negative factor pairs. The only such pair is $-1$, $-1$.

(4) There is only one possibility:

$(3x - 1)(x - 1) = 3x^2 - 4x + 1$

The factorization is $(3x - 1)(x - 1)$.

**39.** $12x^2 - 28x - 24$

(1) We factor out the common factor, 4:

$4(3x^2 - 7x - 6)$

Then we factor the trinomial $3x^2 - 7x - 6$.

(2) Factor $3x^2$. The only possibility is $3x$, $x$. The desired factorization is of the form:

$(3x+\quad)(x+\quad)$

(3) Factor $-6$. The possibilities are 6, $-1$ and $-6$, 1 and 3, $-2$ and $-3$, 2.

(4) We can immediately reject all possibilities in which either factor has a common factor, such as $(3x - 6)$ or $(3x + 3)$, because we factored out the largest common factor at the outset. We try some of the remaining possibilities:

$(3x - 1)(x + 6) = 3x^2 + 17x - 6$

$(3x - 2)(x + 3) = 3x^2 + 7x - 6$

$(3x + 2)(x - 3) = 3x^2 - 7x - 6$

Then $3x^2 - 7x - 6 = (3x + 2)(x - 3)$, so $12x^2 - 28x - 24 = 4(3x + 2)(x - 3)$.

**41.** $-1 + 2x^2 - x = 2x^2 - x - 1$

(1) There is no common factor (other than 1 or $-1$).

(2) Factor the first term, $2x^2$. The only possibility is $2x$, $x$. The desired factorization is of the form:

$(2x+\quad)(x+\quad)$

(3) Factor $-1$. The only possibility is 1, $-1$.

(4) We try some possibilities:

$(2x + 1)(x - 1) = 2x^2 - x - 1$

The factorization is $(2x + 1)(x - 1)$.

**43.** $9x^2 - 18x - 16$

(1) There is no common factor (other than 1 or $-1$).

(2) Factor the first term, $9x^2$. The possibilities are $9x$, $x$ and $3x$, $3x$. We have these as possibilities for factorizations:

$(9x+\quad)(x+\quad)$ and $(3x+\quad)(3x+\quad)$

(3) Factor the last term, $-16$. The possibilities are 16, $-1$ and $-16$, 1 and 8, $-2$ and $-8$, 2 and 4, $-4$.

(4) We try some possibilities:

$(9x + 16)(x - 1) = 9x^2 + 7x - 16$

$(3x + 16)(3x - 1) = 9x^2 + 45x - 16$

$(9x + 8)(x - 2) = 9x^2 - 10x - 16$

$(3x + 8)(3x - 2) = 9x^2 + 18x - 16$

$(3x - 8)(3x + 2) = 9x^2 - 18x - 16$

The factorization is $(3x - 8)(3x + 2)$.

**45.** $15x^2 - 25x - 10$

(1) Factor out the common factor, 5:

$5(3x^2 - 5x - 2)$

Then we factor the trinomial $3x^2 - 5x - 2$. This was done in Excercise 15. We know that $3x^2 - 5x - 2 = (3x + 1)(x - 2)$, so $15x^2 - 25x - 10 = 5(3x + 1)(x - 2)$.

**47.** $12x^3 + 31x^2 + 20x$

(1) We factor out the common factor, $x$:

$x(12x^2 + 31x + 20)$

Then we factor the trinomial $12x^2 + 31x + 20$. This was done in Excercise 17. We know that $12x^2 + 31x + 20 = (3x + 4)(4x + 5)$, so $12x^3 + 31x^2 + 20x = x(3x + 4)(4x + 5)$.

**49.** $14x^4 + 19x^3 - 3x^2$

   (1) Factor out the common factor, $x^2$:

   $x^2(14x^2 + 19x - 3)$

   Then we factor the trinomial $14x^2 + 19x - 3$. This was done in Excercise 19. We know that $14x^2 + 19x - 3 = (7x - 1)(2x + 3)$, so $14x^4 + 19x^3 - 3x^2 = x^2(7x - 1)(2x + 3)$.

**51.** $168x^3 - 45x^2 + 3x$

   (1) Factor out the common factor, $3x$:

   $3x(56x^2 - 15x + 1)$

   Then we factor the trinomial $56x^2 - 15x + 1$.

   (2) Factor $56x^2$. The possibilities are $56x$, $x$ and $28x$, $2x$ and $14x$, $4x$ and $7x$, $8x$. We have these as possibilities for factorizations:

   $(56x+\ \ )(x+\ \ )$ and $(28x+\ \ )(2x+\ \ )$ and $(14x+\ \ )(4x+\ \ )$ and $(7x+\ \ )(8x+\ \ )$

   (3) Factor 1. Since 1 is positive and the middle term is negative we need consider only the negative factor pair $-1, -1$.

   (4) We try some possibilities:

   $(56x - 1)(x - 1) = 56x^2 - 57x + 1$

   $(28x - 1)(2x - 1) = 56x^2 - 30x + 1$

   $(14x - 1)(4x - 1) = 56x^2 - 18x + 1$

   $(7x - 1)(8x - 1) = 56x^2 - 15x + 1$

   Then $56x^2 - 15x + 1 = (7x - 1)(8x - 1)$, so $168x^3 - 45x^2 + 3x = 3x(7x - 1)(8x - 1)$.

**53.** $15x^4 - 19x^2 + 6 = 15(x^2)^2 - 19x^2 + 6$

   (1) There is no common factor (other than 1 or $-1$).

   (2) Factor the first term, $15x^4$. The possibilities are $15x^2$, $x^2$ and $5x^2$, $3x^2$. We have these as possibilities for factorizations:

   $(15x^2+\ \ )(x^2+\ \ )$ and $(5x^2+\ \ )(3x^2+\ \ )$

   (3) Factor 6. Since 6 is positive and the middle term is negative, we need consider only negative factor pairs. Those pairs are $-6, -1$ and $-3, -2$.

   (4) We can immediately reject all possibilities in which either factor has a common factor, such as $(15x^2 - 6)$ or $(3x^2 - 3)$, because we determined at the outset that there is no common factor. We try some of the remaining possibilities:

   $(15x^2 - 1)(x^2 - 6) = 15x^4 - 91x^2 + 6$

   $(15x^2 - 2)(x^2 - 3) = 15x^4 - 47x^2 + 6$

   $(5x^2 - 6)(3x^2 - 1) = 15x^4 - 23x^2 + 6$

   $(5x^2 - 3)(3x^2 - 2) = 15x^4 - 19x^2 + 6$

   The factorization is $(5x^2 - 3)(3x^2 - 2)$.

**55.** $25t^2 + 80t + 64$

   (1) There is no common factor (other than 1 or $-1$).

   (2) Factor the first term, $25t^2$. The possibilities are $25t$, $t$ and $5t$, $5t$. We have these as possibilities for factorizations:

   $(25t+\ \ )(t+\ \ )$ and $(5t+\ \ )(5t+\ \ )$

   (3) Factor the last term, 64. Since all signs are positive, we need consider only positive pairs of factors. Those factor pairs are 64, 1 and 32, 2 and 16, 4 and 8, 8.

   (4) We try some possibilities:

   $(25t + 64)(t + 1) = 25t^2 + 89t + 64$

   $(5t + 32)(5t + 2) = 25t^2 + 170t + 64$

   $(25t + 16)(t + 4) = 25t^2 + 116t + 64$

   $(5t + 8)(5t + 8) = 25t^2 + 80t + 64$

   The factorization is $(5t + 8)(5t + 8)$ or $(5t + 8)^2$.

**57.** $6x^3 + 4x^2 - 10x$

   (1) Factor out the common factor, $2x$:

   $2x(3x^2 + 2x - 5)$

   Then we factor the trinomial $3x^2 + 2x - 5$. We did this in Exercise 35 (after we factored 2 out of the original trinomial). We know that $3x^2 + 2x - 5 = (3x + 5)(x - 1)$, so $6x^3 + 4x^2 - 10x = 2x(3x + 5)(x - 1)$.

**59.** $25x^2 + 79x + 64$

   We follow the same procedure as in Exercise 55. None of the possibilities works. Thus, $25x^2 + 79x + 64$ is not factorable.

**61.** $x^2 + 15x - 11$

   (1) There is no common factor (other than 1 or $-1$).

   (2) Factor the first term, $x^2$. The possibility is $x$, $x$. The desired factorization is of the form:

   $(x+\ \ )(x+\ \ )$

   (3) Factor the last term, $-11$. The possibilities are 11, $-1$, and $-11$ and 1.

   (4) We try the possibilities:

   $(x + 11)(x - 1) = x^2 + 10x - 11$

   $(x - 11)(x + 1) = x^2 - 10x - 11$

   Neither possibility works. Thus, $x^2 + 15x - 11$ is not factorable. (Note that we could also have used the method in Section 4.2 since the leading coefficient of this trinomial is 1.)

**63.** $12m^2 - mn - 20n^2$

(1) There is no common factor (other than 1 or $-1$).

(2) Factor the first term, $12m^2$. The possibilities are $12m$, $m$ and $6m$, $2m$ and $3m$, $4m$. We have these as possibilities for factorizations:

$(12m+\quad)(m+\quad)$ and $(6m+\quad)(2m+\quad)$

and $(3m+\quad)(4m+\quad)$

(3) Factor the last term, $-20n^2$. The possibilities are $20n$, $-n$ and $-20n$, $n$ and $10n$, $-2n$ and $-10n$, $2n$ and $5n$, $-4n$ and $-5n$, $4n$.

(4) We can immediately reject all possibilities in which either factor has a common factor, such as $(12m+20n)$ or $(4m-2n)$, because we determined at the outset that there is no common factor. We try some of the remaining possibilities:

$(12m - n)(m + 20n) = 12m^2 + 239mn - 20n^2$
$(12m + 5n)(m - 4n) = 12m^2 - 43mn - 20n^2$
$(3m - 20n)(4m + n) = 12m^2 - 77mn - 20n^2$
$(3m - 4n)(4m + 5n) = 12m^2 - mn - 20n^2$

The factorization is $(3m - 4n)(4m + 5n)$.

**65.** $6a^2 - ab - 15b^2$

(1) There is no common factor (other than 1 or $-1$).

(2) Factor the first term, $6a^2$. The possibilities are $6a$, $a$ and $3a$, $2a$. We have these as possibilities for factorizations:

$(6a+\quad)(a+\quad)$ and $(3a+\quad)(2a+\quad)$

(3) Factor the last term, $-15b^2$. The possibilities are $15b$, $-b$ and $-15b$, $b$ and $5b$, $-3b$ and $-5b$, $3b$.

(4) We can immediately reject all possibilities in which either factor has a common factor, such as $(6a + 15b)$ or $(3a - 3b)$, because we determined at the outset that there is no common factor. We try some of the remaining possibilities:

$(6a - b)(a + 15b) = 6a^2 + 89ab - 15b^2$
$(3a - b)(2a + 15b) = 6a^2 + 43ab - 15b^2$
$(6a + 5b)(a - 3b) = 6a^2 - 13ab - 15b^2$
$(3a + 5b)(2a - 3b) = 6a^2 + ab - 15b^2$
$(3a - 5b)(2a + 3b) = 6a^2 - ab - 15b^2$

The factorization is $(3a - 5b)(2a + 3b)$.

**67.** $9a^2 + 18ab + 8b^2$

(1) There is no common factor (other than 1 or $-1$).

(2) Factor the first term, $9a^2$. The possibilities are $9a$, $a$ and $3a$, $3a$. We have these as possibilities for factorizations:

$(9a+\quad)(a+\quad)$ and $(3a+\quad)(3a+\quad)$

(3) Factor $8b^2$. Since all signs are positive, we need consider only pairs of factors with positive coefficients. Those factor pairs are $8b$, $b$ and $4b$, $2b$.

(4) We try some possibilities:

$(9a + 8b)(a + b) = 9a^2 + 17ab + 8b^2$
$(3a + 8b)(3a + b) = 9a^2 + 27ab + 8b^2$
$(9a + 4b)(a + 2b) = 9a^2 + 22ab + 8b^2$
$(3a + 4b)(3a + 2b) = 9a^2 + 18ab + 8b^2$

The factorization is $(3a + 4b)(3a + 2b)$.

**69.** $35p^2 + 34pq + 8q^2$

(1) There is no common factor (other than 1 or $-1$).

(2) Factor the first term, $35p^2$. The possibilities are $35p$, $p$ and $7p$, $5p$. We have these as possibilities for factorizations:

$(35p+\quad)(p+\quad)$ and $(7p+\quad)(5p+\quad)$

(3) Factor $8q^2$. Since all signs are positive, we need consider only pairs of factors with positive coefficients. Those factor pairs are $8q$, $q$ and $4q$, $2q$.

(4) We try some possibilities:

$(35p + 8q)(p + q) = 35p^2 + 43pq + 8q^2$
$(7p + 8q)(5p + q) = 35p^2 + 47pq + 8q^2$
$(35p + 4q)(p + 2q) = 35p^2 + 74pq + 8q^2$
$(7p + 4q)(5p + 2q) = 35p^2 + 34pq + 8p^2$

The factorization is $(7p + 4q)(5p + 2q)$.

**71.** $18x^2 - 6xy - 24y^2$

(1) Factor out the common factor, 6:

$6(3x^2 - xy - 4y^2)$

Then we factor the trinomial $3x^2 - xy - 4y^2$.

(2) Factor $3x^2$. The only possibility is $3x$, $x$. The desired factorization is of the form:

$(3x+\quad)(x+\quad)$

(3) Factor $-4y^2$. The possibilities are $4y$, $-y$ and $-4y$, $y$ and $2y$, $-2y$.

(4) We try some possibilities:

$(3x + 4y)(x - y) = 3x^2 + xy - 4y^2$
$(3x - 4y)(x + y) = 3x^2 - xy - 4y^2$

Then $3x^2 - xy - 4y^2 = (3x - 4y)(x + y)$, so
$18x^2 - 6xy - 24y^2 = 6(3x - 4y)(x + y)$.

**73.**
$$A = pq - 7$$
$$A + 7 = pq \qquad \text{Adding 7}$$
$$\frac{A + 7}{p} = q \qquad \text{Dividing by } p$$

**75.** $3x + 2y = 6$
$$2y = 6 - 3x \quad \text{Subtracting } 3x$$
$$y = \frac{6 - 3x}{2} \quad \text{Dividing by 2}$$

**77.** $5 - 4x < -11$
$$-4x < -16 \quad \text{Subtracting 5}$$
$$x > 4 \qquad \text{Dividing by } -4 \text{ and reversing the}$$
$$\text{inequality symbol}$$
The solution set is $\{x | x > 4\}$.

**79.** $20x^{2n} + 16x^n + 3 = 20(x^n)^2 + 16x^n + 3$

(1) There is no common factor (other than 1 and $-1$).

(2) Factor the first term, $20x^{2n}$. The possibilities are $20x^n$, $x^n$ and $10x^n$, $2x^n$ and $5x^n$, $4x^n$. We have these as possibilities for factorizations:
$$(20x^n + \quad)(x^n + \quad) \text{ and } (10x^n + \quad)(2x^n + \quad)$$
$$\text{and } (5x^n + \quad)(4x^n + \quad)$$

(3) Factor the last term, 3. Since all signs are positive, we need consider only the positive factor pair 3, 1.

(4) We try some possibilities:
$$(20x^n + 3)(x^n + 1) = 20x^{2n} + 23x^n + 3$$
$$(10x^n + 3)(2x^n + 1) = 20x^{2n} + 16x^n + 3$$

The factorization is $(10x^n + 3)(2x^n + 1)$.

**81.** $3x^{6a} - 2x^{3a} - 1 = 3(x^{3a})^2 - 2x^{3a} - 1$

(1) There is no common factor (other than 1 or $-1$).

(2) Factor the first term, $3x^{6a}$. The only possibility is $3x^{3a}$, $x^{3a}$. The desired factorization is of the form:
$$(3x^{3a} + \quad)(x^{3a} + \quad)$$

(3) Factor the last term, $-1$. The only possibility is $-1$, 1.

(4) We try the possibilities:
$$(3x^{3a} - 1)(x^{3a} + 1) = 3x^{6a} + 2x^{3a} - 1$$
$$(3x^{3a} + 1)(x^{3a} - 1) = 3x^{6a} - 2x^{3a} - 1$$

The factorization is $(3x^{3a} + 1)(x^{3a} - 1)$.

## Exercise Set 4.4

**1.** $x^2 + 2x + 7x + 14 = x(x + 2) + 7(x + 2)$
$$= (x + 7)(x + 2)$$

**3.** $x^2 - 4x - x + 4 = x(x - 4) - 1(x - 4)$
$$= (x - 1)(x - 4)$$

**5.** $6x^2 + 4x + 9x + 6 = 2x(3x + 2) + 3(3x + 2)$
$$= (2x + 3)(3x + 2)$$

**7.** $3x^2 - 4x - 12x + 16 = x(3x - 4) - 4(3x - 4)$
$$= (x - 4)(3x - 4)$$

**9.** $35x^2 - 40x + 21x - 24 = 5x(7x - 8) + 3(7x - 8)$
$$= (5x + 3)(7x - 8)$$

**11.** $4x^2 + 6x - 6x - 9 = 2x(2x + 3) - 3(2x + 3)$
$$= (2x - 3)(2x + 3)$$

**13.** $2x^4 + 6x^2 + 5x^2 + 15 = 2x^2(x^2 + 3) + 5(x^2 + 3)$
$$= (2x^2 + 5)(x^2 + 3)$$

**15.** $2x^2 - 7x - 4$

(a) First factor out a common factor, if any. There is none (other than 1 or $-1$).

(b) Multiply the leading coefficient, 2 and the constant, $-4$: $2(-4) = -8$.

(c) Look for a factorization of $-8$ in which the sum of the factors is the coefficient of the middle term, $-7$.

| Pairs of factors | Sums of factors |
|------------------|-----------------|
| $-1$,  8         | 7               |
| 1, $-8$          | $-7$            |
| $-2$, 4          | 2               |
| 2, $-4$          | $-2$            |

(d) Split the middle term: $-7x = 1x - 8x$

(e) Factor by grouping:
$$2x^2 - 7x - 4 = 2x^2 + x - 8x - 4$$
$$= x(2x + 1) - 4(2x + 1)$$
$$= (x - 4)(2x + 1)$$

**17.** $3x^2 + 4x - 15$

(a) First factor out a common factor, if any. There is none (other than 1 or $-1$).

(b) Multiply the leading coefficient, 3, and the constant, $-15$: $3(-15) = -45$.

(c) Look for a factorization of $-45$ in which the sum of the factors is the coefficient of the middle term, 4.

| Pairs of factors | Sums of factors |
|:---:|:---:|
| −1,  45 | 44 |
| 1, −45 | −44 |
| −3,  15 | 12 |
| 3, −15 | −12 |
| −5,   9 | 4 |
| 5,  −9 | −4 |

(d) Split the middle term: $4x = -5x + 9x$

(e) Factor by grouping:
$$3x^2 + 4x - 15 = 3x^2 - 5x + 9x - 15$$
$$= x(3x - 5) + 3(3x - 5)$$
$$= (x + 3)(3x - 5)$$

**19.** $6x^2 + 23x + 7$

(a) First factor out a common factor, if any. There is none (other than 1 or −1).

(b) Multiply the leading coefficient, 6, and the constant, 7: $6 \cdot 7 = 42$.

(c) Look for a factorization of 42 in which the sum of the factors is the coefficient of the middle term, 23. We only need to consider positive factors.

| Pairs of factors | Sums of factors |
|:---:|:---:|
| 1,  42 | 43 |
| 2,  21 | 23 |
| 3,  14 | 17 |
| 6,   7 | 13 |

(d) Split the middle term: $23x = 2x + 21x$

(e) Factor by grouping:
$$6x^2 + 23x + 7 = 6x^2 + 2x + 21x + 7$$
$$= 2x(3x + 1) + 7(3x + 1)$$
$$= (2x + 7)(3x + 1)$$

**21.** $3x^2 + 4x + 1$

(a) First factor out a common factor, if any. There is none (other than 1 or −1).

(b) Multiply the leading coefficient, 3, and the constant, 1: $3 \cdot 1 = 3$.

(c) Look for a factorization of 3 in which the sum of the factors is the coefficient of the middle term, 4. The numbers we wand are 1 and 3: $1 \cdot 3 = 3$ amd $1 + 3 = 4$.

(d) Split the middle term: $4x = 1x + 3x$

(e) Factor by grouping:
$$3x^2 + 4x + 1 = 3x^2 + x + 3x + 1$$
$$= x(3x + 1) + 1(3x + 1)$$
$$= (x + 1)(3x + 1)$$

**23.** $4x^2 + 4x - 15$

(a) First factor out a common factor, if any. There is none (other than 1 or −1).

(b) Multiply the leading coefficient, 4, and the constant, −15: $4(-15) = -60$.

(c) Look for a factorization of −60 in which the sum of the factors is the coefficient of the middle term, 4.

| Pairs of factors | Sums of factors |
|:---:|:---:|
| −1,  60 | 59 |
| 1, −60 | −59 |
| −2,  30 | 28 |
| 2, −30 | −28 |
| −3,  20 | 17 |
| 3, −20 | −17 |
| −4,  15 | 11 |
| 4, −15 | −11 |
| −5,  12 | 7 |
| 5, −12 | −7 |
| −6,  10 | 4 |
| 6, −10 | −4 |

(d) Split the middle term: $4x = -6x + 10x$

(e) Factor by grouping:
$$4x^2 + 4x - 15 = 4x^2 - 6x + 10x - 15$$
$$= 2x(2x - 3) + 5(2x - 3)$$
$$= (2x + 5)(2x - 3)$$

**25.** $2x^2 + x - 1$

(a) First factor out a common factor, if any. There is none (other than 1 or −1).

(b) Multiply the leading coefficient, 2, and the constant, −1: $2(-1) = -2$.

(c) Look for a factorization of −2 in which the sum of the factors is the coefficient of the middle term, 1. The numbers we wand are 2 and −1: $2(-1) = -2$ amd $2 - 1 = 1$.

(d) Split the middle term: $x = 2x - 1x$

(e) Factor by grouping:
$$2x^2 + x - 1 = 2x^2 + 2x - x - 1$$
$$= 2x(x + 1) - 1(x + 1)$$
$$= (2x - 1)(x + 1)$$

**27.** $9x^2 - 18x - 16$

(a) First factor out a common factor, if any. There is none (other than 1 or −1).

(b) Multiply the leading coefficient, 9, and the constant, −16: $9(-16) = -144$.

(c) Look for a factorization of −144, so the sum of the factors is the coefficient of the middle term, −18.

| Pairs of factors | Sums of factors |
|:---:|:---:|
| −1,  144 | 143 |
| 1,−144 | −143 |
| −2,  72 | 70 |
| 2, −72 | −70 |
| −3,  48 | 45 |
| 3, −48 | −45 |
| −4,  36 | 32 |
| 4, −36 | −32 |
| −6,  24 | 18 |
| 6, −24 | −18 |
| −8,  18 | 10 |
| 8, −18 | −10 |
| −9,  16 | 7 |
| 9, −16 | −7 |
| −12,  12 | 0 |

(d) Split the middle term: $-18x = 6x - 24x$

(e) Factor by grouping:
$$9x^2 - 18x - 16 = 9x^2 + 6x - 24x - 16$$
$$= 3x(3x + 2) - 8(3x + 2)$$
$$= (3x - 8)(3x + 2)$$

29. $3x^2 + 5x - 2$

(a) First factor out a common factor, if any. There is none (other than 1 or −1).

(b) Multiply the leading coefficient, 3, and the constant, −2: $3(-2) = -6$.

(c) Look for a factorization of −6 in which the sum of the factors is the coefficient of the middle term, 5. The numbers we wand are −1 and 6: $-1(6) = -6$ amd $-1 + 6 = 5$.

(d) Split the middle term: $5x = -1x + 6x$

(e) Factor by grouping:
$$3x^2 + 5x - 2 = 3x^2 - x + 6x - 2$$
$$= x(3x - 1) + 2(3x - 1)$$
$$= (x + 2)(3x - 1)$$

31. $12x^2 - 31x + 20$

(a) First factor out a common factor, if any. There is none (other than 1 or −1).

(b) Multiply the leading coefficient, 12, and the constant, 20: $12 \cdot 20 = 240$.

(c) Look for a factorization of 240 in which the sum of the factors is the coefficient of the middle term, −31. We only need to consider negative factors.

| Pairs of factors | Sums of factors |
|:---:|:---:|
| −1,−240 | −241 |
| −2,−120 | −122 |
| −3,  −8 | −83 |
| −4, −60 | −64 |
| −5, −48 | −53 |
| −6, −40 | −46 |
| −8, −30 | −38 |
| −10, −24 | −34 |
| −12, −20 | −32 |
| −15, −16 | −31 |

(d) Split the middle term: $-31x = -15x - 16x$

(e) Factor by grouping:
$$12x^2 - 31x + 20 = 12x^2 - 15x - 16x + 20$$
$$= 3x(4x - 5) - 4(4x - 5)$$
$$= (3x - 4)(4x - 5)$$

33. $14x^2 + 19x - 3$

(a) First factor out a common factor, if any. There is none (other than 1 or −1).

(b) Multiply the leading coefficient, 14, and the constant, −3: $14(-3) = -42$.

(c) Look for a factorization of −42 so that the sum of the factors is the coefficient of the middle term, 19.

| Pairs of factors | Sums of factors |
|:---:|:---:|
| −1,   42 | 41 |
| 1, −42 | −41 |
| −2,  21 | 19 |
| 2, −21 | −19 |
| −3,  14 | 11 |
| 3, −14 | −11 |
| −6,   7 | 1 |
| 6,  −7 | −1 |

(d) Split the middle term: $19x = -2x + 21x$

(e) Factor by grouping:
$$14x^2 + 19x - 3 = 14x^2 - 2x + 21x - 3$$
$$= 2x(7x - 1) + 3(7x - 1)$$
$$= (2x + 3)(7x - 1)$$

35. $9x^2 + 18x + 8$

(a) First factor out a common factor, if any. There is none (other than 1 or −1).

(b) Multiply the leading coefficient, 9, and the constant, 8: $9 \cdot 8 = 72$.

(c) Look for a factorization of 72 in which the sum of the factors is the coefficient of the middle term, 18. We only need to consider positive factors.

| Pairs of factors | Sums of factors |
|---|---|
| 1, 72 | 73 |
| 2, 36 | 38 |
| 3, 24 | 27 |
| 4, 18 | 22 |
| 6, 12 | 18 |
| 8, 9 | 17 |

(d) Split the middle term: $18x = 6x + 12x$

(e) Factor by grouping:
$$9x^2 + 18x + 8 = 9x^2 + 6x + 12x + 8$$
$$= 3x(3x + 2) + 4(3x + 2)$$
$$= (3x + 4)(3x + 2)$$

**37.** $49 - 42x + 9x^2 = 9x^2 - 42x + 49$

(a) First factor out a common factor, if any. There is none (other than 1 or $-1$).

(b) Multiply the leading coefficient, 9, and the constant, 49: $9 \cdot 49 = 441$.

(c) Look for a factorization of 441 in which the sum of the factors is the coefficient of the middle term, $-42$. We only need to consider negative factors.

| Pairs of factors | Sums of factors |
|---|---|
| $-1, -441$ | $-442$ |
| $-3, -147$ | $-150$ |
| $-7, -63$ | $-70$ |
| $-9, -49$ | $-58$ |
| $-21, -21$ | $-42$ |

(d) Split the middle term: $-42x = -21x - 21x$

(e) Factor by grouping:
$$9x^2 - 42x + 49 = 9x^2 - 21x - 21x + 49$$
$$= 3x(3x - 7) - 7(3x - 7)$$
$$= (3x - 7)(3x - 7), \text{ or}$$
$$(3x - 7)^2$$

**39.** $24x^2 + 47x - 2$

(a) First factor out a common factor, if any. There is none (other than 1 or $-1$).

(b) Multiply the leading coefficient, 24, and the constant, $-2$: $24(-2) = -48$.

(c) Look for a factorization of $-48$ in which the sum of the factors is the coefficient of the middle term, 47. The numbers we want are 48 and $-1$: $48(-1) = -48$ and $48 + (-1) = 47$.

(d) Split the middle term: $47x = 48x - 1x$

(e) Factor by grouping:
$$24x^2 + 47x - 2 = 24x^2 + 48x - x - 2$$
$$= 24x(x + 2) - 1(x + 2)$$
$$= (24x - 1)(x + 2)$$

**41.** $35x^5 - 57x^4 - 44x^3$

(a) We first factor out the common factor, $x^3$.
$$x^3(35x^2 - 57x - 44)$$

(b) Now we factor the trinomial $35x^2 - 57x - 44$. Multiply the leading coefficient, 35, and the constant, $-44$: $35(-44) = -1540$.

(c) Look for a factorization of $-1540$ in which the sum of the factors is the coefficient of the middle term, $-57$.

| Pairs of factors | Sums of factors |
|---|---|
| $7, -220$ | $-213$ |
| $10, -154$ | $-144$ |
| $11, -140$ | $-129$ |
| $14, -110$ | $-96$ |
| $20, -77$ | $-57$ |

(d) Split the middle term: $-57x = 20x - 77x$

(e) Factor by grouping:
$$35x^2 - 57x - 44 = 35x^2 + 20x - 77x - 44$$
$$= 5x(7x + 4) - 11(7x + 4)$$
$$= (5x - 11)(7x + 4)$$

We must include the common factor to get a factorization of the original trinomial.
$$35x^5 - 57x^4 - 44x^3 = x^3(5x - 11)(7x + 4)$$

**43.** $60x + 18x^2 - 6x^3$

(a) We first factor out the common factor, $6x$.
$$60x + 18x^2 - 6x^3 = 6x(10 + 3x - x^2)$$

(b) Now we factor the trinomial $10 + 3x - x^2$. Multiply the leading coefficient, $-1$, and the constant, 10: $-1(10) = -10$.

(c) Look for a factorization of $-10$ in which the sum of the factors is the coefficient of the middle term, 3. The numbers we want are 5 and $-2$: $5(-2) = -10$ and $5 + (-2) = 3$.

(d) Split the middle term: $3x = 5x - 2x$

(e) Factor by grouping:
$$10 + 3x - x^2 = 10 + 5x - 2x - x^2$$
$$= 5(2 + x) - x(2 + x)$$
$$= (5 - x)(2 + x)$$

We must include the common factor to get a factorization of the original trinomial.
$$60x + 18x^2 - 6x^3 = 6x(5 - x)(2 + x)$$

**45.** $6 - 3x \geq -18$

$\quad\quad -3x \geq -24 \quad$ Subtracting 6

$\quad\quad\quad x \leq 8 \quad$ Dividing by $-3$ and reversing the inequality symbol

The solution set is $\{x|x \leq 8\}$.

**47.** $\quad \dfrac{1}{2}x - 6x + 10 \leq x - 5x$

$2\left(\dfrac{1}{2}x - 6x + 10\right) \leq 2(x - 5x) \quad$ Multiplying by 2 to clear the fraction

$\quad\quad x - 12x + 20 \leq 2x - 10x$

$\quad\quad\quad -11x + 20 \leq -8x \quad$ Collecting like terms

$\quad\quad\quad\quad\quad 20 \leq 3x \quad$ Adding $11x$

$\quad\quad\quad\quad\quad \dfrac{20}{3} \leq x \quad$ Dividing by 3

The solution set is $\left\{x|x \geq \dfrac{20}{3}\right\}$.

**49.** $3x - 6x + 2(x - 4) > 2(9 - 4x)$

$\quad 3x - 6x + 2x - 8 > 18 - 8x \quad$ Removing parentheses

$\quad\quad\quad -x - 8 > 18 - 8x \quad$ Collecting like terms

$\quad\quad\quad\quad\quad 7x > 26 \quad$ Adding $8x$ and 8

$\quad\quad\quad\quad\quad x > \dfrac{26}{7} \quad$ Dividing by 7

The solution set is $\left\{x|x > \dfrac{26}{7}\right\}$.

**51.** $9x^{10} - 12x^5 + 4$

(a) First factor out a common factor, if any. There is none (other than 1 or $-1$).

(b) Multiply the leading coefficient, 9, and the constant, 4: $9 \cdot 4 = 36$.

(c) Look for a factorization of 36 in which the sum of the factors is the coefficient of the middle term, $-12$. The factors we want are $-6$ and $-6$.

(d) Split the middle term: $-12x^5 = -6x^5 - 6x^5$

(e) Factor by grouping:

$9x^{10} - 12x^5 + 4 = 9x^{10} - 6x^5 - 6x^5 + 4$

$\quad\quad\quad\quad = 3x^5(3x^5 - 2) - 2(3x^5 - 2)$

$\quad\quad\quad\quad = (3x^5 - 2)(3x^5 - 2), \text{ or}$

$\quad\quad\quad\quad = (3x^5 - 2)^2$

**53.** $16x^{10} + 8x^5 + 1$

(a) First factor out a common factor, if any. There is none (other than 1 or $-1$).

(b) Multiply the leading coefficient, 16, and the constant, 1: $16 \cdot 1 = 16$.

(c) Look for a factorization of 16 in which the sum of the factors is the coefficient of the middle term, 8. The factors we want are 4 and 4.

(d) Split the middle term: $8x^5 = 4x^5 + 4x^5$

(e) Factor by grouping:

$16x^{10} + 8x^5 + 1 = 16x^{10} + 4x^5 + 4x^5 + 1$

$\quad\quad\quad\quad = 4x^5(4x^5 + 1) + 1(4x^5 + 1)$

$\quad\quad\quad\quad = (4x^5 + 1)(4x^5 + 1), \text{ or}$

$\quad\quad\quad\quad = (4x^5 + 1)^2$

## Exercise Set 4.5

**1.** $x^2 - 14x + 49$

(a) We know that $x^2$ and 49 are squares.

(b) There is no minus sign before either $x^2$ or 49.

(c) If we multiply the square roots, $x$ and 7, and double the product, we get $2 \cdot x \cdot 7 = 14x$. This is the opposite of the remaining term, $-14x$.

Thus, $x^2 - 14x + 49$ is a trinomial square.

**3.** $x^2 + 16x - 64$

Both $x^2$ and 64 are squares, but there is a minus sign before 64. Thus, $x^2 + 16x - 64$ is not a trinomial square.

**5.** $x^2 - 2x + 4$

(a) Both $x^2$ and 4 are squares.

(b) There is no minus sign before either $x^2$ or 4.

(c) If we multiply the square roots, $x$ and 2, and double the product, we get $2 \cdot x \cdot 2 = 4x$. This is neither the remaining term nor its opposite.

Thus, $x^2 - 2x + 4$ is not a trinomial square.

**7.** $9x^2 - 36x + 24$

Only one term is a square. Thus, $9x^2 - 36x + 24$ is not a trinomial square.

**9.** $x^2 - 14x + 49 = x^2 - 2 \cdot x \cdot 7 + 7^2 = (x - 7)^2$

$\quad\quad\quad\quad\quad\quad = A^2 - 2 \ A \ B + B^2 = (A - B)^2$

**11.** $x^2 + 16x + 64 = x^2 + 2 \cdot x \cdot 8 + 8^2 = (x + 8)^2$

$\quad\quad\quad\quad\quad\quad = A^2 + 2 \ A \ B + B^2 = (A + B)^2$

**13.** $x^2 - 2x + 1 = x^2 - 2 \cdot x \cdot 1 + 1^2 = (x - 1)^2$

**15.** $4 + 4x + x^2 = x^2 + 4x + 4 \quad\quad$ Changing the order

$\quad\quad\quad\quad = x^2 + 2 \cdot x \cdot 2 + 2^2$

$\quad\quad\quad\quad = (x + 2)^2$

**17.** $q^4 - 6q^2 + 9 = (q^2)^2 - 2 \cdot q^2 \cdot 3 + 3^2 = (q^2 - 3)^2$

**19.** $49 + 56y + 16y^2 = 16y^2 + 56y + 49$

$\quad\quad\quad\quad = (4y)^2 + 2 \cdot 4y \cdot 7 + 7^2$

$\quad\quad\quad\quad = (4y + 7)^2$

**21.** $2x^2 - 4x + 2 = 2(x^2 - 2x + 1)$
$$= 2(x^2 - 2 \cdot x \cdot 1 + 1^2)$$
$$= 2(x - 1)^2$$

**23.** $x^3 - 18x^2 + 81x = x(x^2 - 18x + 81)$
$$= x(x^2 - 2 \cdot x \cdot 9 + 9^2)$$
$$= x(x - 9)^2$$

**25.** $12q^2 - 36q + 27 = 3(4q^2 - 12q + 9)$
$$= 3[(2q)^2 - 2 \cdot 2q \cdot 3 + 3^2]$$
$$= 3(2q - 3)^2$$

**27.** $49 - 42x + 9x^2 = 7^2 - 2 \cdot 7 \cdot 3x + (3x)^2$
$$= (7 - 3x)^2$$

**29.** $5y^4 + 10y^2 + 5 = 5(y^4 + 2y^2 + 1)$
$$= 5[(y^2)^2 + 2 \cdot y^2 \cdot 1 + 1^2]$$
$$= 5(y^2 + 1)^2$$

**31.** $1 + 4x^4 + 4x^2 = 1^2 + 2 \cdot 1 \cdot 2x^2 + (2x^2)^2$
$$= (1 + 2x^2)^2$$

**33.** $4p^2 + 12pq + 9q^2 = (2p)^2 + 2 \cdot 2p \cdot 3q + (3q)^2$
$$= (2p + 3q)^2$$

**35.** $a^2 - 6ab + 9b^2 = a^2 - 2 \cdot a \cdot 3b + (3b)^2$
$$= (a - 3b)^2$$

**37.** $81a^2 - 18ab + b^2 = (9a)^2 - 2 \cdot 9a \cdot b + b^2$
$$= (9a - b)^2$$

**39.** $36a^2 + 96ab + 64b^2 = 4(9a^2 + 24ab + 16b^2)$
$$= 4[(3a)^2 + 2 \cdot 3a \cdot 4b + (4b)^2]$$
$$= 4(3a + 4b)^2$$

**41.** $x^2 - 4$

  (a) The first expression is a square: $x^2$

    The second expression is a square: $4 = 2^2$

  (b) The terms have different signs.

    $x^2 - 4$ is a difference of squares.

**43.** $x^2 + 25$

The terms do not have different signs.

$x^2 + 25$ is not a difference of squares.

**45.** $x^2 - 45$

The number 45 is not a square.

$x^2 - 45$ is not a difference of squares.

**47.** $16x^2 - 25y^2$

  (a) The first expression is a square: $16x^2 = (4x)^2$

    The second expression is a square: $25y^2 = (5y)^2$

  (b) The terms have different signs.

    $16x^2 - 25y^2$ is a difference of squares.

**49.** $y^2 - 4 = y^2 - 2^2 = (y + 2)(y - 2)$

**51.** $p^2 - 9 = p^2 - 3^2 = (p + 3)(p - 3)$

**53.** $-49 + t^2 = t^2 - 49 = t^2 - 7^2 = (t + 7)(t - 7)$

**55.** $a^2 - b^2 = (a + b)(a - b)$

**57.** $25t^2 - m^2 = (5t)^2 - m^2 = (5t + m)(5t - m)$

**59.** $100 - k^2 = 10^2 - k^2 = (10 + k)(10 - k)$

**61.** $16a^2 - 9 = (4a)^2 - 3^2 = (4a + 3)(4a - 3)$

**63.** $4x^2 - 25y^2 = (2x)^2 - (5y)^2 = (2x + 5y)(2x - 5y)$

**65.** $8x^2 - 98 = 2(4x^2 - 49) = 2[(2x)^2 - 7^2] =$
$2(2x + 7)(2x - 7)$

**67.** $36x - 49x^3 = x(36 - 49x^2) = x[6^2 - (7x)^2] =$
$x(6 + 7x)(6 - 7x)$

**69.** $49a^4 - 81 = (7a^2)^2 - 9^2 = (7a^2 + 9)(7a^2 - 9)$

**71.**   $a^4 - 16$

  $= (a^2)^2 - 4^2$

  $= (a^2 + 4)(a^2 - 4)$    Factoring a difference of squares

  $= (a^2 + 4)(a + 2)(a - 2)$  Factoring further: $a^2 - 4$ is a difference of squares.

**73.**   $5x^4 - 405$

  $5(x^4 - 81)$

  $= 5[(x^2)^2 - 9^2]$

  $= 5(x^2 + 9)(x^2 - 9)$

  $= 5(x^2 + 9)(x + 3)(x - 3)$  Factoring $x^2 - 9$

**75.**   $1 - y^8$

  $= 1^2 - (y^4)^2$

  $= (1 + y^4)(1 - y^4)$

  $= (1 + y^4)(1 + y^2)(1 - y^2)$    Factoring $1 - y^4$

  $= (1 + y^4)(1 + y^2)(1 + y)(1 - y)$  Factoring $1 - y^2$

**77.**   $x^{12} - 16$

  $= (x^6)^2 - 4^2$

  $= (x^6 + 4)(x^6 - 4)$

  $= (x^6 + 4)(x^3 + 2)(x^3 - 2)$  Factoring $x^6 - 4$

**79.** $y^2 - \dfrac{1}{16} = y^2 - \left(\dfrac{1}{4}\right)^2$
$$= \left(y + \dfrac{1}{4}\right)\left(y - \dfrac{1}{4}\right)$$

**81.** $25 - \dfrac{1}{49}x^2 = 5^2 - \left(\dfrac{1}{7}x\right)^2$
$$= \left(5 + \dfrac{1}{7}x\right)\left(5 - \dfrac{1}{7}x\right)$$

**83.**     $16m^4 - t^4$
$$= (4m^2)^2 - (t^2)^2$$
$$= (4m^2 + t^2)(4m^2 - t^2)$$
$$= (4m^2 + t^2)(2m + t)(2m - t) \quad \text{Factoring } 4m^2 - t^2$$

**85.** $-110 \div 10$    The quotient of a negative number and a positive number is negative.
$$-110 \div 10 = -11$$

**87.** $-\dfrac{2}{3} \div \dfrac{4}{5} = -\dfrac{2}{3} \cdot \dfrac{5}{4} = -\dfrac{10}{12} = -\dfrac{2 \cdot 5}{2 \cdot 6} = -\dfrac{\cancel{2} \cdot 5}{\cancel{2} \cdot 6} = -\dfrac{5}{6}$

**89.** $-64 \div (-32)$    The quotient of two negative numbers is a positive number.
$$-64 \div (-32) = 2$$

**91.** $49x^2 - 216$

There is no common factor. Also, $49x^2$ is a square, but 216 is not so this expression is not a difference of squares. It is not factorable.

**93.** $x^2 + 22x + 121 = x^2 + 2 \cdot x \cdot 11 + 11^2$
$$= (x + 11)^2$$

**95.** $18x^3 + 12x^2 + 2x = 2x(9x^2 + 6x + 1)$
$$= 2x[(3x)^2 + 2 \cdot 3x \cdot 1 + 1^2]$$
$$= 2x(3x + 1)^2$$

**97.**     $x^8 - 2^8$
$$= (x^4 + 2^4)(x^4 - 2^4)$$
$$= (x^4 + 2^4)(x^2 + 2^2)(x^2 - 2^2)$$
$$= (x^4 + 2^4)(x^2 + 2^2)(x + 2)(x - 2), \text{ or}$$
$$= (x^4 + 16)(x^2 + 4)(x + 2)(x - 2)$$

**99.** $3x^5 - 12x^3 = 3x^3(x^2 - 4) = 3x^3(x + 2)(x - 2)$

**101.** $18x^3 - \dfrac{8}{25}x = 2x\left(9x^2 - \dfrac{4}{25}\right) = 2x\left(3x + \dfrac{2}{5}\right)\left(3x - \dfrac{2}{5}\right)$

**103.** $0.49p - p^3 = p(0.49 - p^2) = p(0.7 + p)(0.7 - p)$

**105.** $0.64x^2 - 1.21 = (0.8x)^2 - (1.1)^2 = (0.8x + 1.1)(0.8x - 1.1)$

**107.** $(x+3)^2 - 9 = [(x+3)+3][(x+3)-3] = (x+6)x$, or $x(x+6)$

**109.** $x^2 - \left(\dfrac{1}{x}\right)^2 = \left(x + \dfrac{1}{x}\right)\left(x - \dfrac{1}{x}\right)$

**111.**     $81 - b^{4k} = 9^2 - (b^{2k})^2$
$$= (9 + b^{2k})(9 - b^{2k})$$
$$= (9 + b^{2k})[3^2 - (b^k)^2]$$
$$= (9 + b^{2k})(3 + b^k)(3 - b^k)$$

**113.** $9b^{2n} + 12b^n + 4 = (3b^n)^2 + 2 \cdot 3b^n \cdot 2 + 2^2 = $
$(3b^n + 2)^2$

**115.**     $(y+3)^2 + 2(y+3) + 1$
$$= (y+3)^2 + 2 \cdot (y+3) \cdot 1 + 1^2$$
$$= [(y+3) + 1]^2$$
$$= (y + 4)^2$$

**117.** If $cy^2 + 6y + 1$ is the square of a binomial, then $2 \cdot a \cdot 1 = 6$ where $a^2 = c$. Then $a = 3$, so $c = a^2 = 3^2 = 9$. (The polynomial is $9y^2 + 6y + 1$.)

## Exercise Set 4.6

**1.** $3x^2 - 192 = 3(x^2 - 64)$        3 is a common factor
$$= 3(x^2 - 8^2) \qquad \text{Difference of squares}$$
$$= 3(x + 8)(x - 8)$$

**3.** $a^2 + 25 - 10a = a^2 - 10a + 25$
$$= a^2 - 2 \cdot a \cdot 5 + 5^2 \quad \text{Trinomial square}$$
$$= (a - 5)^2$$

**5.** $2x^2 - 11x + 12$

There is no common factor (other than 1). This polynomial has three terms, but it is not a trinomial square. Multiply the leading coefficient and the constant, 2 and 12: $2 \cdot 12 = 24$. Try to factor 24 so that the sum of the factors is $-11$. The numbers we want are $-3$ and $-8$: $-3(-8) = 24$ and $-3 + (-8) = -11$. Split the middle term and factor by grouping.
$$2x^2 - 11x + 12 = 2x^2 - 3x - 8x + 12$$
$$= x(2x - 3) - 4(2x - 3)$$
$$= (x - 4)(2x - 3)$$

**7.**     $x^3 + 24x^2 + 144x$
$$= x(x^2 + 24x + 144) \qquad x \text{ is a common factor}$$
$$= x(x^2 + 2 \cdot x \cdot 12 + 12^2) \quad \text{Trinomial square}$$
$$= x(x + 12)^2$$

**9.**     $x^3 + 3x^2 - 4x - 12$
$$= x^2(x + 3) - 4(x + 3) \quad \text{Factoring by grouping}$$
$$= (x^2 - 4)(x + 3)$$
$$= (x + 2)(x - 2)(x + 3) \quad \text{Factoring the difference of squares}$$

**11.** $48x^2 - 3 = 3(16x^2 - 1)$        3 is a common factor
$$= 3[(4x)^2 - 1^2] \qquad \text{Difference of squares}$$
$$= 3(4x + 1)(4x - 1)$$

**13.**     $9x^3 + 12x^2 - 45x$
$$= 3x(3x^2 + 4x - 15) \quad 3x \text{ is a common factor}$$
$$= 3x(3x - 5)(x + 3) \quad \text{Factoring the trinomial}$$

**15.** $x^2 + 4$ is a <u>sum</u> of squares with no common factor. It cannot be factored.

**17.** $x^4 + 7x^2 - 3x^3 - 21x = x(x^3 + 7x - 3x^2 - 21)$
$$= x[x(x^2 + 7) - 3(x^2 + 7)]$$
$$= x[(x - 3)(x^2 + 7)]$$
$$= x(x - 3)(x^2 + 7)$$

**19.**     $x^5 - 14x^4 + 49x^3$
$$= x^3(x^2 - 14x + 49) \quad x^3 \text{ is a common factor}$$
$$= x^3(x^2 - 2 \cdot x \cdot 7 + 7^2) \quad \text{Trinomial square}$$
$$= x^3(x - 7)^2$$

**21.**     $20 - 6x - 2x^2$
$$= -2(-10 + 3x + x^2) \quad -2 \text{ is a common factor}$$
$$= -2(x^2 + 3x - 10) \quad \text{Writing in descending order}$$
$$= -2(x + 5)(x - 2), \quad \text{Using trial and error}$$
$$\text{or } 2(5 + x)(2 - x)$$

**23.** $x^2 - 6x + 1$

There is no common factor (other than 1 or $-1$). This is not a trinomial square, because $-6x \neq 2 \cdot x \cdot 1$ and $-6x \neq -2 \cdot x \cdot 1$. We try factoring using the refined trial and error procedure. We look for two factors of 1 whose sum is $-6$. There are none. The polynomial cannot be factored.

**25.**     $4x^4 - 64$
$$= 4(x^4 - 16) \quad 4 \text{ is a common factor}$$
$$= 4[(x^2)^2 - 4^2] \quad \text{Difference of squares}$$
$$= 4(x^2 + 4)(x^2 - 4) \quad \text{Difference of squares}$$
$$= 4(x^2 + 4)(x + 2)(x - 2)$$

**27.**     $1 - y^8 \quad\quad\quad\quad \text{Difference of squares}$
$$= (1 + y^4)(1 - y^4) \quad\quad \text{Difference of squares}$$
$$= (1 + y^4)(1 + y^2)(1 - y^2) \quad \text{Difference of squares}$$
$$= 1(1 + y^4)(1 + y^2)(1 + y)(1 - y)$$

**29.**     $x^5 - 4x^4 + 3x^3$
$$= x^3(x^2 - 4x + 3) \quad x^3 \text{ is a common factor}$$
$$= x^3(x - 3)(x - 1) \quad \text{Factoring the trinomial using trial and error}$$

**31.**     $\dfrac{1}{81}x^6 - \dfrac{8}{27}x^3 + \dfrac{16}{9}$
$$= \dfrac{1}{9}\left(\dfrac{1}{9}x^6 - \dfrac{8}{3}x^3 + 16\right) \quad \dfrac{1}{9} \text{ is a common factor}$$
$$= \dfrac{1}{9}\left[\left(\dfrac{1}{3}x^3\right)^2 - 2 \cdot \dfrac{1}{3}x^3 \cdot 4 + 4^2\right] \quad \text{Trinomial square}$$
$$= \dfrac{1}{9}\left(\dfrac{1}{3}x^3 - 4\right)^2$$

**33.**     $mx^2 + my^2$
$$= m(x^2 + y^2) \quad m \text{ is a common factor}$$
The factor with more than one term connot be factored further, so we have factored completely.

**35.** $9x^2y^2 - 36xy = 9xy(xy - 4)$

**37.** $2\pi rh + 2\pi r^2 = 2\pi r(h + r)$

**39.**     $(a + b)(x - 3) + (a + b)(x + 4)$
$$= (a + b)[(x - 3) + (x + 4)] \quad (a + b) \text{ is a common factor}$$
$$= (a + b)(2x + 1)$$

**41.** $(x - 1)(x + 1) - y(x + 1) = (x + 1)(x - 1 - y)$
$$(x + 1) \text{ is a common factor}$$

**43.**     $n^2 + 2n + np + 2p$
$$= n(n + 2) + p(n + 2) \quad \text{Factoring by grouping}$$
$$= (n + p)(n + 2)$$

**45.**     $6q^2 - 3q + 2pq - p$
$$= 3q(2q - 1) + p(2q - 1) \quad \text{Factoring by grouping}$$
$$= (3q + p)(2q - 1)$$

**47.**     $4b^2 + a^2 - 4ab$
$$= a^2 - 4ab + 4b^2 \quad \text{Rearranging}$$
$$= a^2 - 2 \cdot a \cdot 2b + (2b)^2 \quad \text{Trinomial square}$$
$$= (a - 2b)^2$$

(Note that if we had rewritten the polynomial as $4b^2 - 4ab + a^2$, we might have written the result as $(4b - a)^2$. The two factorizations are equivalent.)

**49.**     $16x^2 + 24xy + 9y^2$
$$= (4x)^2 + 2 \cdot 4x \cdot 3y + (3y)^2 \quad \text{Trinomial square}$$
$$= (4x + 3y)^2$$

**51.**     $49m^4 - 112m^2n + 64n^2$
$$= (7m^2)^2 - 2 \cdot 7m^2 \cdot 8n + (8n)^2 \quad \text{Trinomial square}$$
$$= (7m^2 - 8n)^2$$

**53.**     $y^4 + 10y^2z^2 + 25z^4$
$$= (y^2)^2 + 2 \cdot y^2 \cdot 5z^2 + (5z^2)^2 \quad \text{Trinomial square}$$
$$= (y^2 + 5z^2)^2$$

**55.**     $\dfrac{1}{4}a^2 + \dfrac{1}{3}ab + \dfrac{1}{9}b^2$
$$= \left(\dfrac{1}{2}a\right)^2 + 2 \cdot \dfrac{1}{2}a \cdot \dfrac{1}{3}b + \left(\dfrac{1}{3}b\right)^2$$
$$= \left(\dfrac{1}{2}a + \dfrac{1}{3}b\right)^2$$

**57.** $a^2 - ab - 2b^2 = (a - 2b)(a + b) \quad \text{Using trial and error}$

**59.**     $2mn - 360n^2 + m^2$
$$= m^2 + 2mn - 360n^2 \quad \text{Rewriting}$$
$$= (m + 20n)(m - 18n) \quad \text{Using trial and error}$$

**61.** $m^2n^2 - 4mn - 32 = (mn - 8)(mn + 4) \quad \text{Using trial and error}$

**63.**  $a^2b^6 + 4ab^5 - 32b^4$

$= b^4(a^2b^2 + 4ab - 32)$   $b^4$ is a common factor

$= b^4(ab + 8)(ab - 4)$    Using trial and error

**65.**  $a^5 + 4a^4b - 5a^3b^2$

$= a^3(a^2 + 4ab - 5b^2)$   $a^3$ is a common factor

$= a^3(a + 5b)(a - b)$    Factoring the trinomial

**67.**  $a^2 - \dfrac{1}{25}b^2$

$= a^2 - \left(\dfrac{1}{5}b\right)^2$    Difference of squares

$= \left(a + \dfrac{1}{5}b\right)\left(a - \dfrac{1}{5}b\right)$

**69.**  $a^4 + a^2bc - 2b^2c^2$

$= (a^2 + 2bc)(a^2 - bc)$  Using trial and error

**71.**  $x^2 - y^2 = (x + y)(x - y)$    Difference of squares

**73.**  $16 - p^4q^4$

$= 4^2 - (p^2q^2)^2$   Difference of squares

$= (4 + p^2q^2)(4 - p^2q^2)$     $4 - p^2q^2$ is a difference of squares

$= (4 + p^2q^2)(2 + pq)(2 - pq)$

**75.**  $1 - 16x^{12}y^{12}$

$= 1^2 - (4x^6y^6)^2$     Difference of squares

$= (1 + 4x^6y^6)(1 - 4x^6y^6)$     $1 - 4x^6y^6$ is a difference of squares

$= (1 + 4x^6y^6)(1 + 2x^3y^3)(1 - 2x^3y^3)$

**77.**  $q^3 + 8q^2 - q - 8$

$= q^2(q + 8) - (q + 8)$    Factoring by grouping

$= (q^2 - 1)(q + 8)$

$= (q + 1)(q - 1)(q + 8)$  Factoring the difference of squares

**79.**  $\dfrac{7}{5} \div \left(-\dfrac{11}{10}\right)$

$= \dfrac{7}{5} \cdot \left(-\dfrac{10}{11}\right)$    Multiplying by the reciprocal of the divisor

$= -\dfrac{7 \cdot 10}{5 \cdot 11}$

$= -\dfrac{7 \cdot 5 \cdot 2}{5 \cdot 11} = -\dfrac{7 \cdot 2}{11} \cdot \dfrac{5}{5}$

$= -\dfrac{14}{11}$

**81.**  $A = aX + bX - 7$

$A + 7 = aX + bX$

$A + 7 = X(a + b)$

$\dfrac{A + 7}{a + b} = X$

**83.**  $a^4 - 2a^2 + 1 = (a^2)^2 - 2 \cdot a^2 \cdot 1 + 1^2$

$= (a^2 - 1)^2$

$= [(a + 1)(a - 1)]^2$

$= (a + 1)^2(a - 1)^2$

**85.**  $12.25x^2 - 7x + 1 = (3.5x)^2 - 2 \cdot (3.5x) \cdot 1 + 1^2$

$= (3.5x - 1)^2$

**87.**  $5x^2 + 13x + 7.2$

Multiply the leading coefficient and the constant, 5 and 7.2: $5(7.2) = 36$. Try to factor 36 so that the sum of the factors is 13. The numbers we want are 9 and 4. Split the middle term and factor by grouping:

$5x^2 + 13x + 7.2 = 5x^2 + 9x + 4x + 7.2$

$= 5x(x + 1.8) + 4(x + 1.8)$

$= (5x + 4)(x + 1.8)$

**89.**   $18 + y^3 - 9y - 2y^2$

$= y^3 - 2y^2 - 9y + 18$

$= y^2(y - 2) - 9(y - 2)$

$= (y^2 - 9)(y - 2)$

$= (y + 3)(y - 3)(y - 2)$

**91.**  $a^3 + 4a^2 + a + 4 = a^2(a + 4) + 1(a + 4)$

$= (a^2 + 1)(a + 4)$

**93.**  $x^4 - 7x^2 - 18 = (x^2 - 9)(x^2 + 2)$

$= (x + 3)(x - 3)(x^2 + 2)$

**95.**  $x^3 - x^2 - 4x + 4 = x^2(x - 1) - 4(x - 1)$

$= (x^2 - 4)(x - 1)$

$= (x + 2)(x - 2)(x - 1)$

**97.**   $y^2(y - 1) - 2y(y - 1) + (y - 1)$

$= (y - 1)(y^2 - 2y + 1)$

$= (y - 1)(y - 1)^2$

$= (y - 1)^3$

**99.**   $(y + 4)^2 + 2x(y + 4) + x^2$

$= (y + 4)^2 + 2 \cdot (y + 4) \cdot x + x^2$  Trinomial square

$= (y + 4 + x)^2$

---

## Exercise Set 4.7

**1.**  $(x + 4)(x + 9) = 0$

$x + 4 = 0$   or   $x + 9 = 0$    Using the principle of zero products

$x = -4$   or     $x = -9$   Solving the two equations separately

Check:

For $-4$

$$\frac{(x+4)(x+9)=0}{(-4+4)(-4+9)\quad|\quad 0}$$
$$\quad\quad 0\cdot 5$$
$$\quad\quad\quad 0\quad|\quad\text{TRUE}$$

For $-9$

$$\frac{(x+4)(x+9)=0}{(9+4)(-9+9)\quad|\quad 0}$$
$$\quad\quad 13\cdot 0$$
$$\quad\quad\quad 0\quad|\quad\text{TRUE}$$

The solutions are $-4$ and $-9$.

**3.** $(x+3)(x-8)=0$

$\quad x+3=0\quad$ or $\quad x-8=0\quad$ Using the principle of zero products

$\quad\quad x=-3\quad$ or $\quad\quad x=8$

Check:

For $-3$

$$\frac{(x+3)(x-8)=0}{(-3+3)(-3-8)\quad|\quad 0}$$
$$\quad\quad 0(-11)$$
$$\quad\quad\quad 0\quad|\quad\text{TRUE}$$

For $8$

$$\frac{(x+3)(x-8)=0}{(8+3)(8-8)\quad|\quad 0}$$
$$\quad\quad 11\cdot 0$$
$$\quad\quad\quad 0\quad|\quad\text{TRUE}$$

The solutions are $-3$ and $8$.

**5.** $(x+12)(x-11)=0$

$\quad x+12=0\quad$ or $\quad x-11=0$

$\quad\quad x=-12\quad$ or $\quad\quad x=11$

The solutions are $-12$ and $11$.

**7.** $x(x+3)=0$

$\quad x=0\quad$ or $\quad x+3=0$

$\quad x=0\quad$ or $\quad\quad x=-3$

The solutions are $0$ and $-3$.

**9.** $0=y(y+18)$

$\quad y=0\quad$ or $\quad y+18=0$

$\quad y=0\quad$ or $\quad\quad y=-18$

The solutions are $0$ and $-18$.

**11.** $(2x+5)(x+4)=0$

$\quad 2x+5=0\quad$ or $\quad x+4=0$

$\quad\quad 2x=-5\quad$ or $\quad\quad x=-4$

$\quad\quad x=-\dfrac{5}{2}\quad$ or $\quad\quad x=-4$

The solutions are $-\dfrac{5}{2}$ and $-4$.

**13.** $(5x+1)(4x-12)=0$

$\quad 5x+1=0\quad$ or $\quad 4x-12=0$

$\quad\quad 5x=-1\quad$ or $\quad\quad 4x=12$

$\quad\quad x=-\dfrac{1}{5}\quad$ or $\quad\quad x=3$

The solutions are $-\dfrac{1}{5}$ and $3$.

**15.** $(7x-28)(28x-7)=0$

$\quad 7x-28=0\quad$ or $\quad 28x-7=0$

$\quad\quad 7x=28\quad$ or $\quad\quad 28x=7$

$\quad\quad x=4\quad$ or $\quad\quad x=\dfrac{7}{28}=\dfrac{1}{4}$

The solutions are $4$ and $\dfrac{1}{4}$.

**17.** $2x(3x-2)=0$

$\quad 2x=0\quad$ or $\quad 3x-2=0$

$\quad x=0\quad$ or $\quad\quad 3x=2$

$\quad x=0\quad$ or $\quad\quad x=\dfrac{2}{3}$

The solutions are $0$ and $\dfrac{2}{3}$.

**19.** $\dfrac{1}{2}x\left(\dfrac{2}{3}x-12\right)=0$

$\quad \dfrac{1}{2}x=0\quad$ or $\quad \dfrac{2}{3}x-12=0$

$\quad x=0\quad$ or $\quad\quad \dfrac{2}{3}x=12$

$\quad x=0\quad$ or $\quad\quad x=\dfrac{3}{2}\cdot 12=18$

The solutions are $0$ and $18$.

**21.** $\left(\dfrac{1}{5}+2x\right)\left(\dfrac{1}{9}-3x\right)=0$

$\quad \dfrac{1}{5}+2x=0\quad$ or $\quad \dfrac{1}{9}-3x=0$

$\quad\quad 2x=-\dfrac{1}{5}\quad$ or $\quad -3x=-\dfrac{1}{9}$

$\quad\quad x=-\dfrac{1}{10}\quad$ or $\quad\quad x=\dfrac{1}{27}$

The solutions are $-\dfrac{1}{10}$ and $\dfrac{1}{27}$.

**23.** $(0.3x-0.1)(0.05x+1)=0$

$\quad 0.3x-0.1=0\quad$ or $\quad 0.05x+1=0$

$\quad\quad 0.3x=0.1\quad$ or $\quad\quad 0.05x=-1$

$\quad\quad x=\dfrac{0.1}{0.3}\quad$ or $\quad\quad x=-\dfrac{1}{0.05}$

$\quad\quad x=\dfrac{1}{3}\quad$ or $\quad\quad x=-20$

The solutions are $\dfrac{1}{3}$ and $-20$.

**25.** $9x(3x - 2)(2x - 1) = 0$

$\quad 9x = 0 \quad$ or $\quad 3x - 2 = 0 \quad$ or $\quad 2x - 1 = 0$

$\quad x = 0 \quad$ or $\qquad 3x = 2 \quad$ or $\qquad 2x = 1$

$\quad x = 0 \quad$ or $\qquad x = \dfrac{2}{3} \quad$ or $\qquad x = \dfrac{1}{2}$

The solutions are $0$, $\dfrac{2}{3}$, and $\dfrac{1}{2}$.

**27.** $\quad x^2 + 6x + 5 = 0$

$\quad (x + 5)(x + 1) = 0 \qquad$ Factoring

$\quad x + 5 = 0 \quad$ or $\quad x + 1 = 0 \qquad$ Using the principle of
$\qquad\qquad\qquad\qquad\qquad\qquad\qquad$ zero products

$\qquad x = -5 \quad$ or $\qquad x = -1$

The solutions are $-5$ and $-1$.

**29.** $\quad x^2 + 7x - 18 = 0$

$\quad (x + 9)(x - 2) = 0 \qquad$ Factoring

$\quad x + 9 = 0 \quad$ or $\quad x - 2 = 0 \quad$ Using the principle of
$\qquad\qquad\qquad\qquad\qquad\qquad\qquad$ zero products

$\qquad x = -9 \quad$ or $\qquad x = 2$

The solutions are $-9$ and $2$.

**31.** $\quad x^2 - 8x + 15 = 0$

$\quad (x - 5)(x - 3) = 0$

$\quad x - 5 = 0 \quad$ or $\quad x - 3 = 0$

$\qquad x = 5 \quad$ or $\qquad x = 3$

The solutions are $5$ and $3$.

**33.** $\quad x^2 - 8x = 0$

$\quad x(x - 8) = 0$

$\quad x = 0 \quad$ or $\quad x - 8 = 0$

$\quad x = 0 \quad$ or $\qquad x = 8$

The solutions are $0$ and $8$.

**35.** $\quad x^2 + 18x = 0$

$\quad x(x + 18) = 0$

$\quad x = 0 \quad$ or $\quad x + 18 = 0$

$\quad x = 0 \quad$ or $\qquad x = -18$

The solutions are $0$ and $-18$.

**37.** $\qquad x^2 = 16$

$\qquad x^2 - 16 = 0 \qquad$ Subtracting 16

$\quad (x - 4)(x + 4) = 0$

$\quad x - 4 = 0 \quad$ or $\quad x + 4 = 0$

$\qquad x = 4 \quad$ or $\qquad x = -4$

The solutions are $4$ and $-4$.

**39.** $\qquad\quad 9x^2 - 4 = 0$

$\quad (3x - 2)(3x + 2) = 0$

$\quad 3x - 2 = 0 \quad$ or $\quad 3x + 2 = 0$

$\qquad 3x = 2 \quad$ or $\qquad 3x = -2$

$\qquad x = \dfrac{2}{3} \quad$ or $\qquad x = -\dfrac{2}{3}$

The solutions are $\dfrac{2}{3}$ and $-\dfrac{2}{3}$.

**41.** $0 = 6x + x^2 + 9$

$0 = x^2 + 6x + 9 \qquad$ Writing in descending order

$0 = (x + 3)(x + 3)$

$\quad x + 3 = 0 \quad$ or $\quad x + 3 = 0$

$\qquad x = -3 \quad$ or $\qquad x = -3$

There is only one solution, $-3$.

**43.** $\qquad x^2 + 16 = 8x$

$\quad x^2 - 8x + 16 = 0 \qquad$ Subtracting $8x$

$\quad (x - 4)(x - 4) = 0$

$\quad x - 4 = 0 \quad$ or $\quad x - 4 = 0$

$\qquad x = 4 \quad$ or $\qquad x = 4$

There is only one solution, $4$.

**45.** $\qquad 5x^2 = 6x$

$\quad 5x^2 - 6x = 0$

$\quad x(5x - 6) = 0$

$\quad x = 0 \quad$ or $\quad 5x - 6 = 0$

$\quad x = 0 \quad$ or $\qquad 5x = 6$

$\quad x = 0 \quad$ or $\qquad x = \dfrac{6}{5}$

The solutions are $0$ and $\dfrac{6}{5}$.

**47.** $\qquad 6x^2 - 4x = 10$

$\quad 6x^2 - 4x - 10 = 0$

$\quad 2(3x^2 - 2x - 5) = 0$

$\quad 2(3x - 5)(x + 1) = 0$

$\quad 3x - 5 = 0 \quad$ or $\quad x + 1 = 0$

$\qquad 3x = 5 \quad$ or $\qquad x = -1$

$\qquad x = \dfrac{5}{3} \quad$ or $\qquad x = -1$

The solutions are $\dfrac{5}{3}$ and $-1$.

**49.** $\qquad 12y^2 - 5y = 2$

$\quad 12y^2 - 5y - 2 = 0$

$\quad (4y + 1)(3y - 2) = 0$

$\quad 4y + 1 = 0 \quad$ or $\quad 3y - 2 = 0$

$\qquad 4y = -1 \quad$ or $\qquad 3y = 2$

$\qquad y = -\dfrac{1}{4} \quad$ or $\qquad y = \dfrac{2}{3}$

The solutions are $-\dfrac{1}{4}$ and $\dfrac{2}{3}$.

**51.**     $t(3t + 1) = 2$

$3t^2 + t = 2$     Multiplying on the left

$3t^2 + t - 2 = 0$     Subtracting 2

$(3t - 2)(t + 1) = 0$

$3t - 2 = 0$   or   $t + 1 = 0$

$3t = 2$   or        $t = -1$

$t = \dfrac{2}{3}$   or        $t = -1$

The solutions are $\dfrac{2}{3}$ and $-1$.

**53.**          $100y^2 = 49$

$100y^2 - 49 = 0$

$(10y + 7)(10y - 7) = 0$

$10y + 7 = 0$     or   $10y - 7 = 0$

$10y = -7$   or        $10y = 7$

$y = -\dfrac{7}{10}$   or        $y = \dfrac{7}{10}$

The solutions are $-\dfrac{7}{10}$ and $\dfrac{7}{10}$.

**55.**          $x^2 - 5x = 18 + 2x$

$x^2 - 5x - 18 - 2x = 0$          Subtracting 18 and $2x$

$x^2 - 7x - 18 = 0$

$(x - 9)(x + 2) = 0$

$x - 9 = 0$   or   $x + 2 = 0$

$x = 9$   or        $x = -2$

The solutions are 9 and $-2$.

**57.**   $10x^2 - 23x + 12 = 0$

$(5x - 4)(2x - 3) = 0$

$5x - 4 = 0$   or   $2x - 3 = 0$

$5x = 4$   or        $2x = 3$

$x = \dfrac{4}{5}$   or        $x = \dfrac{3}{2}$

The solutions are $\dfrac{4}{5}$ and $\dfrac{3}{2}$.

**59.** $(a + b)^2$

**61.** $144 \div -9 = -16$

The two numbers have different signs, so their quotient is negative.

**63.**   $-\dfrac{5}{8} \div \dfrac{3}{16} = -\dfrac{5}{8} \cdot \dfrac{16}{3}$

$= -\dfrac{5 \cdot 16}{8 \cdot 3}$

$= -\dfrac{5 \cdot 8 \cdot 2}{8 \cdot 3}$

$= -\dfrac{10}{3}$

**65.**          $b(b + 9) = 4(5 + 2b)$

$b^2 + 9b = 20 + 8b$

$b^2 + 9b - 8b - 20 = 0$

$b^2 + b - 20 = 0$

$(b + 5)(b - 4) = 0$

$b + 5 = 0$     or   $b - 4 = 0$

$b = -5$   or        $b = 4$

The solutions are $-5$ and 4.

**67.**          $(t - 3)^2 = 36$

$t^2 - 6t + 9 = 36$

$t^2 - 6t - 27 = 0$

$(t - 9)(t + 3) = 0$

$t - 9 = 0$   or   $t + 3 = 0$

$t = 9$   or        $t = -3$

The solutions are 9 and $-3$.

**69.**          $x^2 - \dfrac{1}{64} = 0$

$\left(x - \dfrac{1}{8}\right)\left(x + \dfrac{1}{8}\right) = 0$

$x - \dfrac{1}{8} = 0$   or   $x + \dfrac{1}{8} = 0$

$x = \dfrac{1}{8}$   or        $x = -\dfrac{1}{8}$

The solutions are $\dfrac{1}{8}$ and $-\dfrac{1}{8}$.

**71.**          $\dfrac{5}{16}x^2 = 5$

$\dfrac{5}{16}x^2 - 5 = 0$

$5\left(\dfrac{1}{16}x^2 - 1\right) = 0$

$5\left(\dfrac{1}{4}x - 1\right)\left(\dfrac{1}{4}x + 1\right) = 0$

$\dfrac{1}{4}x - 1 = 0$   or   $\dfrac{1}{4}x + 1 = 0$

$\dfrac{1}{4}x = 1$   or        $\dfrac{1}{4}x = -1$

$x = 4$   or        $x = -4$

The solutions are 4 and $-4$.

**73.** (a)          $x = -3$   or        $x = 4$

$x + 3 = 0$     or   $x - 4 = 0$

$(x + 3)(x - 4) = 0$   Principle of zero products

$x^2 - x - 12 = 0$   Multiplying

(b)          $x = -3$   or        $x = -4$

$x + 3 = 0$     or   $x + 4 = 0$

$(x + 3)(x + 4) = 0$

$x^2 + 7x + 12 = 0$

(c)           $x = \dfrac{1}{2}$   or     $x = \dfrac{1}{2}$

$x - \dfrac{1}{2} = 0$   or   $x - \dfrac{1}{2} = 0$

$\left(x - \dfrac{1}{2}\right)\left(x - \dfrac{1}{2}\right) = 0$

$x^2 - x + \dfrac{1}{4} = 0,$   or

$4x^2 - 4x + 1 = 0$   Multiplying by 4

(d)      $(x - 5)(x + 5) = 0$

$x^2 - 25 = 0$

(e)   $(x - 0)(x - 0.1)\left(x - \dfrac{1}{4}\right) = 0$

$x\left(x - \dfrac{1}{10}\right)\left(x - \dfrac{1}{4}\right) = 0$

$x\left(x^2 - \dfrac{7}{20}x + \dfrac{1}{40}\right) = 0$

$x^3 - \dfrac{7}{20}x^2 + \dfrac{1}{40}x = 0,$   or

$40x^3 - 14x^2 + x = 0$   Multiplying by 40

## Exercise Set 4.8

1. **Familiarize**. Let $x$ = the number, or numbers.

**Translate**. We reword the problem.

7 plus   the square of a number   is 32.

$\downarrow$ $\downarrow$   $\downarrow$   $\downarrow$ $\downarrow$
7 $+$   $x^2$   $= 32$

**Solve**. We solve the equation.

$7 + x^2 = 32$

$7 + x^2 - 32 = 32 - 32$   Subtracting 32

$x^2 - 25 = 0$   Simplifying

$(x + 5)(x - 5) = 0$   Factoring

$x + 5 = 0$   or   $x - 5 = 0$   Using the principle of
zero products

$x = -5$   or     $x = 5$

**Check**. The square of both $-5$ and 5 is 25, and 7 added to 25 is 32. Both numbers check.

**State**. There are two such numbers, $-5$ and 5.

3. **Familiarize**. Let $x$ = the number, or numbers.

**Translate**. We reword the problem.

15 plus   the square of a number   is eight times   the number.

$\downarrow$ $\downarrow$   $\downarrow$   $\downarrow$ $\downarrow$ $\downarrow$   $\downarrow$
15 $+$   $x^2$   $= 8$ $\cdot$   $x$

**Solve**. We solve the equation.

$15 + x^2 = 8x$

$15 + x^2 - 8x = 8x - 8x$   Subtracting $8x$

$15 + x^2 - 8x = 0$   Simplifying

$x^2 - 8x + 15 = 0$   Writing in descending order

$(x - 5)(x - 3) = 0$   Factoring

$x - 5 = 0$   or   $x - 3 = 0$   Using the principle of
zero products

$x = 5$   or     $x = 3$

**Check**. The square of 5 is 25. Fifteen more than 25 is 40 and $8 \cdot 5 = 40$, so 5 checks. The square of 3 is 9. Fifteen more than 9 is 24 and $8 \cdot 3 = 24$, so 3 also checks.

**State**. There are two such numbers, 5 and 3.

5. **Familiarize**. The page numbers on facing pages are consecutive integers. Let $x$ = the smaller integer. Then $x + 1$ = the larger integer.

**Translate**. We reword the problem.

Smaller integer   times   larger integer   is 210.

$\downarrow$   $\downarrow$   $\downarrow$   $\downarrow$ $\downarrow$
$x$   $\cdot$   $(x + 1)$   $= 210$

**Solve**. We solve the equation.

$x(x + 1) = 210$

$x^2 + x = 210$

$x^2 + x - 210 = 0$

$(x + 15)(x - 14) = 0$

$x + 15 = 0$   or   $x - 14 = 0$

$x = -15$   or     $x = 14$

**Check**. The solutions of the equation are $-15$ and 14. Since a page number cannot be negative, $-15$ cannot be a solution of the original problem. We only need to check 14. When $x = 14$, then $x + 1 = 15$, and $14 \cdot 15 = 210$. This checks.

**State**. The page numbers are 14 and 15.

7. **Familiarize**. Let $x$ = the smaller even integer. Then $x + 2$ = the larger even integer.

**Translate**. We reword the problem.

Smaller even integer   times   larger even integer   is 168.

$\downarrow$   $\downarrow$   $\downarrow$   $\downarrow$ $\downarrow$
$x$   $\cdot$   $(x + 2)$   $= 168$

**Solve**.

$x(x + 2) = 168$

$x^2 + 2x = 168$

$x^2 + 2x - 168 = 0$

$(x + 14)(x - 12) = 0$

$x + 14 = 0$   or   $x - 12 = 0$

$x = -14$   or     $x = 12$

**Check**. The solutions of the equation are $-14$ and 12. When $x$ is $-14$, then $x + 2$ is $-12$ and $-14(-12) = 168$.

The numbers $-14$ and $-12$ are consecutive even integers which are solutions of the problem. When $x$ is 12, then $x + 2$ is 14 and $12 \cdot 14 = 168$. The numbers 12 and 14 are also consecutive even integers which are solutions of the problem.

*State.* We have two solutions, each of which consists of a pair of numbers: $-14$ and $-12$, and 12 and 14.

9. *Familiarize.* Let $x =$ the smaller odd integer. Then $x + 2 =$ the larger odd integer.

   *Translate.* We reword the problem.

   $$\underbrace{\text{Smaller odd integer}}_{x} \quad \underset{\downarrow}{\text{times}} \quad \underbrace{\text{larger odd integer}}_{(x+2)} \quad \underset{= \ 255}{\text{is 255.}}$$

   $$x \quad \cdot \quad (x+2) \quad = 255$$

   *Solve.*
   $$x(x+2) = 255$$
   $$x^2 + 2x = 255$$
   $$x^2 + 2x - 255 = 0$$
   $$(x-15)(x+17) = 0$$
   $$x - 15 = 0 \quad \text{or} \quad x + 17 = 0$$
   $$x = 15 \quad \text{or} \quad x = -17$$

   *Check.* The solutions of the equation are 15 and $-17$. When $x$ is 15, then $x + 2$ is 17 and $15 \cdot 17 = 255$. The numbers 15 and 17 are consecutive odd integers which are solutions to the problem. When $x$ is $-17$, then $x + 2$ is $-15$ and $-17(-15) = 255$. The numbers $-17$ and $-15$ are also consecutive odd integers which are solutions to the problem.

   *State.* We have two solutions, each of which consists of a pair of numbers: 15 and 17, and $-17$ and $-15$.

11. *Familiarize.* Using the labels shown on the drawing in the text, we let $w =$ the width and $w + 5 =$ the length. Recall that the area of a rectangle is length times width.

    *Translate.* We reword the problem.

    $$\underbrace{\text{Length}}_{(w+5)} \quad \underset{\downarrow}{\text{times}} \quad \underbrace{\text{width}}_{w} \quad \underset{\downarrow}{\text{is}} \quad \underbrace{\text{area.}}_{84}$$

    $$(w+5) \quad \cdot \quad w \quad = \quad 84$$

    *Solve.* We solve the equation.
    $$(w+5)w = 84$$
    $$w^2 + 5w = 84$$
    $$w^2 + 5w - 84 = 84 - 84$$
    $$w^2 + 5w - 84 = 0$$
    $$(w+12)(w-7) = 0$$
    $$w + 12 = 0 \quad \text{or} \quad w - 7 = 0$$
    $$w = -12 \quad \text{or} \quad w = 7$$

    *Check.* The width of a rectangle cannot have a negative measure, so $-12$ cannot be a solution. Suppose the width is 7 cm. The length is 5 cm greater than the width, so the length is 12 cm and the area is $12 \cdot 7$, or 84 cm². The numbers check in the original problem.

    *State.* The length is 12 cm, and the width is 7 cm.

13. *Familiarize.* First draw a picture. Let $x =$ the length of a side of the square.

    The area of the square is $x \cdot x$, or $x^2$. The perimeter of the square is $x + x + x + x$, or $4x$.

    *Translate.*

    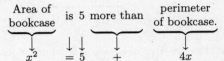

    $$\underbrace{\text{Area of bookcase}}_{x^2} \quad \underset{= \ 5}{\text{is 5}} \quad \underbrace{\text{more than}}_{+} \quad \underbrace{\text{perimeter of bookcase.}}_{4x}$$

    *Solve.*
    $$x^2 = 5 + 4x$$
    $$x^2 - 4x - 5 = 0$$
    $$(x-5)(x+1) = 0$$
    $$x - 5 = 0 \quad \text{or} \quad x + 1 = 0$$
    $$x = 5 \quad \text{or} \quad x = -1$$

    *Check.* The solutions of the equation are 5 and $-1$. The length of a side cannot be negative, so we only check 5. The area is $5 \cdot 5$, or 25. The perimeter is $5 + 5 + 5 + 5$, or 20. The area, 25, is 5 more than the perimeter, 20. This checks.

    *State.* The length of a side is 5.

15. *Familiarize.* Using the labels shown on the drawing in the text, we let $b =$ the base and $b + 1 =$ the height. Recall that the formula for the area of a triangle is $\frac{1}{2} \cdot (\text{base}) \cdot (\text{height})$.

    *Translate.*

    $$\underset{\frac{1}{2}}{\frac{1}{2}} \cdot \underbrace{\text{base}}_{b} \cdot \underbrace{\text{height}}_{(b+1)} \underset{=}{\text{is}} \underbrace{\text{area.}}_{15}$$

    *Solve.* We solve the equation.
    $$\frac{1}{2}b(b+1) = 15$$
    $$2 \cdot \frac{1}{2}b(b+1) = 2 \cdot 15 \quad \text{Multiplying by 2}$$
    $$b(b+1) = 30 \quad \text{Simplifying}$$
    $$b^2 + b = 30$$
    $$b^2 + b - 30 = 0$$
    $$(b+6)(b-5) = 0$$
    $$b + 6 = 0 \quad \text{or} \quad b - 5 = 0$$
    $$b = -6 \quad \text{or} \quad b = 5$$

    *Check.* The base of a triangle cannot have a negative length, so $-6$ cannot be a solution. Suppose the base is 5 cm. The height is 1 cm greater than the base, so the height is 6 cm and the area is $\frac{1}{2} \cdot 5 \cdot 6$, or 15 cm². These numbers check.

*State*. The height is 6 cm, and the base is 5 cm.

17. *Familiarize*. We make a drawing. Let $x =$ the length of a side of the original square. Then $x + 3 =$ the length of a side of the enlarged square.

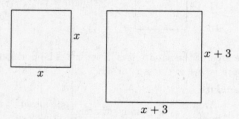

Recall that the area of a square is found by squaring the length of a side.

*Translate*.

$$
\underbrace{\text{Area of}}_{81} \quad \underbrace{\text{is}}_{=} \quad \underbrace{\text{the square of the}}_{(x+3)^2}
$$

Area of enlarged square | is | the square of the lengthened side.

$$81 \quad = \quad (x+3)^2$$

*Solve*.

$$81 = (x+3)^2$$
$$81 = x^2 + 6x + 9$$
$$0 = x^2 + 6x - 72$$
$$0 = (x+12)(x-6)$$
$$x + 12 = 0 \quad \text{or} \quad x - 6 = 0$$
$$x = -12 \quad \text{or} \quad x = 6$$

*Check*. The solutions of the equation are $-12$ and $6$. The length of a side cannot be negative, so $-12$ cannot be a solution. Suppose the length of a side of the original square is 6 km. Then the length of a side of the new square is $6 + 3$, or 9 km. Its area is $9^2$, or 81 km². The numbers check.

*State*. The length of a side of the original square is 6 km.

19. *Familiarize*. Let $x =$ the smaller odd positive integer. Then $x + 2 =$ the larger odd positive integer.

*Translate*.

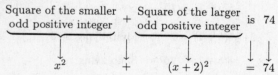

Square of the smaller odd positive integer | + | Square of the larger odd positive integer | is | 74

$$x^2 \quad + \quad (x+2)^2 \quad = \quad 74$$

*Solve*.

$$x^2 + (x+2)^2 = 74$$
$$x^2 + x^2 + 4x + 4 = 74$$
$$2x^2 + 4x - 70 = 0$$
$$2(x^2 + 2x - 35) = 0$$
$$2(x+7)(x-5) = 0$$
$$x + 7 = 0 \quad \text{or} \quad x - 5 = 0$$
$$x = -7 \quad \text{or} \quad x = 5$$

*Check*. The solutions of the equation are $-7$ and $5$. The problem asks for odd positive integers, so $-7$ cannot be a solution. When $x$ is 5, $x + 2$ is 7. The numbers 5 and 7 are consecutive odd positive integers. The sum of their squares, $25 + 49$, is 74. The numbers check.

*State*. The integers are 5 and 7.

21. *Familiarize*. Reread Example 4 in Section 3.3.

*Translate*. Substitute 14 for $n$.

$$14^2 - 14 = N$$

*Solve*. We do the computation on the left.

$$14^2 - 14 = N$$
$$196 - 14 = N$$
$$182 = N$$

*Check*. We can redo the computation, or we can solve the equation $n^2 - n = 182$. The answer checks.

*State*. 182 games will be played.

23. *Familiarize*. Reread Example 4 in Section 3.3.

*Translate*. Substitute 132 for $N$.

$$n^2 - n = 132$$

*Solve*.

$$n^2 - n = 132$$
$$n^2 - n - 132 = 0$$
$$(n-12)(n+11) = 0$$
$$n - 12 = 0 \quad \text{or} \quad n + 11 = 0$$
$$n = 12 \quad \text{or} \quad n = -11$$

*Check*. The solutions of the equation are 12 and $-11$. Since the number of teams cannot be negative, $-11$ cannot be a solution. But 12 checks since $12^2 - 12 = 144 - 12 = 132$.

*State*. There are 12 teams in the league.

25. *Familiarize*. We will use the formula $N = \frac{1}{2}(n^2 - n)$.

*Translate*. Substitute 100 for $n$.

$$N = \frac{1}{2}(100^2 - 100)$$

*Solve*. We do the computation on the right.

$$N = \frac{1}{2}(10,000 - 100)$$
$$N = \frac{1}{2}(9900)$$
$$N = 4950$$

*Check*. We can redo the computation, or we can solve the equation $4950 = \frac{1}{2}(n^2 - n)$. The answer checks.

*State*. 4950 handshakes are possible.

27. *Familiarize*. We will use the formula $N = \frac{1}{2}(n^2 - n)$.

*Translate*. Substitute 300 for $N$.

$$300 = \frac{1}{2}(n^2 - n)$$

*Solve*. We solve the equation.

$$2 \cdot 300 = 2 \cdot \frac{1}{2}(n^2 - n) \qquad \text{Multiplying by 2}$$

$$600 = n^2 - n$$

$$0 = n^2 - n - 600$$

$$0 = (n + 24)(n - 25)$$

$$n + 24 = 0 \quad \text{or} \quad n - 25 = 0$$

$$n = -24 \quad \text{or} \qquad n = 25$$

*Check*. The number of people at a meeting cannot be negative, so $-24$ cannot be a solution. But 25 checks since $\frac{1}{2}(25^2 - 25) = \frac{1}{2}(625 - 25) = \frac{1}{2} \cdot 600 = 300$.

*State*. There were 25 people at the party.

29. *Familiarize*. We make a drawing. Let $x =$ the length of the unknown leg. Then $x + 2 =$ the length of the hypotenuse.

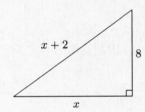

*Translate*. Use the Pythagorean theorem.

$$a^2 + b^2 = c^2$$

$$8^2 + x^2 = (x + 2)^2$$

*Solve*. We solve the equation.

$$8^2 + x^2 = (x + 2)^2$$

$$64 + x^2 = x^2 + 4x + 4$$

$$60 = 4x \qquad \text{Subtracting } x^2 \text{ and } 4$$

$$15 = x$$

*Check*. When $x = 15$, then $x + 2 = 17$ and $8^2 + 15^2 = 17^2$. Thus, 15 and 17 check.

*State*. The lengths of the hypotenuse and the other leg are 17 ft and 15 ft, respectively.

31. *Familiarize*. Using the labels shown on the drawing in the text, we let $x =$ the width of the walk. Then the length and width of the rectangle formed by the pool and walk together are $40 + 2x$ and $20 + 2x$, respectively.

*Translate*.

$$\underbrace{\text{Area}}_{1500} \; \underbrace{\text{is}}_{=} \; \underbrace{\text{length}}_{(40 + 2x)} \; \underbrace{\text{times}}_{\cdot} \; \underbrace{\text{width.}}_{(20 + 2x)}$$

*Solve*. We solve the equation.

$$1500 = (40 + 2x)(20 + 2x)$$

$$1500 = 2(20 + x) \cdot 2(10 + x) \quad \begin{array}{l}\text{Factoring 2 out of each} \\ \text{factor on the right}\end{array}$$

$$1500 = 4 \cdot (20 + x)(10 + x)$$

$$375 = (20 + x)(10 + x) \qquad \text{Dividing by 4}$$

$$375 = 200 + 30x + x^2$$

$$0 = x^2 + 30x - 175$$

$$0 = (x + 35)(x - 5)$$

$$x + 35 = 0 \quad \text{or} \quad x - 5 = 0$$

$$x = -35 \quad \text{or} \qquad x = 5$$

*Check*. The solutions of the equation are $-35$ and 5. Since the width of the walk cannot be negative, $-35$ is not a solution. When $x = 5$, $40 + 2x = 40 + 2 \cdot 5$, or 50 and $20 + 2x = 20 + 2 \cdot 5$, or 30. The total area of the pool and walk is $50 \cdot 30$, or 1500 ft$^2$. This checks.

*State*. The width of the walk is 5 ft.

33. *Familiarize*. Let $y =$ the ten's digit. Then $y + 4 =$ the one's digit and $10y + y + 4$, or $11y + 4$, represents the number.

*Translate*.

$$\underbrace{\text{The number}}_{11y + 4} \; \underbrace{\text{plus}}_{+} \; \underbrace{\begin{array}{c}\text{the product} \\ \text{of the digits}\end{array}}_{y(y + 4)} \; \underbrace{\text{is 58.}}_{= \quad 58}$$

*Solve*. We solve the equation.

$$11y + 4 + y(y + 4) = 58$$

$$11y + 4 + y^2 + 4y = 58$$

$$y^2 + 15y + 4 = 58$$

$$y^2 + 15y - 54 = 0$$

$$(y + 18)(y - 3) = 0$$

$$y + 18 = 0 \quad \text{or} \quad y - 3 = 0$$

$$y = -18 \quad \text{or} \qquad y = 3$$

*Check*. Since $-18$ cannot be a digit of the number, we only need to check 3. When $y = 3$, then $y + 4 = 7$ and the number is 37. We see that $37 + 3 \cdot 7 = 37 + 21$, or 58. The result checks.

*State*. The number is 37.

35. *Familiarize*. We make a drawing. Let $w =$ the width of the piece of cardboard. Then $2w =$ the length.

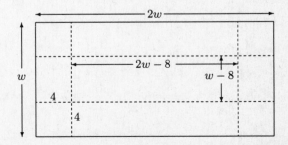

The box will have length $2w - 8$, width $w - 8$, and height 4. Recall that the formula for volume is $V = $ length $\times$ width $\times$ height.

**Translate**.

$$\underbrace{\text{The volume}}\quad \text{is}\quad \underbrace{616\text{cm}^3}.$$

$$\qquad\downarrow\qquad\qquad\downarrow\qquad\downarrow$$

$$(2w - 8)(w - 8)(4) = \quad 616$$

**Solve**. We solve the equation.

$$(2w - 8)(w - 8)(4) = 616$$
$$(2w^2 - 24w + 64)(4) = 616$$
$$8w^2 - 96w + 256 = 616$$
$$8w^2 - 96w - 360 = 0$$
$$8(w^2 - 12w - 45) = 0$$
$$w^2 - 12w - 45 = 0 \qquad \text{Dividing by 8}$$
$$(w - 15)(w + 3) = 0$$

$$w - 15 = 0 \quad \text{or} \quad w + 3 = 0$$
$$w = 15 \quad \text{or} \qquad w = -3$$

**Check**. The width cannot be negative, so we only need to check 15. When $w = 15$, then $2w = 30$ and the dimensions of the box are $30 - 8$ by $15 - 8$ by 4, or 22 by 7 by 4. The volume is $22 \cdot 7 \cdot 4$, or 616.

**State**. The cardboard is 30 cm by 15 cm.

# Chapter 5

# Rational Expressions and Equations

**1.** $\dfrac{-3}{2x}$

To determine which numbers make the rational expression undefined, we set the denominator equal to 0 and solve:

$$2x = 0$$
$$x = 0$$

The expression is undefined for the replacement number 0.

**3.** $\dfrac{5}{x-8}$

To determine which numbers make the rational expression undefined, we set the denominator equal to 0 and solve:

$$x - 8 = 0$$
$$x = 8$$

The expression is undefined for the replacement number 8.

**5.** $\dfrac{3}{2y+5}$

Set the denominator equal to 0 and solve:

$$2y + 5 = 0$$
$$2y = -5$$
$$y = -\frac{5}{2}$$

The expression is undefined for the replacement number $-\dfrac{5}{2}$.

**7.** $\dfrac{x^2+11}{x^2-3x-28}$

Set the denominator equal to 0 and solve:

$$x^2 - 3x - 28 = 0$$
$$(x-7)(x+4) = 0$$
$$x - 7 = 0 \quad \text{or} \quad x + 4 = 0$$
$$x = 7 \quad \text{or} \quad x = -4$$

The expression is undefined for the replacement numbers 7 and −4.

**9.** $\dfrac{m^3-2m}{m^2-25}$

Set the denominator equal to 0 and solve:

$$m^2 - 25 = 0$$
$$(m+5)(m-5) = 0$$
$$m + 5 = 0 \quad \text{or} \quad m - 5 = 0$$
$$m = -5 \quad \text{or} \quad m = 5$$

The expression is undefined for the replacement numbers −5 and 5.

**11.** $\dfrac{x-4}{3}$

Since the denominator is the constant 3, there are no replacement numbers for which the expression is undefined.

**13.** $\dfrac{4x}{4x} \cdot \dfrac{3x^2}{5y} = \dfrac{(4x)(3x^2)}{(4x)(5y)}$  Multiplying the numerators and the denominators

**15.** $\dfrac{2x}{2x} \cdot \dfrac{x-1}{x+4} = \dfrac{2x(x-1)}{2x(x+4)}$  Multiplying the numerators and the denominators

**17.** $\dfrac{3-x}{4-x} \cdot \dfrac{-1}{-1} = \dfrac{(3-x)(-1)}{(4-x)(-1)}$, or $\dfrac{-1(3-x)}{-1(4-x)}$

**19.** $\dfrac{y+6}{y+6} \cdot \dfrac{y-7}{y+2} = \dfrac{(y+6)(y-7)}{(y+6)(y+2)}$

**21.** $\dfrac{8x^3}{32x} = \dfrac{8 \cdot x \cdot x^2}{8 \cdot 4 \cdot x}$  Factoring numerator and denominator

$= \dfrac{8x}{8x} \cdot \dfrac{x^2}{4}$  Factoring the rational expression

$= 1 \cdot \dfrac{x^2}{4}$  $\left(\dfrac{8x}{8x} = 1\right)$

$= \dfrac{x^2}{4}$  We removed a factor of 1.

**23.** $\dfrac{48p^7q^5}{18p^5q^4} = \dfrac{8 \cdot 6 \cdot p^5 \cdot p^2 \cdot q^4 \cdot q}{6 \cdot 3 \cdot p^5 \cdot q^4}$  Factoring numerator and denominator

$= \dfrac{6p^5q^4}{6p^5q^4} \cdot \dfrac{8p^2q}{3}$  Factoring the rational expression

$= 1 \cdot \dfrac{8p^2q}{3}$  $\left(\dfrac{6p^5q^4}{6p^5q^4} = 1\right)$

$= \dfrac{8p^2q}{3}$  Removing a factor of 1

**25.** $\dfrac{4x-12}{4x} = \dfrac{4(x-3)}{4 \cdot x}$

$= \dfrac{4}{4} \cdot \dfrac{x-3}{x}$

$= 1 \cdot \dfrac{x-3}{x}$

$= \dfrac{x-3}{x}$

**27.** $\dfrac{3m^2+3m}{6m^2+9m} = \dfrac{3m(m+1)}{3m(2m+3)}$

$= \dfrac{3m}{3m} \cdot \dfrac{m+1}{2m+3}$

$= 1 \cdot \dfrac{m+1}{2m+3}$

$= \dfrac{m+1}{2m+3}$

**29.** $\dfrac{a^2 - 9}{a^2 + 5a + 6} = \dfrac{(a-3)(a+3)}{(a+2)(a+3)}$

$\qquad = \dfrac{a-3}{a+2} \cdot \dfrac{a+3}{a+3}$

$\qquad = \dfrac{a-3}{a+2} \cdot 1$

$\qquad = \dfrac{a-3}{a+2}$

**31.** $\dfrac{a^2 - 10a + 21}{a^2 - 11a + 28} = \dfrac{(a-7)(a-3)}{(a-7)(a-4)}$

$\qquad = \dfrac{a-7}{a-7} \cdot \dfrac{a-3}{a-4}$

$\qquad = 1 \cdot \dfrac{a-3}{a-4}$

$\qquad = \dfrac{a-3}{a-4}$

**33.** $\dfrac{x^2 - 25}{x^2 - 10x + 25} = \dfrac{(x-5)(x+5)}{(x-5)(x-5)}$

$\qquad = \dfrac{x-5}{x-5} \cdot \dfrac{x+5}{x-5}$

$\qquad = 1 \cdot \dfrac{x+5}{x-5}$

$\qquad = \dfrac{x+5}{x-5}$

**35.** $\dfrac{a^2 - 1}{a - 1} = \dfrac{(a-1)(a+1)}{a-1}$

$\qquad = \dfrac{a-1}{a-1} \cdot \dfrac{a+1}{1}$

$\qquad = 1 \cdot \dfrac{a+1}{1}$

$\qquad = a+1$

**37.** $\dfrac{x^2 + 1}{x + 1}$ cannot be simplified.

Neither the numerator nor the denominator can be factored.

**39.** $\dfrac{6x^2 - 54}{4x^2 - 36} = \dfrac{2 \cdot 3(x^2 - 9)}{2 \cdot 2(x^2 - 9)}$

$\qquad = \dfrac{2(x^2-9)}{2(x^2-9)} \cdot \dfrac{3}{2}$

$\qquad = 1 \cdot \dfrac{3}{2}$

$\qquad = \dfrac{3}{2}$

**41.** $\dfrac{6t + 12}{t^2 - t - 6} = \dfrac{6(t+2)}{(t-3)(t+2)}$

$\qquad = \dfrac{6}{t-3} \cdot \dfrac{t+2}{t+2}$

$\qquad = \dfrac{6}{t-3} \cdot 1$

$\qquad = \dfrac{6}{t-3}$

**43.** $\dfrac{2t^2 + 6t + 4}{4t^2 - 12t - 16} = \dfrac{2(t^2 + 3t + 2)}{4(t^2 - 3t - 4)}$

$\qquad = \dfrac{2(t+2)(t+1)}{2 \cdot 2(t-4)(t+1)}$

$\qquad = \dfrac{2(t+1)}{2(t+1)} \cdot \dfrac{t+2}{2(t-4)}$

$\qquad = 1 \cdot \dfrac{t+2}{2(t-4)}$

$\qquad = \dfrac{t+2}{2(t-4)}$

**45.** $\dfrac{t^2 - 4}{(t+2)^2} = \dfrac{(t-2)(t+2)}{(t+2)(t+2)}$

$\qquad = \dfrac{t-2}{t+2} \cdot \dfrac{t+2}{t+2}$

$\qquad = \dfrac{t-2}{t+2} \cdot 1$

$\qquad = \dfrac{t-2}{t+2}$

**47.** $\dfrac{6 - x}{x - 6} = \dfrac{-(-6 + x)}{x - 6}$

$\qquad = \dfrac{-1(x-6)}{x-6}$

$\qquad = -1 \cdot \dfrac{x-6}{x-6}$

$\qquad = -1 \cdot 1$

$\qquad = -1$

**49.** $\dfrac{a - b}{b - a} = \dfrac{-(-a + b)}{b - a}$

$\qquad = \dfrac{-1(b-a)}{b-a}$

$\qquad = -1 \cdot \dfrac{b-a}{b-a}$

$\qquad = -1 \cdot 1$

$\qquad = -1$

**51.** $\dfrac{6t - 12}{2 - t} = \dfrac{-6(-t + 2)}{2 - t}$

$\qquad = \dfrac{-6(2-t)}{2-t}$

$\qquad = \dfrac{-6(2-t)}{2-t}$

$\qquad = -6$

**53.** $\dfrac{x^2 - 1}{1 - x} = \dfrac{(x+1)(x-1)}{-1(-1 + x)}$

$\qquad = \dfrac{(x+1)(x-1)}{-1(x-1)}$

$\qquad = \dfrac{(x+1)(x\!-\!1)}{-1(x\!-\!1)}$

$\qquad = -(x+1)$

$\qquad = -x - 1$

**55.** $\dfrac{4x^3}{3x} \cdot \dfrac{14}{x} = \dfrac{4x^3 \cdot 14}{3x \cdot x}$ Multiplying the numerators and the denominators

$\phantom{\dfrac{4x^3}{3x} \cdot \dfrac{14}{x}} = \dfrac{4 \cdot x \cdot x \cdot x \cdot 14}{3 \cdot x \cdot x}$ Factoring the numerator and the denominator

$\phantom{\dfrac{4x^3}{3x} \cdot \dfrac{14}{x}} = \dfrac{4 \cdot \not{x} \cdot \not{x} \cdot x \cdot 14}{3 \cdot \not{x} \cdot \not{x}}$ Removing a factor of 1

$\phantom{\dfrac{4x^3}{3x} \cdot \dfrac{14}{x}} = \dfrac{56x}{3}$ Simplifying

**57.** $\dfrac{3c}{d^2} \cdot \dfrac{4d}{6c^3} = \dfrac{3c \cdot 4d}{d^2 \cdot 6c^3}$ Multiplying the numerators and the denominators

$\phantom{\dfrac{3c}{d^2} \cdot \dfrac{4d}{6c^3}} = \dfrac{3 \cdot c \cdot 2 \cdot 2 \cdot d}{d \cdot d \cdot 3 \cdot 2 \cdot c \cdot c \cdot c}$ Factoring the numerator and the denominator

$\phantom{\dfrac{3c}{d^2} \cdot \dfrac{4d}{6c^3}} = \dfrac{\not{3} \cdot \not{c} \cdot \not{2} \cdot 2 \cdot \not{d}}{\not{d} \cdot d \cdot \not{3} \cdot \not{2} \cdot \not{c} \cdot c \cdot c}$

$\phantom{\dfrac{3c}{d^2} \cdot \dfrac{4d}{6c^3}} = \dfrac{2}{dc^2}$

**59.** $\dfrac{x^2 - 3x - 10}{(x - 2)^2} \cdot \dfrac{x - 2}{x - 5} = \dfrac{(x^2 - 3x - 10)(x - 2)}{(x - 2)^2(x - 5)}$

$\phantom{\dfrac{x^2 - 3x - 10}{(x - 2)^2}} = \dfrac{(x - 5)(x + 2)(x - 2)}{(x - 2)(x - 2)(x - 5)}$

$\phantom{\dfrac{x^2 - 3x - 10}{(x - 2)^2}} = \dfrac{(\cancel{x - 5})(x + 2)(\cancel{x - 2})}{(\cancel{x - 2})(x - 2)(\cancel{x - 5})}$

$\phantom{\dfrac{x^2 - 3x - 10}{(x - 2)^2}} = \dfrac{x + 2}{x - 2}$

**61.** $\dfrac{a^2 - 9}{a^2} \cdot \dfrac{a^2 - 3a}{a^2 + a - 12} = \dfrac{(a - 3)(a + 3)(a)(a - 3)}{a \cdot a(a + 4)(a - 3)}$

$\phantom{\dfrac{a^2 - 9}{a^2}} = \dfrac{(\cancel{a - 3})(a + 3)(\not{a})(a - 3)}{\not{a} \cdot a(a + 4)(\cancel{a - 3})}$

$\phantom{\dfrac{a^2 - 9}{a^2}} = \dfrac{(a - 3)(a + 3)}{a(a + 4)}$

**63.** $\dfrac{4a^2}{3a^2 - 12a + 12} \cdot \dfrac{3a - 6}{2a} = \dfrac{4a^2(3a - 6)}{(3a^2 - 12a + 12)2a}$

$\phantom{\dfrac{4a^2}{3a^2 - 12a + 12}} = \dfrac{2 \cdot 2 \cdot a \cdot a \cdot 3 \cdot (a - 2)}{3 \cdot (a - 2) \cdot (a - 2) \cdot 2 \cdot a}$

$\phantom{\dfrac{4a^2}{3a^2 - 12a + 12}} = \dfrac{\not{2} \cdot 2 \cdot \not{a} \cdot a \cdot \not{3} \cdot (\cancel{a - 2})}{\not{3} \cdot (\cancel{a - 2}) \cdot (a - 2) \cdot \not{2} \cdot \not{a}}$

$\phantom{\dfrac{4a^2}{3a^2 - 12a + 12}} = \dfrac{2a}{a - 2}$

**65.** $\dfrac{t^4 - 16}{t^4 - 1} \cdot \dfrac{t^2 + 1}{t^2 + 4}$

$= \dfrac{(t^4 - 16)(t^2 + 1)}{(t^4 - 1)(t^2 + 4)}$

$= \dfrac{(t^2 + 4)(t + 2)(t - 2)(t^2 + 1)}{(t^2 + 1)(t + 1)(t - 1)(t^2 + 4)}$

$= \dfrac{(\cancel{t^2 + 4})(t + 2)(t - 2)(\cancel{t^2 + 1})}{(\cancel{t^2 + 1})(t + 1)(t - 1)(\cancel{t^2 + 4})}$

$= \dfrac{(t + 2)(t - 2)}{(t + 1)(t - 1)}$

**67.** $\dfrac{(x + 4)^3}{(x + 2)^3} \cdot \dfrac{x^2 + 4x + 4}{x^2 + 8x + 16}$

$= \dfrac{(x + 4)^3(x^2 + 4x + 4)}{(x + 2)^3(x^2 + 8x + 16)}$

$= \dfrac{(x + 4)(x + 4)(x + 4)(x + 2)(x + 2)}{(x + 2)(x + 2)(x + 2)(x + 4)(x + 4)}$

$= \dfrac{(\cancel{x + 4})(\cancel{x + 4})(x + 4)(\cancel{x + 2})(\cancel{x + 2})}{(\cancel{x + 2})(\cancel{x + 2})(x + 2)(\cancel{x + 4})(\cancel{x + 4})}$

$= \dfrac{x + 4}{x + 2}$

**69.** $\dfrac{5a^2 - 180}{10a^2 - 10} \cdot \dfrac{20a + 20}{2a - 12} = \dfrac{(5a^2 - 180)(20a + 20)}{(10a^2 - 10)(2a - 12)}$

$\phantom{\dfrac{5a^2 - 180}{10a^2 - 10}} = \dfrac{5(a + 6)(a - 6)(2)(10)(a + 1)}{10(a + 1)(a - 1)(2)(a - 6)}$

$\phantom{\dfrac{5a^2 - 180}{10a^2 - 10}} = \dfrac{5(a + 6)(\cancel{a - 6})(\cancel{2})(\cancel{10})(\cancel{a + 1})}{\cancel{10}(\cancel{a + 1})(a - 1)(\cancel{2})(\cancel{a - 6})}$

$\phantom{\dfrac{5a^2 - 180}{10a^2 - 10}} = \dfrac{5(a + 6)}{a - 1}$

**71. *Familiarize*.** Let $x$ = the smaller even integer. Then $x + 2$ = the larger even integer.

***Translate*.** We reword the problem.

Smaller even integer $\underbrace{\phantom{xxx}}$ times larger even integer $\underbrace{\phantom{xxx}}$ is 360.

$\phantom{xxxxx} x \phantom{xxxxxxx} \cdot \phantom{xxxxxx} (x + 2) \phantom{xxxx} = 360$

***Solve*.**

$x(x + 2) = 360$

$x^2 + 2x = 360$

$x^2 + 2x - 360 = 0$

$(x + 20)(x - 18) = 0$

$x + 20 = 0 \quad \text{or} \quad x - 18 = 0$

$x = -20 \quad \text{or} \qquad x = 18$

***Check*.** The solutions of the equation are $-20$ and $18$. When $x = -20$, then $x + 2 = -18$ and $-20(-18) = 360$. The numbers $-20$ and $-18$ are consecutive even integers which are solutions to the problem. When $x = 18$, then $x + 2 = 20$ and $18 \cdot 20 = 360$. The numbers $18$ and $20$ are also consecutive even integers which are solutions to the problem.

***State*.** We have two solutions, each of which consists of a pair of numbers: $-20$ and $-18$, and $18$ and $20$.

**73.** We factor by grouping.

$2y^3 - 10y^2 + y - 5 = 2y^2(y - 5) + 1(y - 5)$

$\phantom{2y^3 - 10y^2 + y - 5} = (2y^2 + 1)(y - 5)$

**75.** $a^2 - 16a + 64 = a^2 - 2 \cdot a \cdot 8 + 8^2$ Trinomial square

$\phantom{a^2 - 16a + 64} = (a - 8)^2$

**77.**
$$\frac{x^4 - 16y^2}{(x^2 + 4y^2)(x - 2y)}$$

$$= \frac{(x^2 + 4y^2)(x + 2y)(x - 2y)}{(x^2 + 4y^2)(x - 2y)}$$

$$= \frac{(x^2 + 4y^2)\,(x + 2y)(x - 2y)}{(x^2 + 4y^2)\,(x - 2y)(1)}$$

$$= x + 2y$$

**79.**
$$\frac{t^4 - 1}{t^4 - 81} \cdot \frac{t^2 - 9}{t^2 + 1} \cdot \frac{(t - 9)^2}{(t + 1)^2}$$

$$= \frac{(t^2 + 1)(t + 1)(t - 1)(t + 3)(t - 3)(t - 9)(t - 9)}{(t^2 + 9)(t + 3)(t - 3)(t^2 + 1)(t + 1)(t + 1)}$$

$$= \frac{(t^2 + 1)(t + 1)(t - 1)(t + 3)(t - 3)(t - 9)(t - 9)}{(t^2 + 9)(t + 3)(t - 3)(t^2 + 1)(t + 1)(t + 1)}$$

$$= \frac{(t - 1)(t - 9)(t - 9)}{(t^2 + 9)(t + 1)}, \text{ or } \frac{(t - 1)(t - 9)^2}{(t^2 + 9)(t + 1)}$$

**81.**
$$\frac{x^2 - y^2}{(x - y)^2} \cdot \frac{x^2 - 2xy + y^2}{x^2 - 4xy - 5y^2}$$

$$= \frac{(x + y)(x - y)(x - y)(x - y)}{(x - y)(x - y)(x - 5y)(x + y)}$$

$$= \frac{(x + y)(x - y)(x - y)(x - y)}{(x - y)(x - y)(x - 5y)(x + y)}$$

$$= \frac{x - y}{x - 5y}$$

## Exercise Set 5.2

**1.** The reciprocal of $\frac{4}{x}$ is $\frac{x}{4}$ because $\frac{4}{x} \cdot \frac{x}{4} = 1$.

**3.** The reciprocal of $x^2 - y^2$ is $\frac{1}{x^2 - y^2}$ because
$$\frac{x^2 - y^2}{1} \cdot \frac{1}{x^2 - y^2} = 1.$$

**5.** The reciprocal of $\frac{x^2 + 2x - 5}{x^2 - 4x + 7}$ is $\frac{x^2 - 4x + 7}{x^2 + 2x - 5}$ because
$$\frac{x^2 + 2x - 5}{x^2 - 4x + 7} \cdot \frac{x^2 - 4x + 7}{x^2 + 2x - 5} = 1.$$

**7.** $\dfrac{2}{5} \div \dfrac{4}{3} = \dfrac{2}{5} \cdot \dfrac{3}{4}$     Multiplying by the reciprocal of the divisor

$$= \frac{2 \cdot 3}{5 \cdot 4}$$

$$= \frac{2 \cdot 3}{5 \cdot 2 \cdot 2} \quad \text{Factoring the denominator}$$

$$= \frac{2 \cdot 3}{5 \cdot 2 \cdot 2} \quad \text{Removing a factor of 1}$$

$$= \frac{3}{10} \quad \text{Simplifying}$$

**9.** $\dfrac{2}{x} \div \dfrac{8}{x} = \dfrac{2}{x} \cdot \dfrac{x}{8}$     Multiplying by the reciprocal of the divisor

$$= \frac{2 \cdot x}{x \cdot 8}$$

$$= \frac{2 \cdot x \cdot 1}{x \cdot 2 \cdot 4} \quad \begin{array}{l}\text{Factoring the numerator and}\\ \text{the denominator}\end{array}$$

$$= \frac{2 \cdot x \cdot 1}{x \cdot 2 \cdot 4} \quad \text{Removing a factor of 1}$$

$$= \frac{1}{4} \quad \text{Simplifying}$$

**11.** $\dfrac{a}{b^2} \div \dfrac{a^2}{b^3} = \dfrac{a}{b^2} \cdot \dfrac{b^3}{a^2}$     Multiplying by the reciprocal of the divisor

$$= \frac{a \cdot b^3}{b^2 \cdot a^2}$$

$$= \frac{a \cdot b^2 \cdot b}{b^2 \cdot a \cdot a}$$

$$= \frac{a \cdot b^2 \cdot b}{b^2 \cdot a \cdot a}$$

$$= \frac{b}{a}$$

**13.** $\dfrac{a + 2}{a - 3} \div \dfrac{a - 1}{a + 3} = \dfrac{a + 2}{a - 3} \cdot \dfrac{a + 3}{a - 1}$

$$= \frac{(a + 2)(a + 3)}{(a - 3)(a - 1)}$$

**15.** $\dfrac{x^2 - 1}{x} \div \dfrac{x + 1}{x - 1} = \dfrac{x^2 - 1}{x} \cdot \dfrac{x - 1}{x + 1}$

$$= \frac{(x^2 - 1)(x - 1)}{x(x + 1)}$$

$$= \frac{(x - 1)(x + 1)(x - 1)}{x(x + 1)}$$

$$= \frac{(x - 1)(x + 1)(x - 1)}{x(x + 1)}$$

$$= \frac{(x - 1)^2}{x}$$

**17.** $\dfrac{x + 1}{6} \div \dfrac{x + 1}{3} = \dfrac{x + 1}{6} \cdot \dfrac{3}{x + 1}$

$$= \frac{(x + 1) \cdot 3}{6(x + 1)}$$

$$= \frac{3(x + 1)}{2 \cdot 3(x + 1)}$$

$$= \frac{1 \cdot 3(x + 1)}{2 \cdot 3(x + 1)}$$

$$= \frac{1}{2}$$

**19.** $\dfrac{5x-5}{16} \div \dfrac{x-1}{6} = \dfrac{5x-5}{16} \cdot \dfrac{6}{x-1}$

$\qquad = \dfrac{(5x-5)\cdot 6}{16(x-1)}$

$\qquad = \dfrac{5(x-1)\cdot 2 \cdot 3}{2 \cdot 8(x-1)}$

$\qquad = \dfrac{5(x-1)\cdot 2 \cdot 3}{2 \cdot 8(x-1)}$

$\qquad = \dfrac{15}{8}$

**21.** $\dfrac{-6+3x}{5} \div \dfrac{4x-8}{25} = \dfrac{-6+3x}{5} \cdot \dfrac{25}{4x-8}$

$\qquad = \dfrac{(-6+3x)\cdot 25}{5(4x-8)}$

$\qquad = \dfrac{3(x-2)\cdot 5 \cdot 5}{5 \cdot 4(x-2)}$

$\qquad = \dfrac{3(x-2)\cdot 5 \cdot 5}{5 \cdot 4(x-2)}$

$\qquad = \dfrac{15}{4}$

**23.** $\dfrac{a+2}{a-1} \div \dfrac{3a+6}{a-5} = \dfrac{a+2}{a-1} \cdot \dfrac{a-5}{3a+6}$

$\qquad = \dfrac{(a+2)(a-5)}{(a-1)(3a+6)}$

$\qquad = \dfrac{(a+2)(a-5)}{(a-1)\cdot 3 \cdot (a+2)}$

$\qquad = \dfrac{(a+2)(a-5)}{(a-1)\cdot 3 \cdot (a+2)}$

$\qquad = \dfrac{a-5}{3(a-1)}$

**25.** $\dfrac{x^2-4}{x} \div \dfrac{x-2}{x+2} = \dfrac{x^2-4}{x} \cdot \dfrac{x+2}{x-2}$

$\qquad = \dfrac{(x^2-4)(x+2)}{x(x-2)}$

$\qquad = \dfrac{(x-2)(x+2)(x+2)}{x(x-2)}$

$\qquad = \dfrac{(x-2)(x+2)(x+2)}{x(x-2)}$

$\qquad = \dfrac{(x+2)^2}{x}$

**27.** $\dfrac{x^2-9}{4x+12} \div \dfrac{x-3}{6} = \dfrac{x^2-9}{4x+12} \cdot \dfrac{6}{x-3}$

$\qquad = \dfrac{(x^2-9)\cdot 6}{(4x+12)(x-3)}$

$\qquad = \dfrac{(x-3)(x+3)\cdot 3 \cdot 2}{2 \cdot 2(x+3)(x-3)}$

$\qquad = \dfrac{(x-3)(x+3)\cdot 3 \cdot 2}{2 \cdot 2(x+3)(x-3)}$

$\qquad = \dfrac{3}{2}$

**29.** $\dfrac{c^2+3c}{c^2+2c-3} \div \dfrac{c}{c+1} = \dfrac{c^2+3c}{c^2+2c-3} \cdot \dfrac{c+1}{c}$

$\qquad = \dfrac{(c^2+3c)(c+1)}{(c^2+2c-3)c}$

$\qquad = \dfrac{c(c+3)(c+1)}{(c+3)(c-1)c}$

$\qquad = \dfrac{c(c+3)(c+1)}{(c+3)(c-1)c}$

$\qquad = \dfrac{c+1}{c-1}$

**31.** $\dfrac{2y^2-7y+3}{2y^2+3y-2} \div \dfrac{6y^2-5y+1}{3y^2+5y-2}$

$\qquad = \dfrac{2y^2-7y+3}{2y^2+3y-2} \cdot \dfrac{3y^2+5y-2}{6y^2-5y+1}$

$\qquad = \dfrac{(2y^2-7y+3)(3y^2+5y-2)}{(2y^2+3y-2)(6y^2-5y+1)}$

$\qquad = \dfrac{(2y-1)(y-3)(3y-1)(y+2)}{(2y-1)(y+2)(3y-1)(2y-1)}$

$\qquad = \dfrac{(2y-1)(y-3)(3y-1)(y+2)}{(2y-1)(y+2)(3y-1)(2y-1)}$

$\qquad = \dfrac{y-3}{2y-1}$

**33.** $\dfrac{x^2-1}{4x+4} \div \dfrac{2x^2-4x+2}{8x+8} = \dfrac{x^2-1}{4x+4} \cdot \dfrac{8x+8}{2x^2-4x+2}$

$\qquad = \dfrac{(x^2-1)(8x+8)}{(4x+4)(2x^2-4x+2)}$

$\qquad = \dfrac{(x+1)(x-1)(2)(4)(x+1)}{4(x+1)(2)(x-1)(x-1)}$

$\qquad = \dfrac{(x+1)(x-1)(2)(4)(x+1)}{4(x+1)(2)(x-1)(x-1)}$

$\qquad = \dfrac{x+1}{x-1}$

**35.** *Familiarize*. Let $x =$ the number.

*Translate*.

| Sixteen more than | | the square of a number | is | eight times the number. |
|---|---|---|---|---|
| $16$ | $+$ | $x^2$ | $=$ | $8x$ |

*Solve*.

$\qquad 16 + x^2 = 8x$

$\quad x^2 - 8x + 16 = 0$

$\quad (x-4)(x-4) = 0$

$\quad x - 4 = 0 \quad \text{or} \quad x - 4 = 0$

$\qquad x = 4 \quad \text{or} \qquad x = 4$

*Check*. The square of 4, which is 16, plus 16 is 32, and eight times 4 is 32. The number checks.

*State*. The number is 4.

**37.** $(2x^{-3}y^4)^2 = 2^2(x^{-3})^2(y^4)^2$

$\qquad = 2^2x^{-6}y^8$      Multiplying exponents

$\qquad = 4x^{-6}y^8$      $(2^2 = 4)$

$\qquad = \dfrac{4y^8}{x^6}$      $\left(x^{-6} = \dfrac{1}{x^6}\right)$

**39.** $\left(\dfrac{2x^3}{y^5}\right)^2 = \dfrac{2^2(x^3)^2}{(y^5)^2}$

$\qquad = \dfrac{2^2x^6}{y^{10}}$      Multiplying exponents

$\qquad = \dfrac{4x^6}{y^{10}}$      $(2^2 = 4)$

**41.** $\dfrac{3a^2 - 5ab - 12b^2}{3ab + 4b^2} \div (3b^2 - ab)$

$\quad = \dfrac{3a^2 - 5ab - 12b^2}{3ab + 4b^2} \cdot \dfrac{1}{3b^2 - ab}$

$\quad = \dfrac{(3a + 4b)(a - 3b)}{b(3a + 4b) \cdot b(3b - a)}$

$\quad = \dfrac{(3a + 4b)(-1)(3b - a)}{b(3a + 4b) \cdot b(3b - a)}$

$\quad = \dfrac{(3a + 4b)(-1)(3b - a)}{b(3a + 4b) \cdot b(3b - a)}$

$\quad = -\dfrac{1}{b^2}$

**43.** $\dfrac{3x + 3y + 3}{9x} \div \dfrac{x^2 + 2xy + y^2 - 1}{x^4 + x^2}$

$\quad = \dfrac{3x + 3y + 3}{9x} \cdot \dfrac{x^4 + x^2}{x^2 + 2xy + y^2 - 1}$

$\quad = \dfrac{3(x + y + 1)(x^2)(x^2 + 1)}{9x[(x + y) + 1][(x + y) - 1]}$

$\quad = \dfrac{3(x + y + 1)(x)(x)(x^2 + 1)}{3 \cdot 3 \cdot x(x + y + 1)(x + y - 1)}$

$\quad = \dfrac{3x(x + y + 1)}{3x(x + y + 1)} \cdot \dfrac{x(x^2 + 1)}{3(x + y - 1)}$

$\quad = \dfrac{x(x^2 + 1)}{3(x + y - 1)}$

## Exercise Set 5.3

**1.** $12 = 2 \cdot 2 \cdot 3$

$27 = 3 \cdot 3 \cdot 3$

$\text{LCM} = 2 \cdot 2 \cdot 3 \cdot 3 \cdot 3$, or $108$

**3.** $8 = 2 \cdot 2 \cdot 2$

$9 = 3 \cdot 3$

$\text{LCM} = 2 \cdot 2 \cdot 2 \cdot 3 \cdot 3$, or $72$

**5.** $6 = 2 \cdot 3$

$9 = 3 \cdot 3$

$21 = 3 \cdot 7$

$\text{LCM} = 2 \cdot 3 \cdot 3 \cdot 7$, or $126$

**7.** $24 = 2 \cdot 2 \cdot 2 \cdot 3$

$36 = 2 \cdot 2 \cdot 3 \cdot 3$

$40 = 2 \cdot 2 \cdot 2 \cdot 5$

$\text{LCM} = 2 \cdot 2 \cdot 2 \cdot 3 \cdot 3 \cdot 5$, or $360$

**9.** $10 = 2 \cdot 5$

$100 = 2 \cdot 2 \cdot 5 \cdot 5$

$500 = 2 \cdot 2 \cdot 5 \cdot 5 \cdot 5$

$\text{LCM} = 2 \cdot 2 \cdot 5 \cdot 5 \cdot 5$, or $500$

(We might have observed at the outset that both 10 and 100 are factors of 500, so the LCM is 500.)

**11.** $24 = 2 \cdot 2 \cdot 2 \cdot 3$

$18 = 2 \cdot 3 \cdot 3$

$\text{LCD} = 2 \cdot 2 \cdot 2 \cdot 3 \cdot 3$, or $72$

$\dfrac{7}{24} + \dfrac{11}{18} = \dfrac{7}{2 \cdot 2 \cdot 2 \cdot 3} \cdot \dfrac{3}{3} + \dfrac{11}{2 \cdot 3 \cdot 3} \cdot \dfrac{2 \cdot 2}{2 \cdot 2}$

$\qquad = \dfrac{21}{2 \cdot 2 \cdot 2 \cdot 3 \cdot 3} + \dfrac{44}{2 \cdot 2 \cdot 2 \cdot 3 \cdot 3}$

$\qquad = \dfrac{65}{72}$

**13.** $\dfrac{1}{6} + \dfrac{3}{40}$

$= \dfrac{1}{2 \cdot 3} + \dfrac{3}{2 \cdot 2 \cdot 2 \cdot 5}$

     LCD is $2 \cdot 2 \cdot 2 \cdot 3 \cdot 5$, or $120$

$= \dfrac{1}{2 \cdot 3} \cdot \dfrac{2 \cdot 2 \cdot 5}{2 \cdot 2 \cdot 5} + \dfrac{3}{2 \cdot 2 \cdot 2 \cdot 5} \cdot \dfrac{3}{3}$

$= \dfrac{20 + 9}{2 \cdot 2 \cdot 2 \cdot 3 \cdot 5}$

$= \dfrac{29}{120}$

**15.** $\dfrac{1}{20} + \dfrac{1}{30} + \dfrac{2}{45}$

$= \dfrac{1}{2 \cdot 2 \cdot 5} + \dfrac{1}{2 \cdot 3 \cdot 5} + \dfrac{2}{3 \cdot 3 \cdot 5}$

     LCD is $2 \cdot 2 \cdot 3 \cdot 3 \cdot 5$, or $180$

$= \dfrac{1}{2 \cdot 2 \cdot 5} \cdot \dfrac{3 \cdot 3}{3 \cdot 3} + \dfrac{1}{2 \cdot 3 \cdot 5} \cdot \dfrac{2 \cdot 3}{2 \cdot 3} + \dfrac{2}{3 \cdot 3 \cdot 5} \cdot \dfrac{2 \cdot 2}{2 \cdot 2}$

$= \dfrac{9 + 6 + 8}{2 \cdot 2 \cdot 3 \cdot 3 \cdot 5}$

$= \dfrac{23}{180}$

**17.** $6x^2 = 2 \cdot 3 \cdot x \cdot x$

$12x^3 = 2 \cdot 2 \cdot 3 \cdot x \cdot x \cdot x$

$\text{LCM} = 2 \cdot 2 \cdot 3 \cdot x \cdot x \cdot x$, or $12x^3$

**19.** $2x^2 = 2 \cdot x \cdot x$

$6xy = 2 \cdot 3 \cdot x \cdot y$

$18y^2 = 2 \cdot 3 \cdot 3 \cdot y \cdot y$

$\text{LCM} = 2 \cdot 3 \cdot 3 \cdot x \cdot x \cdot y \cdot y$, or $18x^2y^2$

**21.** $2(y-3) = 2 \cdot (y-3)$
$6(y-3) = 2 \cdot 3 \cdot (y-3)$
$\text{LCM} = 2 \cdot 3 \cdot (y-3)$, or $6(y-3)$

**23.** $t, t+2, t-2$
The expressions are not factorable, so the LCM is their product:
$\text{LCM} = t(t+2)(t-2)$

**25.** $x^2 - 4 = (x+2)(x-2)$
$x^2 + 5x + 6 = (x+3)(x+2)$
$\text{LCM} = (x+2)(x-2)(x+3)$

**27.** $t^3 + 4t^2 + 4t = t(t^2 + 4t + 4) = t(t+2)(t+2)$
$t^2 - 4t = t(t-4)$
$\text{LCM} = t(t+2)(t+2)(t-4) = t(t+2)^2(t-4)$

**29.** $a+1 = a+1$
$(a-1)^2 = (a-1)(a-1)$
$a^2 - 1 = (a+1)(a-1)$
$\text{LCM} = (a+1)(a-1)(a-1) = (a+1)(a-1)^2$

**31.** $m^2 - 5m + 6 = (m-3)(m-2)$
$m^2 - 4m + 4 = (m-2)(m-2)$
$\text{LCM} = (m-3)(m-2)(m-2) = (m-3)(m-2)^2$

**33.** $2 + 3x = 2 + 3x$
$4 - 9x^2 = (2+3x)(2-3x)$
$2 - 3x = 2 - 3x$
$\text{LCM} = (2+3x)(2-3x)$

**35.** $10v^2 + 30v = 10v(v+3) = 2 \cdot 5 \cdot v(v+3)$
$5v^2 + 35v + 60 = 5(v^2 + 7v + 12)$
$= 5(v+4)(v+3)$
$\text{LCM} = 2 \cdot 5 \cdot v(v+3)(v+4) = 10v(v+3)(v+4)$

**37.** $9x^3 - 9x^2 - 18x = 9x(x^2 - x - 2)$
$= 3 \cdot 3 \cdot x(x-2)(x+1)$
$6x^5 - 24x^4 + 24x^3 = 6x^3(x^2 - 4x + 4)$
$= 2 \cdot 3 \cdot x \cdot x \cdot x(x-2)(x-2)$
$\text{LCM} = 2 \cdot 3 \cdot 3 \cdot x \cdot x \cdot x(x-2)(x-2)(x+1) = 18x^3(x-2)^2(x+1)$

**39.** $x^5 + 4x^4 + 4x^3 = x^3(x^2 + 4x + 4)$
$= x \cdot x \cdot x(x+2)(x+2)$
$3x^2 - 12 = 3(x^2 - 4) = 3(x+2)(x-2)$
$2x + 4 = 2(x+2)$
$\text{LCM} = 2 \cdot 3 \cdot x \cdot x \cdot x(x+2)(x+2)(x-2)$
$= 6x^3(x+2)^2(x-2)$

**41.** $x^2 - 6x + 9 = x^2 - 2 \cdot x \cdot 3 + 3^2$    Trinomial square
$= (x-3)^2$

**43.** $x^2 - 9 = x^2 - 3^2$    Difference of squares
$= (x+3)(x-3)$

**45.** $x^2 + 6x + 9 = x^2 + 2 \cdot x \cdot 3 + 3^2$    Trinomial square
$= (x+3)^2$

**47.** $72 = 2 \cdot 2 \cdot 2 \cdot 3 \cdot 3$
$90 = 2 \cdot 3 \cdot 3 \cdot 5$
$96 = 2 \cdot 2 \cdot 2 \cdot 2 \cdot 2 \cdot 3$
$\text{LCM} = 2 \cdot 2 \cdot 2 \cdot 2 \cdot 2 \cdot 3 \cdot 3 \cdot 5$, or $1440$

**49.** The time it takes the joggers to meet again at the starting place is the LCM of the times it takes them to complete one round of the course.
$6 = 2 \cdot 3$
$8 = 2 \cdot 2 \cdot 2$
$\text{LCM} = 2 \cdot 2 \cdot 2 \cdot 3$, or $24$
It takes 24 min.

---

## Exercise Set 5.4

**1.** $\dfrac{5}{8} + \dfrac{3}{8} + \dfrac{5+3}{8} = \dfrac{8}{8} = 1$

**3.** $\dfrac{1}{3+x} + \dfrac{5}{3+x} = \dfrac{1+5}{3+x} = \dfrac{6}{3+x}$

**5.** $\dfrac{x^2 + 7x}{x^2 - 5x} + \dfrac{x^2 - 4x}{x^2 - 5x} = \dfrac{(x^2 + 7x) + (x^2 - 4x)}{x^2 - 5x}$
$= \dfrac{2x^2 + 3x}{x^2 - 5x}$
$= \dfrac{x(2x+3)}{x(x-5)}$
$= \dfrac{\not{x}(2x+3)}{\not{x}(x-5)}$
$= \dfrac{2x+3}{x-5}$

**7.** $\dfrac{7}{8} + \dfrac{5}{-8} = \dfrac{7}{8} + \dfrac{5}{-8} \cdot \dfrac{-1}{-1}$
$= \dfrac{7}{8} + \dfrac{-5}{8}$
$= \dfrac{7 + (-5)}{8}$
$= \dfrac{2}{8} = \dfrac{\not{2} \cdot 1}{4 \cdot \not{2}}$
$= \dfrac{1}{4}$

**9.** 
$$\frac{3}{t} + \frac{4}{-t} = \frac{3}{t} + \frac{4}{-t} \cdot \frac{-1}{-1}$$
$$= \frac{3}{t} + \frac{-4}{t}$$
$$= \frac{3 + (-4)}{t}$$
$$= \frac{-1}{t}$$
$$= -\frac{1}{t}$$

**11.** 
$$\frac{2x+7}{x-6} + \frac{3x}{6-x} = \frac{2x+7}{x-6} + \frac{3x}{6-x} \cdot \frac{-1}{-1}$$
$$= \frac{2x+7}{x-6} + \frac{-3x}{x-6}$$
$$= \frac{(2x+7) + (-3x)}{x-6}$$
$$= \frac{-x+7}{x-6}$$

**13.** 
$$\frac{y^2}{y-3} + \frac{9}{3-y} = \frac{y^2}{y-3} + \frac{9}{3-y} \cdot \frac{-1}{-1}$$
$$= \frac{y^2}{y-3} + \frac{-9}{y-3}$$
$$= \frac{y^2 + (-9)}{y-3}$$
$$= \frac{y^2 - 9}{y-3}$$
$$= \frac{(y+3)(y-3)}{y-3}$$
$$= \frac{(y+3)(y-3)}{1(y-3)}$$
$$= y + 3$$

**15.** 
$$\frac{b-7}{b^2-16} + \frac{7-b}{16-b^2} = \frac{b-7}{b^2-16} + \frac{7-b}{16-b^2} \cdot \frac{-1}{-1}$$
$$= \frac{b-7}{b^2-16} + \frac{b-7}{b^2-16}$$
$$= \frac{(b-7) + (b-7)}{b^2-16}$$
$$= \frac{2b-14}{b^2-16}$$

**17.** 
$$\frac{a^2}{a-b} + \frac{b^2}{b-a} = \frac{a^2}{a-b} + \frac{b^2}{b-a} \cdot \frac{-1}{-1}$$
$$= \frac{a^2}{a-b} + \frac{-b^2}{a-b}$$
$$= \frac{a^2 + (-b^2)}{a-b}$$
$$= \frac{a^2 - b^2}{a-b}$$
$$= \frac{(a+b)(a-b)}{a-b}$$
$$= \frac{(a+b)(a-b)}{1(a-b)}$$
$$= a + b$$

**19.** 
$$\frac{x+3}{x-5} + \frac{2x-1}{5-x} + \frac{2(3x-1)}{x-5}$$
$$= \frac{x+3}{x-5} + \frac{2x-1}{5-x} \cdot \frac{-1}{-1} + \frac{2(3x-1)}{x-5}$$
$$= \frac{x+3}{x-5} + \frac{1-2x}{x-5} + \frac{2(3x-1)}{x-5}$$
$$= \frac{(x+3) + (1-2x) + (6x-2)}{x-5}$$
$$= \frac{5x+2}{x-5}$$

**21.** 
$$\frac{2(4x+1)}{5x-7} + \frac{3(x-2)}{7-5x} + \frac{-10x-1}{5x-7}$$
$$= \frac{2(4x+1)}{5x-7} + \frac{3(x-2)}{7-5x} \cdot \frac{-1}{-1} + \frac{-10x-1}{5x-7}$$
$$= \frac{2(4x+1)}{5x-7} + \frac{-3(x-2)}{5x-7} + \frac{-10x-1}{5x-7}$$
$$= \frac{(8x+2) + (-3x+6) + (-10x-1)}{5x-7}$$
$$= \frac{-5x+7}{5x-7}$$
$$= \frac{-1(5x-7)}{5x-7}$$
$$= \frac{-1(5x-7)}{5x-7}$$
$$= -1$$

**23.** 
$$\frac{x+1}{(x+3)(x-3)} + \frac{4(x-3)}{(x-3)(x+3)} + \frac{(x-1)(x-3)}{(3-x)(x+3)}$$
$$= \frac{x+1}{(x+3)(x-3)} + \frac{4(x-3)}{(x-3)(x+3)} + \frac{(x-1)(x-3)}{(3-x)(x+3)} \cdot \frac{-1}{-1}$$
$$= \frac{x+1}{(x+3)(x-3)} + \frac{4(x-3)}{(x-3)(x+3)} + \frac{-1(x^2-4x+3)}{(x-3)(x+3)}$$
$$= \frac{(x+1) + (4x-12) + (-x^2+4x-3)}{(x+3)(x-3)}$$
$$= \frac{-x^2+9x-14}{(x+3)(x-3)}$$

**25.** $\dfrac{2}{x} + \dfrac{5}{x^2} = \dfrac{2}{x} + \dfrac{5}{x \cdot x}$  LCD $= x \cdot x$, or $x^2$

$\qquad = \dfrac{2}{x} \cdot \dfrac{x}{x} + \dfrac{5}{x \cdot x}$

$\qquad = \dfrac{2x + 5}{x^2}$

**27.** $\left. \begin{array}{l} 6r = 2 \cdot 3 \cdot r \\ 8r = 2 \cdot 2 \cdot 2 \cdot r \end{array} \right\}$ LCD $= 2 \cdot 2 \cdot 2 \cdot 3 \cdot r$, or $24r$

$\dfrac{5}{6r} + \dfrac{7}{8r} = \dfrac{5}{6r} \cdot \dfrac{4}{4} + \dfrac{7}{8r} \cdot \dfrac{3}{3}$

$\qquad = \dfrac{20 + 21}{24r}$

$\qquad = \dfrac{41}{24r}$

**29.** $\left. \begin{array}{l} xy^2 = x \cdot y \cdot y \\ x^2y = x \cdot x \cdot y \end{array} \right\}$ LCD $= x \cdot x \cdot y \cdot y$, or $x^2y^2$

$\dfrac{4}{xy^2} + \dfrac{6}{x^2y} = \dfrac{4}{xy^2} \cdot \dfrac{x}{x} + \dfrac{6}{x^2y} \cdot \dfrac{y}{y}$

$\qquad = \dfrac{4x + 6y}{x^2y^2}$

**31.** $\left. \begin{array}{l} 9t^3 = 3 \cdot 3 \cdot t \cdot t \cdot t \\ 6t^2 = 2 \cdot 3 \cdot t \cdot t \end{array} \right\}$ LCD $= 2 \cdot 3 \cdot 3 \cdot t \cdot t \cdot t$, or $18t^3$

$\dfrac{2}{9t^3} + \dfrac{1}{6t^2} = \dfrac{2}{9t^3} \cdot \dfrac{2}{2} + \dfrac{1}{6t^2} \cdot \dfrac{3t}{3t}$

$\qquad = \dfrac{4 + 3t}{18t^3}$

**33.** LCD $= x^2y^2$ (See Exercise 29.)

$\dfrac{x + y}{xy^2} + \dfrac{3x + y}{x^2y} = \dfrac{x + y}{xy^2} \cdot \dfrac{x}{x} + \dfrac{3x + y}{x^2y} \cdot \dfrac{y}{y}$

$\qquad = \dfrac{x(x + y) + y(3x + y)}{x^2y^2}$

$\qquad = \dfrac{x^2 + xy + 3xy + y^2}{x^2y^2}$

$\qquad = \dfrac{x^2 + 4xy + y^2}{x^2y^2}$

**35.** The denominators do not factor, so the LCD is their product, $(x - 2)(x + 2)$.

$\dfrac{3}{x - 2} + \dfrac{3}{x + 2} = \dfrac{3}{x - 2} \cdot \dfrac{x + 2}{x + 2} + \dfrac{3}{x + 2} \cdot \dfrac{x - 2}{x - 2}$

$\qquad = \dfrac{3(x + 2) + 3(x - 2)}{(x - 2)(x + 2)}$

$\qquad = \dfrac{3x + 6 + 3x - 6}{(x - 2)(x + 2)}$

$\qquad = \dfrac{6x}{(x - 2)(x + 2)}$

**37.** $\left. \begin{array}{l} 3x = 3 \cdot x \\ x + 1 = x + 1 \end{array} \right\}$ LCD $= 3x(x + 1)$

$\dfrac{3}{x + 1} + \dfrac{2}{3x} = \dfrac{3}{x + 1} \cdot \dfrac{3x}{3x} + \dfrac{2}{3x} \cdot \dfrac{x + 1}{x + 1}$

$\qquad = \dfrac{9x + 2(x + 1)}{3x(x + 1)}$

$\qquad = \dfrac{9x + 2x + 2}{3x(x + 1)}$

$\qquad = \dfrac{11x + 2}{3x(x + 1)}$

**39.** $\left. \begin{array}{l} x^2 - 16 = (x + 4)(x - 4) \\ x - 4 = x - 4 \end{array} \right\}$ LCD $= (x + 4)(x - 4)$

$\dfrac{2x}{x^2 - 16} + \dfrac{x}{x - 4} = \dfrac{2x}{(x + 4)(x - 4)} + \dfrac{x}{x - 4} \cdot \dfrac{x + 4}{x + 4}$

$\qquad = \dfrac{2x + x(x + 4)}{(x + 4)(x - 4)}$

$\qquad = \dfrac{2x + x^2 + 4x}{(x + 4)(x - 4)}$

$\qquad = \dfrac{x^2 + 6x}{(x + 4)(x - 4)}$

**41.** $\dfrac{5}{z + 4} + \dfrac{3}{3z + 12} = \dfrac{5}{z + 4} + \dfrac{3}{3(z + 4)}$  LCD $= 3(z + 4)$

$\qquad = \dfrac{5}{z + 4} \cdot \dfrac{3}{3} + \dfrac{3}{3(z + 4)}$

$\qquad = \dfrac{15 + 3}{3(z + 4)} = \dfrac{18}{3(z + 4)}$

$\qquad = \dfrac{3 \cdot 6}{3(z + 4)} = \dfrac{\cancel{3} \cdot 6}{\cancel{3}(z + 4)}$

$\qquad = \dfrac{6}{z + 4}$

**43.** $\dfrac{3}{x - 1} + \dfrac{2}{(x - 1)^2}$  LCD $= (x - 1)^2$

$\qquad = \dfrac{3}{x - 1} \cdot \dfrac{x - 1}{x - 1} + \dfrac{2}{(x - 1)^2}$

$\qquad = \dfrac{3(x - 1) + 2}{(x - 1)^2}$

$\qquad = \dfrac{3x - 3 + 2}{(x - 1)^2}$

$\qquad = \dfrac{3x - 1}{(x - 1)^2}$

**45.** $\dfrac{4a}{5a - 10} + \dfrac{3a}{10a - 20} = \dfrac{4a}{5(a - 2)} + \dfrac{3a}{2 \cdot 5(a - 2)}$

$\qquad\qquad\qquad\qquad$ LCD $= 2 \cdot 5(a - 2)$

$\qquad = \dfrac{4a}{5(a - 2)} \cdot \dfrac{2}{2} + \dfrac{3a}{2 \cdot 5(a - 2)}$

$\qquad = \dfrac{8a + 3a}{10(a - 2)}$

$\qquad = \dfrac{11a}{10(a - 2)}$

**47.** $\dfrac{x+4}{x} + \dfrac{x}{x+4}$ $\qquad$ LCD $= x(x+4)$

$= \dfrac{x+4}{x} \cdot \dfrac{x+4}{x+4} + \dfrac{x}{x+4} \cdot \dfrac{x}{x}$

$= \dfrac{(x+4)^2 + x^2}{x(x+4)}$

$= \dfrac{x^2 + 8x + 16 + x^2}{x(x+4)}$

$= \dfrac{2x^2 + 8x + 16}{x(x+4)}$

**49.** $\dfrac{4}{a^2 - a - 2} + \dfrac{3}{a^2 + 4a + 3}$

$= \dfrac{4}{(a-2)(a+1)} + \dfrac{3}{(a+3)(a+1)}$

$\qquad$ LCD $= (a-2)(a+1)(a+3)$

$= \dfrac{4}{(a-2)(a+1)} \cdot \dfrac{a+3}{a+3} + \dfrac{3}{(a+3)(a+1)} \cdot \dfrac{a-2}{a-2}$

$= \dfrac{4(a+3) + 3(a-2)}{(a-2)(a+1)(a+3)}$

$= \dfrac{4a + 12 + 3a - 6}{(a-2)(a+1)(a+3)}$

$= \dfrac{7a + 6}{(a-2)(a+1)(a+3)}$

**51.** $\dfrac{x+3}{x-5} + \dfrac{x-5}{x+3}$ $\qquad$ LCD $= (x-5)(x+3)$

$= \dfrac{x+3}{x-5} \cdot \dfrac{x+3}{x+3} + \dfrac{x-5}{x+3} \cdot \dfrac{x-5}{x-5}$

$= \dfrac{(x+3)^2 + (x-5)^2}{(x-5)(x+3)}$

$= \dfrac{x^2 + 6x + 9 + x^2 - 10x + 25}{(x-5)(x+3)}$

$= \dfrac{2x^2 - 4x + 34}{(x-5)(x+3)}$

**53.** $\dfrac{a}{a^2 - 1} + \dfrac{2a}{a^2 - a}$

$= \dfrac{a}{(a+1)(a-1)} + \dfrac{2a}{a(a-1)}$

$\qquad$ LCD $= a(a+1)(a-1)$

$= \dfrac{a}{(a+1)(a-1)} \cdot \dfrac{a}{a} + \dfrac{2a}{a(a-1)} \cdot \dfrac{a+1}{a+1}$

$= \dfrac{a^2 + 2a(a+1)}{a(a+1)(a-1)} = \dfrac{a^2 + 2a^2 + 2a}{a(a+1)(a-1)}$

$= \dfrac{3a^2 + 2a}{a(a+1)(a-1)} = \dfrac{a(3a+2)}{a(a+1)(a-1)}$

$= \dfrac{\cancel{a}(3a+2)}{\cancel{a}(a+1)(a-1)} = \dfrac{3a+2}{(a+1)(a-1)}$

**55.** $\dfrac{6}{x-y} + \dfrac{4x}{y^2 - x^2}$

$= \dfrac{6}{x-y} + \dfrac{4x}{(y-x)(y+x)}$

$= \dfrac{6}{x-y} + \dfrac{4x}{(y-x)(y+x)} \cdot \dfrac{-1}{-1}$

$= \dfrac{6}{x-y} + \dfrac{-4x}{(x-y)(x+y)}$

$\qquad [-1(y-x) = x-y; \, y+x = x+y]$

$\qquad$ LCD $= (x-y)(x+y)$

$= \dfrac{6}{x-y} \cdot \dfrac{x+y}{x+y} + \dfrac{-4x}{(x-y)(x+y)}$

$= \dfrac{6(x+y) - 4x}{(x-y)(x+y)}$

$= \dfrac{6x + 6y - 4x}{(x-y)(x+y)}$

$= \dfrac{2x + 6y}{(x-y)(x+y)}$

**57.** $\dfrac{4-a}{25-a^2} + \dfrac{a+1}{a-5}$

$= \dfrac{4-a}{25-a^2} \cdot \dfrac{-1}{-1} + \dfrac{a+1}{a-5}$

$= \dfrac{a-4}{a^2 - 25} + \dfrac{a+1}{a-5}$

$= \dfrac{a-4}{(a+5)(a-5)} + \dfrac{a+1}{a-5}$

$\qquad$ LCD $= (a+5)(a-5)$

$= \dfrac{a-4}{(a+5)(a-5)} + \dfrac{a+1}{a-5} \cdot \dfrac{a+5}{a+5}$

$= \dfrac{a-4}{(a+5)(a-5)} + \dfrac{(a+1)(a+5)}{(a+5)(a-5)}$

$= \dfrac{(a-4) + (a+1)(a+5)}{(a+5)(a-5)}$

$= \dfrac{a-4 + a^2 + 6a + 5}{(a+5)(a-5)}$

$= \dfrac{a^2 + 7a + 1}{(a+5)(a-5)}$

**59.** $\dfrac{2}{t^2 + t - 6} + \dfrac{3}{t^2 - 9}$

$= \dfrac{2}{(t+3)(t-2)} + \dfrac{3}{(t+3)(t-3)}$

$\qquad$ LCD $= (t+3)(t-2)(t-3)$

$= \dfrac{2}{(t+3)(t-2)} \cdot \dfrac{t-3}{t-3} + \dfrac{3}{(t+3)(t-3)} \cdot \dfrac{t-2}{t-2}$

$= \dfrac{2(t-3) + 3(t-2)}{(t+3)(t-2)(t-3)}$

$= \dfrac{2t - 6 + 3t - 6}{(t+3)(t-2)(t-3)}$

$= \dfrac{5t - 12}{(t+3)(t-2)(t-3)}$

**61.** $(x^2 + x) - (x + 1) = x^2 + x - x - 1 = x^2 - 1$

**63.** $(2x^4 y^3)^{-3} = \dfrac{1}{(2x^4 y^3)^3} = \dfrac{1}{2^3 (x^4)^3 (y^3)^3} = \dfrac{1}{8x^{12} y^9}$

**65.** $\left(\dfrac{x^{-4}}{y^7}\right)^3 = \dfrac{(x^{-4})^3}{(y^7)^3} = \dfrac{x^{-12}}{y^{21}} = \dfrac{1}{x^{12} y^{21}}$

**67.** To find the perimeter we add the lengths of the sides:

$$\dfrac{y+4}{3} + \dfrac{y+4}{3} + \dfrac{y-2}{5} + \dfrac{y-2}{5} \quad \text{LCD} = 3 \cdot 5$$

$$= \dfrac{y+4}{3} \cdot \dfrac{5}{5} + \dfrac{y+4}{3} \cdot \dfrac{5}{5} + \dfrac{y-2}{5} \cdot \dfrac{3}{3} + \dfrac{y-2}{5} \cdot \dfrac{3}{3}$$

$$= \dfrac{5y + 20 + 5y + 20 + 3y - 6 + 3y - 6}{3 \cdot 5}$$

$$= \dfrac{16y + 28}{15}$$

To find the area we multiply the length and the width:

$$\left(\dfrac{y+4}{3}\right)\left(\dfrac{y-2}{5}\right) = \dfrac{(y+4)(y-2)}{3 \cdot 5} = \dfrac{y^2 + 2y - 8}{15}$$

**69.**

$$\dfrac{5}{z+2} + \dfrac{4z}{z^2 - 4} + 2$$

$$= \dfrac{5}{z+2} + \dfrac{4z}{(z+2)(z-2)} + \dfrac{2}{1} \quad \text{LCD} = (z+2)(z-2)$$

$$= \dfrac{5}{z+2} \cdot \dfrac{z-2}{z-2} + \dfrac{4z}{(z+2)(z-2)} + \dfrac{2}{1} \cdot \dfrac{(z+2)(z-2)}{(z+2)(z-2)}$$

$$= \dfrac{5z - 10 + 4z + 2(z^2 - 4)}{(z+2)(z-2)}$$

$$= \dfrac{5z - 10 + 4z + 2z^2 - 8}{(z+2)(z-2)} = \dfrac{2z^2 + 9z - 18}{(z+2)(z-2)}$$

$$= \dfrac{(2z-3)(z+6)}{(z+2)(z-2)}$$

**71.**

$$\dfrac{3z^2}{z^4 - 4} + \dfrac{5z^2 - 3}{2z^4 + z^2 - 6}$$

$$= \dfrac{3z^2}{(z^2 + 2)(z^2 - 2)} + \dfrac{5z^2 - 3}{(2z^2 - 3)(z^2 + 2)}$$

$$\text{LCD} = (z^2 + 2)(z^2 - 2)(2z^2 - 3)$$

$$= \dfrac{3z^2}{(z^2 + 2)(z^2 - 2)} \cdot \dfrac{2z^2 - 3}{2z^2 - 3} +$$

$$\dfrac{5z^2 - 3}{(2z^2 - 3)(z^2 + 2)} \cdot \dfrac{z^2 - 2}{z^2 - 2}$$

$$= \dfrac{6z^4 - 9z^2 + 5z^4 - 13z^2 + 6}{(z^2 + 2)(z^2 - 2)(2z^2 - 3)}$$

$$= \dfrac{11z^4 - 22z^2 + 6}{(z^2 + 2)(z^2 - 2)(2z^2 - 3)}$$

---

**Exercise Set 5.5**

**1.** $\dfrac{7}{x} - \dfrac{3}{x} = \dfrac{7-3}{x} = \dfrac{4}{x}$

**3.** $\dfrac{y}{y-4} - \dfrac{4}{y-4} = \dfrac{y-4}{y-4} = 1$

**5.**

$$\dfrac{2x-3}{x^2 + 3x - 4} - \dfrac{x-7}{x^2 + 3x - 4}$$

$$= \dfrac{2x - 3 - (x - 7)}{x^2 + 3x - 4}$$

$$= \dfrac{2x - 3 - x + 7}{x^2 + 3x - 4}$$

$$= \dfrac{x + 4}{x^2 + 3x - 4}$$

$$= \dfrac{x + 4}{(x + 4)(x - 1)}$$

$$= \dfrac{(\cancel{x+4}) \cdot 1}{(\cancel{x+4})(x - 1)}$$

$$= \dfrac{1}{x - 1}$$

**7.** $\dfrac{11}{6} - \dfrac{5}{-6} = \dfrac{11}{6} - \dfrac{5}{-6} \cdot \dfrac{-1}{-1}$

$$= \dfrac{11}{6} - \dfrac{-5}{6}$$

$$= \dfrac{11 - (-5)}{6}$$

$$= \dfrac{11 + 5}{6}$$

$$= \dfrac{16}{6}$$

$$= \dfrac{8}{3}$$

**9.** $\dfrac{5}{a} - \dfrac{8}{-a} = \dfrac{5}{a} - \dfrac{8}{-a} \cdot \dfrac{-1}{-1}$

$$= \dfrac{5}{a} - \dfrac{-8}{a}$$

$$= \dfrac{5 - (-8)}{a}$$

$$= \dfrac{5 + 8}{a}$$

$$= \dfrac{13}{a}$$

**11.** $\dfrac{4}{y-1} - \dfrac{4}{1-y} = \dfrac{4}{y-1} - \dfrac{4}{1-y} \cdot \dfrac{-1}{-1}$

$$= \dfrac{4}{y-1} - \dfrac{4(-1)}{(1-y)(-1)}$$

$$= \dfrac{4}{y-1} - \dfrac{-4}{y-1}$$

$$= \dfrac{4 - (-4)}{y-1}$$

$$= \dfrac{4 + 4}{y-1}$$

$$= \dfrac{8}{y-1}$$

**13.** $\dfrac{3-x}{x-7} - \dfrac{2x-5}{7-x} = \dfrac{3-x}{x-7} - \dfrac{2x-5}{7-x} \cdot \dfrac{-1}{-1}$

$\qquad = \dfrac{3-x}{x-7} - \dfrac{(2x-5)(-1)}{(7-x)(-1)}$

$\qquad = \dfrac{3-x}{x-7} - \dfrac{5-2x}{x-7}$

$\qquad = \dfrac{(3-x)-(5-2x)}{x-7}$

$\qquad = \dfrac{3-x-5+2x}{x-7}$

$\qquad = \dfrac{x-2}{x-7}$

**15.** $\dfrac{a-2}{a^2-25} - \dfrac{6-a}{25-a^2} = \dfrac{a-2}{a^2-25} - \dfrac{6-a}{25-a^2} \cdot \dfrac{-1}{-1}$

$\qquad = \dfrac{a-2}{a^2-25} - \dfrac{(6-a)(-1)}{(25-a^2)(-1)}$

$\qquad = \dfrac{a-2}{a^2-25} - \dfrac{a-6}{a^2-25}$

$\qquad = \dfrac{(a-2)-(a-6)}{a^2-25}$

$\qquad = \dfrac{a-2-a+6}{a^2-25}$

$\qquad = \dfrac{4}{a^2-25}$

**17.** $\dfrac{4-x}{x-9} - \dfrac{3x-8}{9-x} = \dfrac{4-x}{x-9} - \dfrac{3x-8}{9-x} \cdot \dfrac{-1}{-1}$

$\qquad = \dfrac{4-x}{x-9} - \dfrac{8-3x}{x-9}$

$\qquad = \dfrac{(4-x)-(8-3x)}{x-9}$

$\qquad = \dfrac{4-x-8+3x}{x-9}$

$\qquad = \dfrac{2x-4}{x-9}$

**19.** $\dfrac{2(x-1)}{2x-3} - \dfrac{3(x+2)}{2x-3} - \dfrac{x-1}{3-2x}$

$= \dfrac{2(x-1)}{2x-3} - \dfrac{3(x+2)}{2x-3} - \dfrac{x-1}{3-2x} \cdot \dfrac{-1}{-1}$

$= \dfrac{2(x-1)}{2x-3} - \dfrac{3(x+2)}{2x-3} - \dfrac{1-x}{2x-3}$

$= \dfrac{(2x-2)-(3x+6)-(1-x)}{2x-3}$

$= \dfrac{2x-2-3x-6-1+x}{2x-3}$

$= \dfrac{-9}{2x-3}$

**21.** $\dfrac{a-2}{10} - \dfrac{a+1}{5} = \dfrac{a-2}{10} - \dfrac{a+1}{5} \cdot \dfrac{2}{2} \qquad \text{LCD} = 10$

$\qquad = \dfrac{a-2}{10} - \dfrac{2(a+1)}{10}$

$\qquad = \dfrac{(a-2)-2(a+1)}{10}$

$\qquad = \dfrac{a-2-2a-2}{10}$

$\qquad = \dfrac{-a-4}{10}$

**23.** $\dfrac{4z-9}{3z} - \dfrac{3z-8}{4z} = \dfrac{4z-9}{3z} \cdot \dfrac{4}{4} - \dfrac{3z-8}{4z} \cdot \dfrac{3}{3}$

$\qquad\qquad \text{LCD} = 3 \cdot 4 \cdot z, \text{ or } 12z$

$\qquad = \dfrac{16z-36}{12z} - \dfrac{9z-24}{12z}$

$\qquad = \dfrac{16z-36-(9z-24)}{12z}$

$\qquad = \dfrac{16z-36-9z+24}{12z}$

$\qquad = \dfrac{7z-12}{12z}$

**25.** $\dfrac{4x+2t}{3xt^2} - \dfrac{5x-3t}{x^2t} \qquad \text{LCD} = 3x^2t^2$

$= \dfrac{4x+2t}{3xt^2} \cdot \dfrac{x}{x} - \dfrac{5x-3t}{x^2t} \cdot \dfrac{3t}{3t}$

$= \dfrac{4x^2+2tx}{3x^2t^2} - \dfrac{15xt-9t^2}{3x^2t^2}$

$= \dfrac{4x^2+2tx-(15xt-9t^2)}{3x^2t^2}$

$= \dfrac{4x^2+2tx-15xt+9t^2}{3x^2t^2}$

$= \dfrac{4x^2-13xt+9t^2}{3x^2t^2}$

**27.** $\dfrac{5}{x+5} - \dfrac{3}{x-5} \qquad \text{LCD} = (x+5)(x-5)$

$= \dfrac{5}{x+5} \cdot \dfrac{x-5}{x-5} - \dfrac{3}{x-5} \cdot \dfrac{x+5}{x+5}$

$= \dfrac{5x-25}{(x+5)(x-5)} - \dfrac{3x+15}{(x+5)(x-5)}$

$= \dfrac{5x-25-(3x+15)}{(x+5)(x-5)}$

$= \dfrac{5x-25-3x-15}{(x+5)(x-5)}$

$= \dfrac{2x-40}{(x+5)(x-5)}$

**29.** $\dfrac{3}{2t^2 - 2t} - \dfrac{5}{2t - 2}$

$= \dfrac{3}{2t(t-1)} - \dfrac{5}{2(t-1)}$ \quad LCD $= 2t(t-1)$

$= \dfrac{3}{2t(t-1)} - \dfrac{5}{2(t-1)} \cdot \dfrac{t}{t}$

$= \dfrac{3}{2t(t-1)} - \dfrac{5t}{2t(t-1)}$

$= \dfrac{3 - 5t}{2t(t-1)}$

**31.** $\dfrac{2s}{t^2 - s^2} - \dfrac{s}{t - s}$ \quad LCD $= (t-s)(t+s)$

$= \dfrac{2s}{(t-s)(t+s)} - \dfrac{s}{t-s} \cdot \dfrac{t+s}{t+s}$

$= \dfrac{2s}{(t-s)(t+s)} - \dfrac{st + s^2}{(t-s)(t+s)}$

$= \dfrac{2s - (st + s^2)}{(t-s)(t+s)}$

$= \dfrac{2s - st - s^2}{(t-s)(t+s)}$

**33.** $\dfrac{y-5}{y} - \dfrac{3y-1}{4y} = \dfrac{y-5}{y} \cdot \dfrac{4}{4} - \dfrac{3y-1}{4y}$ \quad LCD $= 4y$

$= \dfrac{4y - 20}{4y} - \dfrac{3y-1}{4y}$

$= \dfrac{4y - 20 - (3y-1)}{4y}$

$= \dfrac{4y - 20 - 3y + 1}{4y}$

$= \dfrac{y - 19}{4y}$

**35.** $\dfrac{a}{x+a} - \dfrac{a}{x-a}$ \quad LCD $= (x+a)(x-a)$

$= \dfrac{a}{x+a} \cdot \dfrac{x-a}{x-a} - \dfrac{a}{x-a} \cdot \dfrac{x+a}{x+a}$

$= \dfrac{ax - a^2}{(x+a)(x-a)} - \dfrac{ax + a^2}{(x+a)(x-a)}$

$= \dfrac{ax - a^2 - (ax + a^2)}{(x+a)(x-a)}$

$= \dfrac{ax - a^2 - ax - a^2}{(x+a)(x-a)}$

$= \dfrac{-2a^2}{(x+a)(x-a)}$

**37.** $\dfrac{5x}{x^2 - 9} - \dfrac{4}{3 - x}$

$= \dfrac{5x}{(x+3)(x-3)} - \dfrac{4}{3-x}$ \quad $x-3$ and $3-x$ are opposites

$= \dfrac{5x}{(x+3)(x-3)} - \dfrac{4}{3-x} \cdot \dfrac{-1}{-1}$

$= \dfrac{5x}{(x+3)(x-3)} - \dfrac{-4}{x-3}$ \quad LCD $= (x+3)(x-3)$

$= \dfrac{5x}{(x+3)(x-3)} - \dfrac{-4}{x-3} \cdot \dfrac{x+3}{x+3}$

$= \dfrac{5x}{(x+3)(x-3)} - \dfrac{-4x - 12}{(x+3)(x-3)}$

$= \dfrac{5x - (-4x - 12)}{(x+3)(x-3)}$

$= \dfrac{5x + 4x + 12}{(x+3)(x-3)}$

$= \dfrac{9x + 12}{(x+3)(x-3)}$

**39.** $\dfrac{t^2}{2t^2 - 2t} - \dfrac{1}{2t - 2}$

$= \dfrac{t^2}{2t(t-1)} - \dfrac{1}{2(t-1)}$ \quad LCD $= 2t(t-1)$

$= \dfrac{t^2}{2t(t-1)} - \dfrac{1}{2(t-1)} \cdot \dfrac{t}{t}$

$= \dfrac{t^2}{2t(t-1)} - \dfrac{t}{2t(t-1)}$

$= \dfrac{t^2 - t}{2t(t-1)}$

$= \dfrac{t(t-1)}{2t(t-1)}$

$= \dfrac{t(t-1)(1)}{2t(t-1)}$

$= \dfrac{1}{2}$

**41.**

$$\frac{x}{x^2+5x+6}-\frac{2}{x^2+3x+2}$$

$$=\frac{x}{(x+3)(x+2)}-\frac{2}{(x+2)(x+1)}$$

$$\text{LCD}=(x+3)(x+2)(x+1)$$

$$=\frac{x}{(x+3)(x+2)}\cdot\frac{x+1}{x+1}-\frac{2}{(x+2)(x+1)}\cdot\frac{x+3}{x+3}$$

$$=\frac{x^2+x}{(x+3)(x+2)(x+1)}-\frac{2x+6}{(x+3)(x+2)(x+1)}$$

$$=\frac{x^2+x-(2x+6)}{(x+3)(x+2)(x+1)}$$

$$=\frac{x^2+x-2x-6}{(x+3)(x+2)(x+1)}$$

$$=\frac{x^2-x-6}{(x+3)(x+2)(x+1)}$$

$$=\frac{(x-3)(x+2)}{(x+3)(x+2)(x+1)}$$

$$=\frac{(x-3)\cancel{(x+2)}}{(x+3)\cancel{(x+2)}(x+1)}$$

$$=\frac{x-3}{(x+3)(x+1)}$$

**43.**

$$\frac{3(2x+5)}{x-1}-\frac{3(2x-3)}{1-x}+\frac{6x+1}{x-1}$$

$$=\frac{3(2x+5)}{x-1}-\frac{3(2x-3)}{1-x}\cdot\frac{-1}{-1}+\frac{6x-1}{x-1}$$

$$=\frac{3(2x+5)}{x-1}-\frac{-3(2x-3)}{x-1}+\frac{6x-1}{x-1}$$

$$=\frac{(6x+15)-(-6x+9)+(6x-1)}{x-1}$$

$$=\frac{6x+15+6x-9+6x-1}{x-1}$$

$$=\frac{18x+5}{x-1}$$

**45.**

$$\frac{x-y}{x^2-y^2}+\frac{x+y}{x^2-y^2}-\frac{2x}{x^2-y^2}$$

$$=\frac{x-y+x+y-2x}{x^2-y^2}$$

$$=\frac{0}{x^2-y^2}$$

$$=0$$

**47.**

$$\frac{10}{2y-1}-\frac{6}{1-2y}+\frac{y}{2y-1}+\frac{y-4}{1-2y}$$

$$=\frac{10}{2y-1}-\frac{6}{1-2y}\cdot\frac{-1}{-1}+\frac{y}{2y-1}+\frac{y-4}{1-2y}\cdot\frac{-1}{-1}$$

$$=\frac{10}{2y-1}-\frac{-6}{2y-1}+\frac{y}{2y-1}+\frac{4-y}{2y-1}$$

$$=\frac{10-(-6)+y+4-y}{2y-1}$$

$$=\frac{10+6+y+4-y}{2y-1}$$

$$=\frac{20}{2y-1}$$

**49.**

$$\frac{a+6}{4-a^2}-\frac{a+3}{a+2}+\frac{a-3}{2-a}$$

$$=\frac{a+6}{(2+a)(2-a)}-\frac{a+3}{2+a}+\frac{a-3}{2-a}$$

$$a+2=2+a;\ \text{LCD}=(2+a)(2-a)$$

$$=\frac{a+6}{(2+a)(2-a)}-\frac{a+3}{2+a}\cdot\frac{2-a}{2-a}+\frac{a-3}{2-a}\cdot\frac{2+a}{2+a}$$

$$=\frac{(a+6)-(a+3)(2-a)+(a-3)(2+a)}{(2+a)(2-a)}$$

$$=\frac{a+6-(-a^2-a+6)+(a^2-a-6)}{(2+a)(2-a)}$$

$$=\frac{a+6+a^2+a-6+a^2-a-6}{(2+a)(2-a)}$$

$$=\frac{2a^2+a-6}{(2+a)(2-a)}$$

$$=\frac{(2a-3)(a+2)}{(2+a)(2-a)}$$

$$=\frac{(2a-3)\cancel{(2+a)}}{\cancel{(2+a)}(2-a)}$$

$$=\frac{2a-3}{2-a}$$

**51.**

$$\frac{2z}{1-2z}+\frac{3z}{2z+1}-\frac{3}{4z^2-1}$$

$$=\frac{2z}{1-2z}\cdot\frac{-1}{-1}+\frac{3z}{2z+1}-\frac{3}{4z^2-1}$$

$$=\frac{-2z}{2z-1}+\frac{3z}{2z+1}-\frac{3}{(2z-1)(2z+1)}$$

$$\text{LCD}=(2z-1)(2z+1)$$

$$=\frac{-2z}{2z-1}\cdot\frac{2z+1}{2z+1}+\frac{3z}{2z+1}\cdot\frac{2z-1}{2z-1}-$$

$$\frac{3}{(2z-1)(2z+1)}$$

$$=\frac{(-4z^2-2z)+(6z^2-3z)-3}{(2z-1)(2z+1)}$$

$$=\frac{2z^2-5z-3}{(2z-1)(2z+1)}$$

$$=\frac{(z-3)(2z+1)}{(2z-1)(2z+1)}$$

$$=\frac{(z-3)\cancel{(2z+1)}}{(2z-1)\cancel{(2z+1)}}$$

$$=\frac{z-3}{2z-1}$$

**53.**

$$\frac{1}{x+y} - \frac{1}{x-y} + \frac{2x}{x^2-y^2}$$

$$= \frac{1}{x+y} - \frac{1}{x-y} + \frac{2x}{(x+y)(x-y)}$$

$$\qquad\qquad \text{LCD} = (x+y)(x-y)$$

$$= \frac{1}{x+y} \cdot \frac{x-y}{x-y} - \frac{1}{x-y} \cdot \frac{x+y}{x+y} + \frac{x+y}{x+y}$$

$$\qquad\qquad \frac{2x}{(x+y)(x-y)}$$

$$= \frac{x-y-(x+y)+2x}{(x+y)(x-y)}$$

$$= \frac{x-y-x-y+2x}{(x+y)(x-y)}$$

$$= \frac{2x-2y}{(x+y)(x-y)}$$

$$= \frac{2(x-y)}{(x+y)(x-y)}$$

$$= \frac{2(x\!\!\!\!/-y\!\!\!\!/)}{(x+y)(x\!\!\!\!/-y\!\!\!\!/)}$$

$$= \frac{2}{x+y}$$

**55.** $\dfrac{x^8}{x^3} = x^{8-3} = x^5$

**57.** $(a^2 b^{-5})^{-4} = a^{2(-4)} b^{-5(-4)} = a^{-8} b^{20} = \dfrac{b^{20}}{a^8}$

**59.** $\dfrac{66x^2}{11x^5} = \dfrac{6 \cdot \cancel{11} \cdot \cancel{x^2}}{\cancel{11} \cdot \cancel{x^2} \cdot x^3} = \dfrac{6}{x^3}$

**61.**

$$\frac{5}{3-2x} + \frac{3}{2x-3} - \frac{x-3}{2x^2-x-3}$$

$$= \frac{5}{3-2x} \cdot \frac{-1}{-1} + \frac{3}{2x-3} - \frac{x-3}{2x^2-x-3}$$

$$= \frac{-5}{2x-3} + \frac{3}{2x-3} - \frac{x-3}{(2x-3)(x+1)}$$

$$\qquad\qquad \text{LCD} = (2x-3)(x+1)$$

$$= \frac{-5}{2x-3} \cdot \frac{x+1}{x+1} + \frac{3}{2x-3} \cdot \frac{x+1}{x+1} - \frac{x-3}{(2x-3)(x+1)}$$

$$= \frac{(-5x-5)+(3x+3)-(x-3)}{(2x-3)(x+1)}$$

$$= \frac{-5x-5+3x+3-x+3}{(2x-3)(x+1)}$$

$$= \frac{-3x+1}{(2x-3)(x+1)}, \text{ or } \frac{1-3x}{(2x-3)(x+1)}$$

**Exercise Set 5.6**

**1.** $\qquad \dfrac{3}{8} + \dfrac{4}{5} = \dfrac{x}{20}, \text{ LCM} = 40$

$$40\left(\frac{3}{8} + \frac{4}{5}\right) = 40 \cdot \frac{x}{20}$$

$$40 \cdot \frac{3}{8} + 40 \cdot \frac{4}{5} = 40 \cdot \frac{x}{20}$$

$$15 + 32 = 2x$$

$$47 = 2x$$

$$\frac{47}{2} = x$$

Check:

$$\frac{3}{8} + \frac{4}{5} = \frac{x}{20}$$

| $\dfrac{3}{8} + \dfrac{4}{5}$ | $\dfrac{\frac{47}{2}}{20}$ |
|---|---|
| $\dfrac{15}{40} + \dfrac{32}{40}$ | $\dfrac{47}{2} \cdot \dfrac{1}{20}$ |
| $\dfrac{47}{40}$ | $\dfrac{47}{40}$  TRUE |

This checks, so the solution is $\dfrac{47}{2}$.

**3.** $\qquad \dfrac{1}{x} = \dfrac{2}{3} - \dfrac{5}{6}, \text{ LCM} = 6x$

$$6x \cdot \frac{1}{x} = 6x\left(\frac{2}{3} - \frac{5}{6}\right)$$

$$6x \cdot \frac{1}{x} = 6x \cdot \frac{2}{3} - 6x \cdot \frac{5}{6}$$

$$6 = 4x - 5x$$

$$6 = -x$$

$$-6 = x$$

Check:

$$\frac{1}{x} = \frac{2}{3} - \frac{5}{6}$$

| $\dfrac{1}{-6}$ | $\dfrac{2}{3} - \dfrac{5}{6}$ |
|---|---|
| $-\dfrac{1}{6}$ | $\dfrac{4}{6} - \dfrac{5}{6}$ |
|  | $-\dfrac{1}{6}$  TRUE |

This checks, so the solution is $-6$.

**5.** $\qquad \dfrac{1}{6} + \dfrac{1}{8} = \dfrac{1}{t}, \text{ LCM} = 24t$

$$24t\left(\frac{1}{6} + \frac{1}{8}\right) = 24t \cdot \frac{1}{t}$$

$$24t \cdot \frac{1}{6} + 24t \cdot \frac{1}{8} = 24t \cdot \frac{1}{t}$$

$$4t + 3t = 24$$

$$7t = 24$$

$$t = \frac{24}{7}$$

Check:

$$\frac{1}{6} + \frac{1}{8} = \frac{1}{t}$$

| $\frac{1}{6} + \frac{1}{8}$ | $\frac{1}{24/7}$ |
|---|---|
| $\frac{4}{24} + \frac{3}{24}$ | $1 \cdot \frac{7}{24}$ |
| $\frac{7}{24}$ | $\frac{7}{24}$   TRUE |

This checks, so the solution is $\frac{24}{7}$.

**7.**
$$x + \frac{4}{x} = -5, \ \text{LCM} = x$$

$$x\left(x + \frac{4}{x}\right) = x(-5)$$

$$x \cdot x + x \cdot \frac{4}{x} = x(-5)$$

$$x^2 + 4 = -5x$$

$$x^2 + 5x + 4 = 0$$

$$(x + 4)(x + 1) = 0$$

$$x + 4 = 0 \quad \text{or} \quad x + 1 = 0$$

$$x = -4 \quad \text{or} \quad x = -1$$

Check:

| $x + \frac{4}{x} = -5$ | | $x + \frac{4}{x} = -5$ | |
|---|---|---|---|
| $-4 + \frac{4}{-4}$ | $-5$ | $-1 + \frac{4}{-1}$ | $-5$ |
| $-4 - 1$ | | $-1 - 4$ | |
| $-5$ | TRUE | $-5$ | TRUE |

Both of these check, so the two solutions are $-4$ and $-1$.

**9.**
$$\frac{x}{4} - \frac{4}{x} = 0, \ \text{LCM} = 4x$$

$$4x\left(\frac{x}{4} - \frac{4}{x}\right) = 4x \cdot 0$$

$$4x \cdot \frac{x}{4} - 4x \cdot \frac{4}{x} = 4x \cdot 0$$

$$x^2 - 16 = 0$$

$$(x + 4)(x - 4) = 0$$

$$x + 4 = 0 \quad \text{or} \quad x - 4 = 0$$

$$x = -4 \quad \text{or} \quad x = 4$$

Check:

| $\frac{x}{4} - \frac{4}{x} = 0$ | | $\frac{x}{4} - \frac{4}{x} = 0$ | |
|---|---|---|---|
| $\frac{-4}{4} - \frac{4}{-4}$ | $0$ | $\frac{4}{4} - \frac{4}{4}$ | $0$ |
| $-1 - (-1)$ | | $1 - 1$ | |
| $-1 + 1$ | | $0$ | TRUE |
| $0$ | TRUE | | |

Both of these check, so the two solutions are $-4$ and $4$.

**11.**
$$\frac{5}{x} = \frac{6}{x} - \frac{1}{3}, \ \text{LCM} = 3x$$

$$3x \cdot \frac{5}{x} = 3x\left(\frac{6}{x} - \frac{1}{3}\right)$$

$$3x \cdot \frac{5}{x} = 3x \cdot \frac{6}{x} - 3x \cdot \frac{1}{3}$$

$$15 = 18 - x$$

$$-3 = -x$$

$$3 = x$$

Check:

$$\frac{5}{x} = \frac{6}{x} - \frac{1}{3}$$

| $\frac{5}{3}$ | $\frac{6}{3} - \frac{1}{3}$ |
|---|---|
| | $\frac{5}{3}$   TRUE |

This checks, so the solution is 3.

**13.**
$$\frac{5}{3x} + \frac{3}{x} = 1, \ \text{LCM} = 3x$$

$$3x\left(\frac{5}{3x} + \frac{3}{x}\right) = 3x \cdot 1$$

$$3x \cdot \frac{5}{3x} + 3x \cdot \frac{3}{x} = 3x \cdot 1$$

$$5 + 9 = 3x$$

$$14 = 3x$$

$$\frac{14}{3} = x$$

Check:

$$\frac{5}{3x} + \frac{3}{x} = 1$$

| $\frac{5}{3 \cdot (14/3)} + \frac{3}{(14/3)}$ | $1$ |
|---|---|
| $\frac{5}{14} + \frac{9}{14}$ | |
| $\frac{14}{14}$ | |
| $1$ | TRUE |

This checks, so the solution is $\frac{14}{3}$.

**15.** $\dfrac{t-2}{t+3} = \dfrac{3}{8}$, LCM $= 8(t+3)$

$$8(t+3)\left(\dfrac{t-2}{t+3}\right) = 8(t+3)\left(\dfrac{3}{8}\right)$$

$$8(t-2) = 3(t+3)$$

$$8t-16 = 3t+9$$

$$5t = 25$$

$$t = 5$$

Check:

$$\dfrac{t-2}{t+3} = \dfrac{3}{8}$$

| $\dfrac{5-2}{5+3}$ | $\dfrac{3}{8}$ |
|---|---|
| $\dfrac{3}{8}$ | TRUE |

This checks, so the solution is 5.

**17.** $\dfrac{2}{x+1} = \dfrac{1}{x-2}$, LCM $= (x+1)(x-2)$

$$(x+1)(x-2) \cdot \dfrac{2}{x+1} = (x+1)(x-2) \cdot \dfrac{1}{x-2}$$

$$2(x-2) = x+1$$

$$2x-4 = x+1$$

$$x = 5$$

This checks, so the solution is 5.

**19.** $\dfrac{x}{6} - \dfrac{x}{10} = \dfrac{1}{6}$, LCM $= 30$

$$30\left(\dfrac{x}{6} - \dfrac{x}{10}\right) = 30 \cdot \dfrac{1}{6}$$

$$30 \cdot \dfrac{x}{6} - 30 \cdot \dfrac{x}{10} = 30 \cdot \dfrac{1}{6}$$

$$5x - 3x = 5$$

$$2x = 5$$

$$x = \dfrac{5}{2}$$

This checks, so the solution is $\dfrac{5}{2}$.

**21.** $\dfrac{t+2}{5} - \dfrac{t-2}{4} = 1$, LCM $= 20$

$$20\left(\dfrac{t+2}{5} - \dfrac{t-2}{4}\right) = 20 \cdot 1$$

$$20\left(\dfrac{t+2}{5}\right) - 20\left(\dfrac{t-2}{4}\right) = 20 \cdot 1$$

$$4(t+2) - 5(t-2) = 20$$

$$4t+8 - 5t+10 = 20$$

$$-t+18 = 20$$

$$-t = 2$$

$$t = -2$$

This checks, so the solution is $-2$.

**23.** $\dfrac{a-3}{3a+2} = \dfrac{1}{5}$, LCM $= 5(3a+2)$

$$5(3a+2) \cdot \dfrac{a-3}{3a+2} = 5(3a+2) \cdot \dfrac{1}{5}$$

$$5(a-3) = 3a+2$$

$$5a-15 = 3a+2$$

$$2a = 17$$

$$a = \dfrac{17}{2}$$

This checks, so the solution is $\dfrac{17}{2}$.

**25.** $\dfrac{x-1}{x-5} = \dfrac{4}{x-5}$, LCM $= x-5$

$$(x-5) \cdot \dfrac{x-1}{x-5} = (x-5) \cdot \dfrac{4}{x-5}$$

$$x-1 = 4$$

$$x = 5$$

The number 5 is not a solution because it makes a denominator zero. Thus, there is no solution.

**27.** $\dfrac{2}{x+3} = \dfrac{5}{x}$, LCM $= x(x+3)$

$$x(x+3) \cdot \dfrac{2}{x+3} = x(x+3) \cdot \dfrac{5}{x}$$

$$2x = 5(x+3)$$

$$2x = 5x+15$$

$$-15 = 3x$$

$$-5 = x$$

This checks, so the solution is $-5$.

**29.** $\dfrac{x-2}{x-3} = \dfrac{x-1}{x+1}$, LCM $= (x-3)(x+1)$

$$(x-3)(x+1) \cdot \dfrac{x-2}{x-3} = (x-3)(x+1) \cdot \dfrac{x-1}{x+1}$$

$$(x+1)(x-2) = (x-3)(x-1)$$

$$x^2 - x - 2 = x^2 - 4x + 3$$

$$-x - 2 = -4x + 3$$

$$3x = 5$$

$$x = \dfrac{5}{3}$$

This checks, so the solution is $\dfrac{5}{3}$.

**31.** $\dfrac{1}{x+3} + \dfrac{1}{x-3} = \dfrac{1}{x^2-9}$,

$$\text{LCM} = (x+3)(x-3)$$

$$(x+3)(x-3)\left(\dfrac{1}{x+3} + \dfrac{1}{x-3}\right) = (x+3)(x-3) \cdot \dfrac{1}{(x+3)(x-3)}$$

$$(x-3) + (x+3) = 1$$

$$2x = 1$$

$$x = \dfrac{1}{2}$$

This checks, so the solution is $\dfrac{1}{2}$.

**33.**
$$\frac{x}{x+4} - \frac{4}{x-4} = \frac{x^2+16}{x^2-16},$$
$$\text{LCM} = (x+4)(x-4)$$

$$(x+4)(x-4)\left(\frac{x}{x+4} - \frac{x}{x-4}\right) = (x+4)(x-4)\cdot\frac{x^2+16}{(x+4)(x-4)}$$

$$x(x-4) - 4(x+4) = x^2+16$$
$$x^2 - 4x - 4x - 16 = x^2 + 16$$
$$x^2 - 8x - 16 = x^2 + 16$$
$$-8x - 16 = 16$$
$$-8x = 32$$
$$x = -4$$

The number $-4$ is not a solution because it makes a denominator zero. Thus, there is no solution.

**35.**
$$\frac{4-a}{8-a} = \frac{4}{a-8} \qquad \begin{array}{l}8-a \text{ and } a-8\\ \text{are opposites}\end{array}$$

$$\frac{4-a}{8-a}\cdot\frac{-1}{-1} = \frac{4}{a-8}$$

$$\frac{a-4}{a-8} = \frac{4}{a-8},\ \text{LCM} = a-8$$

$$(a-8)\left(\frac{a-4}{a-8}\right) = (a-8)\left(\frac{4}{a-8}\right)$$

$$a - 4 = 4$$
$$a = 8$$

The number 8 is not a solution because it makes a denominator zero. Thus, there is no solution.

**37.** $(a^2b^5)^{-3} = \dfrac{1}{(a^2b^5)^3} = \dfrac{1}{(a^2)^3(b^5)^3} = \dfrac{1}{a^6b^{15}}$

**39.** $\left(\dfrac{2x}{t^2}\right)^4 = \dfrac{(2x)^4}{(t^2)^4} = \dfrac{2^4x^4}{t^8} = \dfrac{16x^4}{t^8}$

**41.** $4x^{-5}\cdot 8x^{11} = 4\cdot 8x^{-5+11} = 32x^6$

**43.**
$$\frac{4}{y-2} - \frac{2y-3}{y^2-4} = \frac{5}{y+2},$$
$$\text{LCM} = (y+2)(y-2)$$

$$(y+2)(y-2)\left(\frac{4}{y-2} - \frac{2y-3}{(y+2)(y-2)}\right) =$$
$$\qquad\qquad\qquad (y+2)(y-2)\cdot\frac{5}{y+2}$$

$$4(y+2) - (2y-3) = 5(y-2)$$
$$4y + 8 - 2y + 3 = 5y - 10$$
$$2y + 11 = 5y - 10$$
$$21 = 3y$$
$$7 = y$$

This checks, so the solution is 7.

**45.**
$$\frac{x+1}{x+2} = \frac{x+3}{x+4},$$
$$\text{LCM} = (x+2)(x+4)$$

$$(x+2)(x+4)\left(\frac{x+1}{x+2}\right) = (x+2)(x+4)\left(\frac{x+3}{x+4}\right)$$
$$(x+4)(x+1) = (x+2)(x+3)$$
$$x^2 + 5x + 4 = x^2 + 5x + 6$$
$$4 = 6 \qquad \text{Subtracting } x^2 \text{ and } 5x$$

We get a false equation, so the original equation has no solution.

**47.**
$$4a - 3 = \frac{a+13}{a+1},\ \text{LCM} = a+1$$

$$(a+1)(4a-3) = (a+1)\cdot\frac{a+13}{a+1}$$

$$4a^2 + a - 3 = a + 13$$
$$4a^2 - 16 = 0$$
$$4(a+2)(a-2) = 0$$
$$a+2 = 0 \quad \text{or} \quad a-2 = 0$$
$$a = -2 \quad \text{or} \qquad a = 2$$

Both of these check, so the two solutions are $-2$ and 2.

**49.**
$$\frac{y^2-4}{y+3} = 2 - \frac{y-2}{y+3},\ \text{LCM} = y+3$$

$$(y+3)\cdot\frac{y^2-4}{y+3} = (y+3)\left(2 - \frac{y-2}{y+3}\right)$$

$$y^2 - 4 = 2(y+3) - (y-2)$$
$$y^2 - 4 = 2y + 6 - y + 2$$
$$y^2 - 4 = y + 8$$
$$y^2 - y - 12 = 0$$
$$(y-4)(y+3) = 0$$
$$y - 4 = 0 \quad \text{or} \quad y + 3 = 0$$
$$y = 4 \quad \text{or} \qquad y = -3$$

The number 4 is a solution, but $-3$ is not because it makes a denominator zero.

## Exercise Set 5.7

**1. Familiarize.** Let $x =$ the number. Then $\dfrac{1}{x}$ is the reciprocal of the number.

**Translate.**

| The reciprocal of 6 | plus | the reciprocal of 8 | is | the reciprocal of the number. |
|:---:|:---:|:---:|:---:|:---:|
| ↓ | ↓ | ↓ | ↓ | ↓ |
| $\dfrac{1}{6}$ | $+$ | $\dfrac{1}{8}$ | $=$ | $\dfrac{1}{x}$ |

*Solve*. We solve the equation.

$$\frac{1}{6} + \frac{1}{8} = \frac{1}{x}, \text{ LCM} = 24x$$

$$24x\left(\frac{1}{6} + \frac{1}{8}\right) = 24x \cdot \frac{1}{x}$$

$$24x \cdot \frac{1}{6} + 24x \cdot \frac{1}{8} = 24x \cdot \frac{1}{x}$$

$$4x + 3x = 24$$

$$7x = 24$$

$$x = \frac{24}{7}$$

*Check*. The reciprocal of $\frac{24}{7}$ is $\frac{7}{24}$. Also,
$\frac{1}{6} + \frac{1}{8} = \frac{4}{24} + \frac{3}{24} = \frac{7}{24}$, so the value checks.

*State*. The number is $\frac{24}{7}$.

**3.** *Familiarize*. Let $x = $ the smaller number. Then $x + 5 = $ the larger number.

*Translate*.

| The larger number | divided by | the smaller number | is | $\frac{4}{3}$. |
|:---:|:---:|:---:|:---:|:---:|
| ↓ | ↓ | ↓ | ↓ | ↓ |
| $(x+5)$ | ÷ | $x$ | = | $\frac{4}{3}$ |

*Solve*. We solve the equation.

$$\frac{x+5}{x} = \frac{4}{3}, \text{ LCM} = 3x$$

$$3x\left(\frac{x+5}{x}\right) = 3x \cdot \frac{4}{3}$$

$$3(x+5) = 4x$$

$$3x + 15 = 4x$$

$$15 = x$$

*Check*. If the smaller number is 15, then the larger is
$15 + 5$, or 20. The quotient of 20 divided by 15 is $\frac{20}{15}$, or
$\frac{4}{3}$. The values check.

*State*. The numbers are 15 and 20.

**5.** *Familiarize*. We complete the table shown in the text.

| $d$ | $=$ | $r$ | $\cdot$ | $t$ | |
|---|---|---|---|---|---|
| | Distance | Speed | Time | | |
| Car | 150 | $r$ | $t$ | $\rightarrow 150 = r(t)$ |
| Truck | 350 | $r + 40$ | $t$ | $\rightarrow 350 = (r+40)t$ |

*Translate*. We apply the formula $d = rt$ along the rows of the table to obtain two equations:

$$150 = rt,$$

$$350 = (r+40)t$$

Then we solve each equation for $t$ and set the results equal:

Solving $150 = rt$ for $t$: $t = \frac{150}{r}$

Solving $350 = (r+40)t$ for $t$: $t = \frac{350}{r+40}$

Thus, we have

$$\frac{150}{r} = \frac{350}{r+40}.$$

*Solve*. We multiply by the LCM, $r(r+40)$.

$$r(r+40) \cdot \frac{150}{r} = r(r+40) \cdot \frac{350}{r+40}$$

$$150(r+40) = 350r$$

$$150r + 6000 = 350r$$

$$6000 = 200r$$

$$30 = r$$

*Check*. If $r$ is 30 km/h, then $r + 40$ is 70 km/h. The time for the car is 150/30, or 5 hr. The time for the truck is 350/70, or 5 hr. The times are the same. The values check.

*State*. The speed of the car is 30 km/h, and the speed of the truck is 70 km/h.

**7.** *Familiarize*. We complete the table shown in the text.

| $d$ | $=$ | $r$ | $\cdot$ | $t$ |
|---|---|---|---|---|
| | Distance | Speed | Time |
| Freight | 330 | $r - 14$ | $t$ |
| Passenger | 400 | $r$ | $t$ |

*Translate*. From the rows of the table we have two equations:

$$330 = (r-14)t,$$

$$400 = rt$$

We solve each equation for $t$ and set the results equal:

Solving $330 = (r-14)t$ for $t$: $t = \frac{330}{r-14}$

Solving $400 = rt$ for $t$: $t = \frac{400}{r}$

Thus, we have

$$\frac{330}{r-14} = \frac{400}{r}.$$

*Solve*. We multiply by the LCM, $r(r-14)$.

$$r(r-14) \cdot \frac{330}{r-14} = r(r-14) \cdot \frac{400}{r}$$

$$330r = 400(r-14)$$

$$330r = 400r - 5600$$

$$-70r = -5600$$

$$r = 80$$

Then substitute 80 for $r$ in either equation to find $t$:

$$t = \frac{400}{r}$$

$$t = \frac{400}{80} \quad \text{Substituting 80 for } r$$

$$t = 5$$

**Check**. If $r = 80$, then $r - 14 = 66$. In 5 hr the freight train travels $66 \cdot 5$, or 330 mi, and the passenger train travels $80 \cdot 5$, or 400 mi. The values check.

**State**. The speed of the passenger train is 80 mph. The speed of the freight train is 66 mph.

9. **Familiarize**. We let $r$ represent the speed going. Then $2r$ is the speed returning. We let $t$ represent the time going. Then $t - 3$ represents the time returning. We organize the information in a table.

$$d \quad = \quad r \quad \cdot \quad t$$

|          | Distance | Speed | Time  |
|----------|----------|-------|-------|
| Going    | 120      | $r$   | $t$   |
| Returning| 120      | $2r$  | $t - 3$ |

**Translate**. The rows of the table give us two equations:
$$120 = rt,$$
$$120 = 2r(t - 3)$$
We can solve each equation for $r$ and set the results equal:

Solving $120 = rt$ for $r$: $r = \dfrac{120}{t}$

Solving $120 = 2r(t - 3)$ for $r$: $r = \dfrac{120}{2(t - 3)}$, or
$$r = \frac{60}{t - 3}$$

Then $\dfrac{120}{t} = \dfrac{60}{t - 3}$.

**Solve**. We multiply on both sides by the LCM, $t(t - 3)$.
$$t(t - 3) \cdot \frac{120}{t} = t(t - 3) \cdot \frac{60}{t - 3}$$
$$120(t - 3) = 60t$$
$$120t - 360 = 60t$$
$$-360 = -60t$$
$$6 = t$$

Then substitute 6 for $t$ in either equation to find $r$, the speed going:
$$r = \frac{120}{t}$$
$$r = \frac{120}{6} \qquad \text{Substituting 6 for } t$$
$$r = 20$$

**Check**. If $r = 20$ and $t = 6$, then $2r = 2 \cdot 20$, or 40 mph and $t - 3 = 6 - 3$, or 3 hr. The distance going is $6 \cdot 20$, or 120 mi. The distance returning is $40 \cdot 3$, or 120 mi. The numbers check.

**State**. The speed going is 20 mph.

11. **Familiarize**. The job takes Juanita 12 hours working alone and Antoine 16 hours working alone. Then in 1 hour Juanita does $\dfrac{1}{12}$ of the job, and Antoine does $\dfrac{1}{16}$ of the job. Working together, they can do $\dfrac{1}{12} + \dfrac{1}{16}$, or $\dfrac{7}{48}$

of the job in 1 hour. In 2 hours Juanita does $2\left(\dfrac{1}{12}\right)$ of the job, and Antoine does $2\left(\dfrac{1}{16}\right)$ of the job. Working together they can do $2\left(\dfrac{1}{12}\right) + 2\left(\dfrac{1}{16}\right)$, or $\dfrac{7}{24}$ of the job in 2 hours. Continuing this reasoning, we find that they do $6\left(\dfrac{1}{12}\right) + 6\left(\dfrac{1}{16}\right)$, or $\dfrac{7}{8}$ of the job in 6 hours working together and $7\left(\dfrac{1}{12}\right) + 7\left(\dfrac{1}{16}\right)$, or $1\dfrac{1}{48}$ of the job in 7 hours working together. Since $1\dfrac{1}{48}$ is more of the job than needs to be done, we see that the answer is somewhere between 6 hr and 7 hr.

**Translate**. Using the work principle, we see that we want some number $t$ such that
$$\frac{t}{12} + \frac{t}{16} = 1.$$

**Solve**. We solve the equation.
$$\frac{t}{12} + \frac{t}{16} = 1, \text{ LCM} = 48$$
$$48\left(\frac{t}{12} + \frac{t}{16}\right) = 48 \cdot 1$$
$$48 \cdot \frac{t}{12} + 48 \cdot \frac{t}{16} = 48 \cdot 1$$
$$4t + 3t = 48$$
$$7t = 48$$
$$t = \frac{48}{7}, \text{ or } 6\frac{6}{7}$$

**Check**. The check can be done by recalculating. We also have another check. In the familiarization step we learned that the time must be between 6 hr and 7 hr. The answer, $6\dfrac{6}{7}$ hr, is between 6 hr and 7 hr and is less than 12 hr, the time it takes Juanita to do the job working alone.

**State**. Working together, it takes them $6\dfrac{6}{7}$ hr to build the garage.

13. **Familiarize**. The job takes Rory 12 hours working alone and Mira 9 hours working alone. Then in 1 hour Rory does $\dfrac{1}{12}$ of the job and Mira does $\dfrac{1}{9}$ of the job. Working together they can do $\dfrac{1}{12} + \dfrac{1}{9}$, or $\dfrac{7}{36}$ of the job in 1 hour. In two hours, Rory does $2\left(\dfrac{1}{12}\right)$ of the job and Mira does $2\left(\dfrac{1}{9}\right)$ of the job. Working together they can do $2\left(\dfrac{1}{12}\right) + 2\left(\dfrac{1}{9}\right)$, or $\dfrac{7}{18}$ of the job in two hours. In 3 hours they can do $3\left(\dfrac{1}{12}\right) + 3\left(\dfrac{1}{9}\right)$, or $\dfrac{7}{12}$ of the job. In 4 hours, they can do $\dfrac{7}{9}$, and in 5 hours they can do $\dfrac{35}{36}$ of the job. In 6 hours, they can do $\dfrac{7}{6}$, or $1\dfrac{1}{6}$ of the job which is more of the job than needs to be done. The answer is somewhere between 5 hr and 6 hr.

**Translate**. Using the work principle, we see that we want some number $t$ such that
$$t\left(\frac{1}{12}\right) + t\left(\frac{1}{9}\right) = 1.$$

*Solve*. We solve the equation.

$$\frac{t}{12} + \frac{t}{9} = 1, \text{ LCM} = 36$$

$$36\left(\frac{t}{12} + \frac{t}{9}\right) = 36 \cdot 1$$

$$3t + 4t = 36$$

$$7t = 36$$

$$t = \frac{36}{7}, \text{ or } 5\frac{1}{7}$$

*Check*. The check can be done by recalculating. We also have another check. In the familiarization step we learned the time must be between 5 hr and 6 hr. The answer, $5\frac{1}{7}$ hr, is between 5 hr and 6 hr and is less than 9 hours, the time it takes Mira alone.

*State*. Working together, it takes them $5\frac{1}{7}$ hr to fit the kitchen.

**15.** $\frac{54 \text{ days}}{6 \text{ days}} = 9$

**17.** $\frac{4.6 \text{ km}}{2 \text{ hr}} = 2.3 \text{ km/h}$

**19.** *Familiarize*. A 120-lb person should eat at least 44 g of protein each day, and we wish to find the minimum protein required for a 180-lb person. We can set up ratios. We let $p =$ the minimum number of grams of protein a 180-lb person should eat each day.

*Translate*. If we assume the rates of protein intake are the same, the ratios are the same and we have an equation.

$$\begin{array}{l} \text{Protein} \rightarrow \\ \text{Weight} \rightarrow \end{array} \frac{44}{120} = \frac{p}{180} \begin{array}{l} \leftarrow \text{Protein} \\ \leftarrow \text{Weight} \end{array}$$

*Solve*. We solve the equation.

$$360 \cdot \frac{44}{120} = 360 \cdot \frac{p}{180} \text{ Multiplying by the LCM, 360}$$

$$3 \cdot 44 = 2 \cdot p$$

$$132 = 2p$$

$$66 = p$$

*Check*. $\frac{44}{120} = \frac{4 \cdot 11}{4 \cdot 30} = \frac{\cancel{4} \cdot 11}{\cancel{4} \cdot 30} = \frac{11}{30}$ and

$\frac{66}{180} = \frac{6 \cdot 11}{6 \cdot 30} = \frac{\cancel{6} \cdot 11}{\cancel{6} \cdot 30} = \frac{11}{30}$. The ratios are the same.

*State*. A 180-lb person should eat a minimum of 66 g of protein each day.

**21.** *Familiarize*. A student travels 234 kilometers in 14 days, and we wish to find how far the student would travel in 42 days. We can set up ratios. We let $K =$ the number of kilometers the student would travel in 42 days.

*Translate*. Assuming the rates are the same, we can translate to a proportion.

$$\begin{array}{l} \text{Kilometers} \rightarrow \\ \text{Days} \rightarrow \end{array} \frac{K}{42} = \frac{234}{14} \begin{array}{l} \leftarrow \text{Kilometers} \\ \leftarrow \text{Days} \end{array}$$

*Solve*. We solve the equation.
We multiply by 42 to get $K$ alone.

$$42 \cdot \frac{K}{42} = 42 \cdot \frac{234}{14}$$

$$K = \frac{9828}{14}$$

$$K = 702$$

*Check*.
$\frac{702}{42} \approx 16.7$ and $\frac{234}{14} \approx 16.7$.
The ratios are the same.

*State*. The student would travel 702 kilometers in 42 days.

**23.** *Familiarize*. 10 cm$^3$ of human blood contains 1.2 grams of heomglobin, and we wish to find how many grams of hemoglobin are contained in 16 cm$^3$ of the same blood. We can set up ratios. We let $H =$ the number of grams of hemoglobin contained in 16 cm$^3$ of the same blood.

*Translate*. Assuming the rates are the same, we can translate to a proportion.

$$\begin{array}{l} \text{Grams} \rightarrow \\ \text{cm}^3 \rightarrow \end{array} \frac{H}{16} = \frac{1.2}{10} \begin{array}{l} \leftarrow \text{Grams} \\ \leftarrow \text{cm}^3 \end{array}$$

*Solve*. We solve the equation.
We multiply by 16 to get $H$ alone.

$$16 \cdot \frac{H}{16} = 16 \cdot \frac{1.2}{10}$$

$$H = \frac{19.2}{10}$$

$$H = 1.92$$

*Check*.
$\frac{1.92}{16} = 0.12$ and $\frac{1.2}{10} = 0.12$.
The ratios are the same.

*State*. Thus 16 cm$^3$ of the same blood would contain 1.92 grams of hemoglobin.

**25.** *Familiarize*. The ratio of blue whales tagged to the total blue whale population, $P$, is $\frac{500}{P}$. Of the 400 blue whales checked later, 20 were tagged. The ratio of blue whales tagged to blue whales checked is $\frac{20}{400}$.

*Translate*. Assuming the two ratios are the same, we can translate to a proportion.

$$\begin{array}{l} \text{Whales tagged} \\ \text{originally} \longrightarrow \\ \text{Whale} \longrightarrow \\ \text{population} \end{array} \frac{500}{P} = \frac{20}{400} \begin{array}{l} \leftarrow \text{Tagged whales} \\ \text{caught later} \\ \leftarrow \text{Whales caught} \\ \text{later} \end{array}$$

*Solve*. We solve the equation.

$$400P \cdot \frac{500}{P} = 400P \cdot \frac{20}{400} \text{ Multiplying by the LCM,}$$
$$400P$$

$$400 \cdot 500 = P \cdot 20$$

$$200,000 = 20P$$

$$10,000 = P$$

**Check.**

$$\frac{500}{10,000} = \frac{1}{20} \quad \text{and} \quad \frac{20}{400} = \frac{1}{20}.$$

The ratios are the same.

**State.** The blue whale population is about 10,000.

**27. Familiarize.** The ratio of the weight of an object on Mars to the weight of an object on earth is 0.4 to 1.

a) We wish to find how much a 12-ton rocket would weigh on Mars.

b) We wish to find how much a 120-lb astronaut would weigh on Mars.

We can set up ratios. We let $r$ = the weight of a 12-ton rocket and $a$ = the weight of a 120-lb astronaut on Mars.

**Translate.** Assuming the ratios are the same, we can translate to proportions.

a)
$$\begin{array}{ll} \text{Weight} & \text{Weight} \\ \text{on Mars} \rightarrow \dfrac{0.4}{1} = \dfrac{r}{12} & \leftarrow \text{on Mars} \\ \text{Weight} \rightarrow & \leftarrow \text{Weight} \\ \text{on earth} & \text{on earth} \end{array}$$

b)
$$\begin{array}{ll} \text{Weight} & \text{Weight} \\ \text{on Mars} \rightarrow \dfrac{0.4}{1} = \dfrac{a}{120} & \leftarrow \text{on Mars} \\ \text{Weight} \rightarrow & \leftarrow \text{Weight} \\ \text{on earth} & \text{on earth} \end{array}$$

**Solve.** We solve each proportion.

a)
$$\frac{0.4}{1} = \frac{r}{12}$$
$$12(0.4) = r$$
$$4.8 = r$$

b)
$$\frac{0.4}{1} = \frac{1}{120}$$
$$120(0.4) = a$$
$$48 = a$$

**Check.** $\dfrac{0.4}{1} = 0.4$, $\dfrac{4.8}{12} = 0.4$, and $\dfrac{48}{120} = 0.4$.

The ratios are the same.

**State.** a) A 12-ton rocket would weigh 4.8 tons on Mars.

b) A 120-lb astronaut would weigh 48 lb on Mars.

**29. Familiarize.** Let $g$ = the number of additional games the team must win in order to finish with a 0.750 record. Then the total number of wins will be $25 + g$, and there will be a total of $36 + 12$, or 48 games in the season. We can set up ratios.

**Translate.** Assuming the ratios are the same, we can set up a proportion.

$$\begin{array}{l} \text{Games won} \rightarrow \dfrac{25 + g}{48} = \dfrac{0.750}{1} \leftarrow \text{Games won} \\ \text{Games played} \rightarrow \quad\quad\quad\quad\quad \leftarrow \text{Games played} \end{array}$$

**Solve.** We solve the proportion.

$$48\left(\frac{25 + g}{48}\right) = 48 \cdot \frac{0.750}{1}$$
$$25 + g = 36$$
$$g = 11$$

**Check.** $\dfrac{25 + 11}{48} = \dfrac{36}{48} = 0.75$ and $\dfrac{0.750}{1} = 0.75$.

The ratios are the same.

**State.** The team must win 11 more games.

**31. Familiarize.** The job takes Larry 8 days working alone and Moe 10 days working alone. Let $x$ represent the number of days it would take Curly working alone. Then in 1 day Larry does $\frac{1}{8}$ of the job, Moe does $\frac{1}{10}$ of the job, and Curly does $\frac{1}{x}$ of the job. In 1 day they would complete $\frac{1}{8} + \frac{1}{10} + \frac{1}{x}$ of the job, and in 3 days they would complete $3\left(\frac{1}{8} + \frac{1}{10} + \frac{1}{x}\right)$, or $\frac{3}{8} + \frac{3}{10} + \frac{3}{x}$.

**Translate.** The amount done in 3 days is one entire job, so we have

$$\frac{3}{8} + \frac{3}{10} + \frac{3}{x} = 1.$$

**Solve.** We solve the equation.

$$\frac{3}{8} + \frac{3}{10} + \frac{3}{x} = 1, \quad \text{LCM} = 40x$$
$$40x\left(\frac{3}{8} + \frac{3}{10} + \frac{3}{x}\right) = 40x \cdot 1$$
$$40x \cdot \frac{3}{8} + 40x \cdot \frac{3}{10} + 40x \cdot \frac{3}{x} = 40x$$
$$15x + 12x + 120 = 40x$$
$$120 = 13x$$
$$\frac{120}{13} = x$$

**Check.** If it takes Curly $\dfrac{120}{13}$, or $9\dfrac{3}{13}$ days, to complete the job, then in one day Curly does $\dfrac{1}{\frac{120}{13}}$, or $\dfrac{13}{120}$, of the job, and in 3 days he does $3\left(\dfrac{13}{120}\right)$, or $\dfrac{13}{40}$, of the job. The portion of the job done by Larry, Moe, and Curly in 3 days is $\dfrac{3}{8} + \dfrac{3}{10} + \dfrac{13}{40} = \dfrac{15}{40} + \dfrac{12}{40} + \dfrac{13}{40} = \dfrac{40}{40} = 1$ entire job. The answer checks.

**State.** It will take Curly $9\dfrac{3}{13}$ days to complete the job working alone.

**33. Familiarize.** Let $x$ represent the numerator and $x + 1$ represent the denominator of the original fraction. The fraction is $\dfrac{x}{x + 1}$. If 2 is subtracted from the numerator and the denominator, the resulting fraction is $\dfrac{x - 2}{x + 1 - 2}$, or $\dfrac{x - 2}{x - 1}$.

**Translate.**

$$\underbrace{\text{The resulting fraction}}_{\displaystyle \frac{x-2}{x-1}} \quad \underset{=}{\text{is}} \quad \underset{\displaystyle \frac{1}{2}}{\frac{1}{2}}$$

*Solve*. We solve the equation.

$$\frac{x-2}{x-1} = \frac{1}{2}, \text{ LCM} = 2(x-1)$$

$$2(x-1) \cdot \frac{x-2}{x-1} = 2(x-1) \cdot \frac{1}{2}$$

$$2(x-2) = x - 1$$

$$2x - 4 = x - 1$$

$$x = 3$$

*Check*. If $x = 3$, then $x + 1 = 4$ and the original fraction is $\frac{3}{4}$. If 2 is subtracted from both numerator and denominator, the resulting fraction is $\frac{3-2}{4-2}$, or $\frac{1}{2}$. The value checks.

*State*. The original fraction was $\frac{3}{4}$.

**35.** 1) Start with the given proportion:

$$\frac{A}{B} = \frac{C}{D}$$

2) Find a second true proportion:

$$\frac{A}{B} = \frac{C}{D}$$

$$\frac{D}{A} \cdot \frac{A}{B} = \frac{D}{A} \cdot \frac{C}{D}$$

$$\frac{D}{B} = \frac{C}{A}$$

3) Find a third true proportion:

$$\frac{A}{B} = \frac{C}{D}$$

$$\frac{B}{C} \cdot \frac{A}{B} = \frac{B}{C} \cdot \frac{C}{D}$$

$$\frac{A}{C} = \frac{B}{D}$$

4) Find a fourth true proportion:

$$\frac{A}{B} = \frac{C}{D}$$

$$\frac{DB}{AC} \cdot \frac{A}{B} = \frac{DB}{AC} \cdot \frac{C}{D}$$

$$\frac{D}{C} = \frac{B}{A}$$

**37.** *Familiarize*. We let $t$ = the time it would take Ann to do the job working alone. Then $t + 6$ = the time it would take Betty to do the job working alone.

*Translate*. Using the work principle we have

$$\frac{4}{t} + \frac{4}{t+6} = 1.$$

*Solve*. We solve the equation.

$$\frac{4}{t} + \frac{4}{t+6} = 1, \text{ LCM} = t(t+6)$$

$$t(t+6)\left(\frac{4}{t} + \frac{4}{t+6}\right) = t(t+6) \cdot 1$$

$$4(t+6) + 4t = t(t+6)$$

$$4t + 24 + 4t = t^2 + 6t$$

$$8t + 24 = t^2 + 6t$$

$$0 = t^2 - 2t - 24$$

$$0 = (t-6)(t+4)$$

$$t - 6 = 0 \quad \text{or} \quad t + 4 = 0$$

$$t = 6 \quad \text{or} \qquad t = -4$$

*Check*. Since the time cannot be negative, we only need to check 6. If Ann can do the job, working alone, in 6 hr, then in 4 hr she does $\frac{4}{6}$, or $\frac{2}{3}$ of the job. If Betsy can do the job, working alone, in $6 + 6$, or 12 hr, then in 4 hr she does $\frac{4}{12}$, or $\frac{1}{3}$ of the job. Together they do $\frac{2}{3} + \frac{1}{3}$, or 1 entire job in 4 hr. Our result checks.

*State*. Working alone, it will take Ann 6 hours and Betty 12 hours to complete the report.

**39.** *Familiarize*. Let $t$ = the number of minutes after 5:00 at which the hands of the clock will first be together. While the minute hand moves through $t$ minutes, the hour hand moves through $t/12$ minutes. At 5:00 the hour hand is on the 25-minute mark. We wish to find when a move of the minute hand through $t$ minutes is equal to $25 + t/12$ minutes.

*Translate*. We use the last sentence of the familiarization step to write an equation.

$$t = 25 + \frac{t}{12}$$

*Solve*. We solve the equation.

$$t = 25 + \frac{t}{12}$$

$$12 \cdot t = 12\left(25 + \frac{t}{12}\right)$$

$$12t = 300 + t \qquad \text{Multiplying by 12}$$

$$11t = 300$$

$$t = \frac{300}{11} \text{ or } 27\frac{3}{11}$$

*Check*. At $27\frac{3}{11}$ minutes after 5:00, the minute hand is at the $27\frac{3}{11}$-minutes mark and the hour hand is at the $25 + \dfrac{27\frac{3}{11}}{12}$-minute mark. Simplifying $25 + \dfrac{27\frac{3}{11}}{12}$, we get

$$25 + \frac{\frac{300}{11}}{12} = 25 + \frac{300}{11} \cdot \frac{1}{12} = 25 + \frac{25}{11} = 25 + 2\frac{3}{11} = 27\frac{3}{11}.$$

Thus, the hands are together.

*State*. The hands are first together $27\frac{3}{11}$ minutes after 5:00.

## Exercise Set 5.8

**1.**     $S = 2\pi rh$

$\dfrac{S}{2\pi h} = \dfrac{2\pi rh}{2\pi h}$     Dividing by $2\pi h$

$\dfrac{S}{2\pi h} = r$     Simplifying

**3.**     $A = \dfrac{1}{2}bh$

$2 \cdot A = 2 \cdot \dfrac{1}{2}bh$     Multiplying by 2

$2A = bh$     Simplifying

$\dfrac{2A}{h} = \dfrac{bh}{h}$     Dividing by $h$

$\dfrac{2a}{h} = b$     Simplifying

**5.**     $S = 180(n - 2)$

$S = 180n - 360$     Removing parentheses

$S + 360 = 180n$     Adding 360

$\dfrac{S + 360}{180} = n$     Dividing by 180

**7.**     $V = \dfrac{1}{3}k(B + b + 4M)$

$3 \cdot V = 3 \cdot \dfrac{1}{3}k(B + b + 4M)$

Multiplying by 3

$3V = k(B + b + 4M)$     Simplifying

$3V = kB + kb + 4kM$     Removing parentheses

$3V - kB - 4kM = kb$     Subtracting $kB$ and $4kM$

$\dfrac{3V - kB - 4kM}{k} = b$     Dividing by $k$

**9.**   $S(r - 1) = rl - a$

$Sr - S = rl - a$     Removing parentheses

$Sr - rl = S - a$     Adding $S$ and subtracting $rl$

$r(S - l) = S - a$     Factoring out $r$

$r = \dfrac{S - a}{S - l}$     Dividing by $S - l$

**11.**     $A = \dfrac{1}{2}h(b_1 + b_2)$

$2A = h(b_1 + b_2)$     Multiplying by 2

$\dfrac{2A}{b_1 + b_2} = h$     Dividing by $b_1 + b_2$

**13.**     $\dfrac{A - B}{AB} = Q$

$AB \cdot \dfrac{A - B}{AB} = AB \cdot Q$     Multiplying by $AB$

$A - B = ABQ$     Simplifying

$A = ABQ + B$     Adding $B$

$A = B(AQ + 1)$     Factoring out $B$

$\dfrac{A}{AQ + 1} = B$     Dividing by $AQ + 1$

**15.**     $\dfrac{1}{p} + \dfrac{1}{q} = \dfrac{1}{f}$,  LCM $= pqf$

$pqf\left(\dfrac{1}{p} + \dfrac{1}{q}\right) = pqf \cdot \dfrac{1}{f}$     Multiplying by $pqf$

$pqf \cdot \dfrac{1}{p} + pqf \cdot \dfrac{1}{q} = pqf \cdot \dfrac{1}{f}$     Removing parentheses

$qf + pf = pq$     Simplifying

$qf = pq - pf$     Subtracting $pf$

$qf = p(q - f)$     Factoring out $p$

$\dfrac{qf}{q - f} = p$     Dividing by $q - f$

**17.**     $\dfrac{A}{P} = 1 + r$

$P \cdot \dfrac{A}{P} = P \cdot (1 + r)$  Multiplying by $P$

$A = P(1 + r)$   Simplifying

**19.**     $\dfrac{1}{R} = \dfrac{1}{r_1} + \dfrac{1}{r_2}$,  LCM $= Rr_1r_2$

$Rr_1r_2 \cdot \dfrac{1}{R} = Rr_1r_2\left(\dfrac{1}{r_1} + \dfrac{1}{r_2}\right)$  Multiplying by $Rr_1r_2$

$Rr_1r_2 \cdot \dfrac{1}{R} = Rr_1r_2 \cdot \dfrac{1}{r_1} + Rr_1r_2 \cdot \dfrac{1}{r_2}$   Removing parentheses

$r_1r_2 = Rr_2 + Rr_1$     Simplifying

$r_1r_2 = R(r_2 + r_1)$     Factoring out $R$

$\dfrac{r_1r_2}{r_2 + r_1} = R$     Dividing by $r_2 + r_1$

**21.**     $\dfrac{A}{B} = \dfrac{C}{D}$,         LCM $= BD$

$BD \cdot \dfrac{A}{B} = BD \cdot \dfrac{C}{D}$     Multiplying by $BD$

$DA = BC$     Simplifying

$D = \dfrac{BC}{A}$     Dividing by $A$

**23.**
$$h_1 = q\left(1 + \frac{h_2}{p}\right)$$

$$h_1 = q + \frac{qh_2}{p} \qquad \text{Removing parentheses}$$

$$h_1 - q = \frac{qh_2}{p} \qquad \text{Subtracting } q$$

$$p(h_1 - q) = qh_2 \qquad \text{Multiplying by } p$$

$$\frac{p(h_1 - q)}{q} = h_2 \qquad \text{Dividing by } q$$

**25.**
$$C = \frac{Ka - b}{a}$$

$$a \cdot C = a \cdot \left(\frac{Ka - b}{a}\right) \qquad \text{Multiplying by } a$$

$$aC = Ka - b \qquad \text{Simplifying}$$

$$b = Ka - aC \qquad \text{Adding } b \text{ and subtracting } aC$$

$$b = a(K - C) \qquad \text{Factoring out } a$$

$$\frac{b}{K - C} = a \qquad \text{Dividing by } K - C$$

**27.** $(5x^4 - 6x^3 + 23x^2 - 79x + 24) -$
$\qquad (-18x^4 - 56x^3 + 84x - 17)$
$= 5x^4 - 6x^3 + 23x^2 - 79x + 24 + 18x^4 + 56x^3 -$
$\qquad 84x + 17$
$= 23x^4 + 50x^3 + 23x^2 - 163x + 41$

**29.** $30y^4 + 9y^2 - 12 = 3(10y^4 + 3y^2 - 4)$
$\qquad\qquad\qquad\quad = 3(5y^2 + 4)(2y^2 - 1)$

**31.** $y^2 + 2y - 35$

The factorization will be of the form $(y + \quad)(y + \quad)$. We look for two factors of 35 whose sum is 2. The factors we need are 7 and $-5$. Then $y^2 + 2y - 35 = (y + 7)(y - 5)$.

**33.**
$$u = -F\left(E - \frac{P}{T}\right)$$

$$u = -EF + \frac{FP}{T} \qquad \text{Removing parentheses}$$

$$T \cdot u = T\left(-EF + \frac{FP}{T}\right) \qquad \text{Multiplying by } T$$

$$Tu = -EFT + FP$$

$$Tu + EFT = FP \qquad \text{Adding } EFT$$

$$T(u + EF) = FP \qquad \text{Factoring out } T$$

$$T = \frac{FP}{u + EF} \qquad \text{Dividing by } u + EF$$

**35.** When $C = F$, we have
$$C = \frac{5}{9}(C - 32)$$

$$9C = 5(C - 32) \qquad \text{Multiplying by 9}$$

$$9C = 5C - 160 \qquad \text{Removing parentheses}$$

$$4C = -160 \qquad \text{Subtracting } 5C$$

$$C = -40 \qquad \text{Dividing by 4}$$

At $-40°$ the Fahrenheit and Celsius readings are the same.

---

## Exercise Set 5.9

**1.**
$$\frac{1 + \dfrac{9}{16}}{1 - \dfrac{3}{4}} \qquad \text{LCM of the denominators is 16.}$$

$$= \frac{1 + \dfrac{9}{16}}{1 - \dfrac{3}{4}} \cdot \frac{16}{16} \qquad \text{Multiplying by 1 using } \frac{16}{16}$$

$$= \frac{\left(1 + \dfrac{9}{16}\right)16}{\left(1 - \dfrac{3}{4}\right)16} \qquad \text{Multiplying numerator and denominator by 16}$$

$$= \frac{1(16) + \dfrac{9}{16}(16)}{1(16) - \dfrac{3}{4}(16)}$$

$$= \frac{16 + 9}{16 - 12}$$

$$= \frac{25}{4}$$

**3.** $\dfrac{1 - \dfrac{3}{5}}{1 + \dfrac{1}{5}}$

$= \dfrac{1 \cdot \dfrac{5}{5} - \dfrac{3}{5}}{1 \cdot \dfrac{5}{5} + \dfrac{1}{5}}$    Getting a common denominator in numerator and in denominator

$= \dfrac{\dfrac{5}{5} - \dfrac{3}{5}}{\dfrac{5}{5} + \dfrac{1}{5}}$

$= \dfrac{\dfrac{2}{5}}{\dfrac{6}{5}}$    Subtracting in numerator; adding in denominator

$= \dfrac{2}{5} \cdot \dfrac{5}{6}$    Multiplying by the reciprocal of the divisor

$= \dfrac{2 \cdot 5}{5 \cdot 2 \cdot 3}$

$= \dfrac{\cancel{2} \cdot \cancel{5} \cdot 1}{\cancel{5} \cdot \cancel{2} \cdot 3}$

$= \dfrac{1}{3}$

**5.** $\dfrac{\dfrac{1}{2} + \dfrac{3}{4}}{\dfrac{5}{8} - \dfrac{5}{6}} = \dfrac{\dfrac{1}{2} \cdot \dfrac{2}{2} + \dfrac{3}{4}}{\dfrac{5}{8} \cdot \dfrac{3}{3} - \dfrac{5}{6} \cdot \dfrac{4}{4}}$    Getting a common denominator in numerator and denominator

$= \dfrac{\dfrac{2}{4} + \dfrac{3}{4}}{\dfrac{15}{24} - \dfrac{20}{24}}$

$= \dfrac{\dfrac{5}{4}}{\dfrac{-5}{24}}$    Adding in numerator; subtracting in denominator

$= \dfrac{5}{4} \cdot \dfrac{24}{-5}$    Multiplying by the reciprocal of the divisor

$= \dfrac{5 \cdot 4 \cdot 6}{4 \cdot (-1) \cdot 5}$

$= \dfrac{\cancel{5} \cdot \cancel{4} \cdot 6}{\cancel{4} \cdot (-1) \cdot \cancel{5}}$

$= -6$

**7.** $\dfrac{\dfrac{1}{x} + 3}{\dfrac{1}{x} - 5}$    LCM of the denominators is $x$.

$= \dfrac{\dfrac{1}{x} + 3}{\dfrac{1}{x} - 5} \cdot \dfrac{x}{x}$    Multiplying by 1 using $\dfrac{x}{x}$

$= \dfrac{\left(\dfrac{1}{x} + 3\right)x}{\left(\dfrac{1}{x} - 5\right)x}$

$= \dfrac{\dfrac{1}{x} \cdot x + 3 \cdot x}{\dfrac{1}{x} \cdot x - 5 \cdot x}$

$= \dfrac{1 + 3x}{1 - 5x}$

**9.** $\dfrac{4 - \dfrac{1}{x^2}}{2 - \dfrac{1}{x}}$    LCM of the denominators is $x^2$.

$= \dfrac{4 - \dfrac{1}{x^2}}{2 - \dfrac{1}{x}} \cdot \dfrac{x^2}{x^2}$

$= \dfrac{\left(4 - \dfrac{1}{x^2}\right)x^2}{\left(2 - \dfrac{1}{x}\right)x^2}$

$= \dfrac{4 \cdot x^2 - \dfrac{1}{x^2} \cdot x^2}{2 \cdot x^2 - \dfrac{1}{x} \cdot x^2}$

$= \dfrac{4x^2 - 1}{2x^2 - x}$

$= \dfrac{(2x + 1)(2x - 1)}{x(2x - 1)}$    Factoring numerator and denominator

$= \dfrac{(2x + 1)\cancel{(2x - 1)}}{x\cancel{(2x - 1)}}$

$= \dfrac{2x + 1}{x}$

**11.** $\dfrac{8+\dfrac{8}{d}}{1+\dfrac{1}{d}} = \dfrac{8\cdot\dfrac{d}{d}+\dfrac{8}{d}}{1\cdot\dfrac{d}{d}+\dfrac{1}{d}}$

$= \dfrac{\dfrac{8d+8}{d}}{\dfrac{d+1}{d}}$

$= \dfrac{8d+8}{d}\cdot\dfrac{d}{d+1}$

$= \dfrac{8(d+1)(d)}{d(d+1)}$

$= \dfrac{8\cancel{(d+1)}\cancel{(d)}}{\cancel{d}\cancel{(d+1)}(1)}$

$= 8$

**13.** $\dfrac{\dfrac{x}{8}-\dfrac{8}{x}}{\dfrac{1}{8}+\dfrac{1}{x}}$     LCM of the denominators is $8x$.

$= \dfrac{\dfrac{x}{8}-\dfrac{8}{x}}{\dfrac{1}{8}+\dfrac{1}{x}}\cdot\dfrac{8x}{8x}$

$= \dfrac{\left(\dfrac{x}{8}-\dfrac{8}{x}\right)8x}{\left(\dfrac{1}{8}+\dfrac{1}{x}\right)8x}$

$= \dfrac{\dfrac{x}{8}(8x)-\dfrac{8}{x}(8x)}{\dfrac{1}{8}(8x)+\dfrac{1}{x}(8x)}$

$= \dfrac{x^2-64}{x+8}$

$= \dfrac{(x+8)(x-8)}{x+8}$

$= \dfrac{\cancel{(x+8)}(x-8)}{1\cancel{(x+8)}}$

$= x-8$

**15.** $\dfrac{1+\dfrac{1}{y}}{1-\dfrac{1}{y^2}} = \dfrac{1\cdot\dfrac{y}{y}+\dfrac{1}{y}}{1\cdot\dfrac{y^2}{y^2}-\dfrac{1}{y^2}}$

$= \dfrac{\dfrac{y+1}{y}}{\dfrac{y^2-1}{y^2}}$

$= \dfrac{y+1}{y}\cdot\dfrac{y^2}{y^2-1}$

$= \dfrac{(y+1)y\cdot y}{y(y+1)(y-1)}$

$= \dfrac{\cancel{(y+1)}\cancel{y}\cdot y}{\cancel{y}\cancel{(y+1)}(y-1)}$

$= \dfrac{y}{y-1}$

**17.** $\dfrac{\dfrac{1}{5}-\dfrac{1}{a}}{\dfrac{5-a}{5}}$     LCM of the denominators is $5a$.

$= \dfrac{\dfrac{1}{5}-\dfrac{1}{a}}{\dfrac{5-a}{5}}\cdot\dfrac{5a}{5a}$

$= \dfrac{\left(\dfrac{1}{5}-\dfrac{1}{a}\right)5a}{\left(\dfrac{5-a}{5}\right)5a}$

$= \dfrac{\dfrac{1}{5}(5a)-\dfrac{1}{a}(5a)}{a(5-a)}$

$= \dfrac{a-5}{5a-a^2}$

$= \dfrac{a-5}{-a(-5+a)}$

$= \dfrac{1\cancel{(a-5)}}{-a\cancel{(a-5)}}$

$= -\dfrac{1}{a}$

**19.** $\dfrac{\dfrac{1}{a}+\dfrac{1}{b}}{\dfrac{1}{a^2}-\dfrac{1}{b^2}}$     LCM of the denominators is $a^2b^2$.

$= \dfrac{\dfrac{1}{a}+\dfrac{1}{b}}{\dfrac{1}{a^2}-\dfrac{1}{b^2}}\cdot\dfrac{a^2b^2}{a^2b^2}\cdot$

$= \dfrac{\left(\dfrac{1}{a}+\dfrac{1}{b}\right)\cdot a^2b^2}{\left(\dfrac{1}{a^2}-\dfrac{1}{b^2}\right)\cdot a^2b^2}$

$= \dfrac{\dfrac{1}{a}\cdot a^2b^2+\dfrac{1}{b}\cdot a^2b^2}{\dfrac{1}{a^2}\cdot a^2b^2-\dfrac{1}{b^2}\cdot a^2b^2}$

$= \dfrac{ab^2+a^2b}{b^2-a^2}$

$= \dfrac{ab(b+a)}{(b+a)(b-a)}$

$= \dfrac{ab\cancel{(b+a)}}{\cancel{(b+a)}(b-a)}$

$= \dfrac{ab}{b-a}$

**21.**

$$\dfrac{\dfrac{p}{q}+\dfrac{q}{p}}{\dfrac{1}{p}+\dfrac{1}{q}} \qquad \text{LCM of the denominators is } pq.$$

$$=\dfrac{\left(\dfrac{p}{q}+\dfrac{q}{p}\right)\cdot pq}{\left(\dfrac{1}{p}+\dfrac{1}{q}\right)\cdot pq}$$

$$=\dfrac{\dfrac{p}{q}\cdot pq+\dfrac{q}{p}\cdot pq}{\dfrac{1}{p}\cdot pq+\dfrac{1}{q}\cdot pq}$$

$$=\dfrac{p^2+q^2}{q+p}$$

**23.**
$$(2x^3-4x^2+x-7)+(4x^4+x^3+4x^2+x)$$
$$=4x^4+3x^3+2x-7$$

**25.**  $p^2-10p+25=p^2-2\cdot p\cdot 5+5^2$ \qquad Trinomial square
$$=(p-5)^2$$

**27.**  $50p^2-100=50(p^2-2)$ \qquad Factoring out the common factor

Since $p^2-2$ cannot be factored, we have factored completely.

**29.**

$$\dfrac{1}{\dfrac{2}{x-1}-\dfrac{1}{3x-2}}$$

$$=\dfrac{1}{\dfrac{2}{x-1}-\dfrac{1}{3x-2}}\cdot\dfrac{(x-1)(3x-2)}{(x-1)(3x-2)}$$

$$=\dfrac{(x-1)(3x-2)}{\left(\dfrac{2}{x-1}-\dfrac{1}{3x-2}\right)(x-1)(3x-2)}$$

$$=\dfrac{(x-1)(3x-2)}{\dfrac{2}{x-1}(x-1)(3x-2)-\dfrac{1}{3x-2}(x-1)(3x-2)}$$

$$=\dfrac{(x-1)(3x-2)}{2(3x-2)-(x-1)}$$

$$=\dfrac{(x-1)(3x-2)}{6x-4-x+1}$$

$$=\dfrac{(x-1)(3x-2)}{5x-3}$$

**31.**

$$\dfrac{\dfrac{a}{b}-\dfrac{c}{d}}{\dfrac{b}{a}-\dfrac{d}{c}}=\dfrac{\dfrac{a}{b}\cdot\dfrac{d}{d}-\dfrac{c}{d}\cdot\dfrac{b}{b}}{\dfrac{b}{a}\cdot\dfrac{c}{c}-\dfrac{d}{c}\cdot\dfrac{a}{a}}$$

$$=\dfrac{\dfrac{ad-bc}{bd}}{\dfrac{bc-ad}{ac}}$$

$$=\dfrac{ad-bc}{bd}\cdot\dfrac{ac}{bc-ad}$$

$$=\dfrac{-1(bc-ad)(ac)}{bd(bc-ad)}$$

$$=\dfrac{-1(bc-ad)\,(ac)}{bd(bc-ad)}$$

$$=-\dfrac{ac}{bd}$$

**33.**  $1+\cfrac{1}{1+\cfrac{1}{1+\cfrac{1}{1+\cfrac{1}{x}}}}=1+\cfrac{1}{1+\cfrac{1}{1+\cfrac{1}{\frac{x+1}{x}}}}$

$$=1+\cfrac{1}{1+\cfrac{1}{1+\cfrac{x}{x+1}}}$$

$$=1+\cfrac{1}{1+\cfrac{1}{\frac{x+1+x}{x+1}}}$$

$$=1+\cfrac{1}{1+\cfrac{1}{\frac{2x+1}{x+1}}}$$

$$=1+\cfrac{1}{1+\cfrac{x+1}{2x+1}}$$

$$=1+\cfrac{1}{\frac{2x+1+x+1}{2x+1}}$$

$$=1+\cfrac{1}{\frac{3x+2}{2x+1}}$$

$$=1+\cfrac{2x+1}{3x+2}$$

$$=\dfrac{3x+2+2x+1}{3x+2}$$

$$=\dfrac{5x+3}{3x+2}$$

# Chapter 6
# Graphs of Equations and Inequalities

## Exercise Set 6.1

**1.** We find the portion of the graph labeled medical care and find that 12% of income is spent on medical expenses.

**3.** We find the portion of the graph labeled Jazz and read that 3.7% of all recordings sold are jazz.

**5.** *Familiarize.* Let $p$ = the percent of all recordings sold that are either soul or pop/rock. We will use the graph to find the percent for each type of music and then find their sum.

*Translate.* We reword the problem.

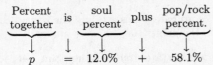

| Percent together | is | soul percent | plus | pop/rock percent. |
|---|---|---|---|---|
| $p$ | $=$ | $12.0\%$ | $+$ | $58.1\%$ |

*Solve.* We do the computation.

$$p = 12.0\% + 58.1\% = 70.1\%$$

*Check.* We go over the computation. The answer checks.

*State.* Together, 70.1% of all recordings sold are either soul or pop/rock.

**7.** *Familiarize.* The graph tells us that 58.1% of all recordings sold are pop/rock, 12.0% are soul, and 9.0% are country. Let $p$, $s$, and $c$ represent the number of pop/rock, soul, and country recordings sold, respectively.

*Translate.* We reword the problem and write an equation for each type of music.

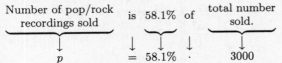

| Number of pop/rock recordings sold | is | 58.1% | of | total number sold. |
|---|---|---|---|---|
| $p$ | $=$ | $58.1\%$ | $\cdot$ | $3000$ |

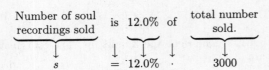

| Number of soul recordings sold | is | 12.0% | of | total number sold. |
|---|---|---|---|---|
| $s$ | $=$ | $12.0\%$ | $\cdot$ | $3000$ |

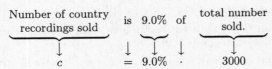

| Number of country recordings sold | is | 9.0% | of | total number sold. |
|---|---|---|---|---|
| $c$ | $=$ | $9.0\%$ | $\cdot$ | $3000$ |

*Solve.* We do the three computations.

$$p = 0.581 \cdot 3000 = 1743$$
$$s = 0.12 \cdot 3000 = 360$$
$$c = 0.09 \cdot 3000 = 270$$

*Check.* We go over the computations. The answers check.

*State.* The store sells 1743 pop/rock, 360 soul, and 270 country recordings.

**9.** *Familiarize.* To find the income that can be expected if 5 yr of college are completed we go to the right, across the bottom scale, to the bar representing income for a person with 5 + (5 or more) yr of college. We then go up to the top of the bar and, from there, back to the left to read approximately $27,500 on the income scale. Doing the same for the bar representing income for a person with 4 yr of college we read approximately $22,500 on the income scale. We let $a$ = the amount by which the income for a person with 5 yr of college completed exceeds the income for a person with 4 yr of college.

*Translate.* We restate the problem and translate to an equation.

| 4 yr income | plus | excess 5 yr income | is | 5 yr income |
|---|---|---|---|---|
| $\$22,500$ | $+$ | $a$ | $=$ | $\$27,000$ |

*Solve.* We solve the equation.

$$22,500 + a = 27,500$$
$$a = 5000 \quad \text{Subtracting 22,500}$$

*Check.* Since $\$22,500 + \$5000 = \$27,500$, the answer checks.

*State.* Approximately $5000 more income can be expected if 5 yr, rather than 4 yr, of college are completed.

**11.** *Familiarize.* To find the income that can be expected if you complete 4 yr of college we go to the right, across the bottom scale, to the bar representing income for a person with 4 yr of college. We then go up to the top of the bar and, from there, back to the left to read approximately $22,500 on the income scale. Doing the same for the bar representing income for a person with 1-3 yr of college we read approximately $18,500. We let $a$ = the amount by which the income for a person with 4 yr of college completed exceeds the income for a person with 1-3 yr of college.

*Translate.* We restate the problem and translate to an equation.

| 1-3 yr income | plus | excess 4 yr income | is | 4 yr income |
|---|---|---|---|---|
| $\$18,500$ | $+$ | $a$ | $=$ | $\$22,500$ |

*Solve.* We solve the equation.

$$18,500 + a = 22,500$$
$$a = 4000 \quad \text{Subtracting 18,500}$$

**Check**.  Since $18,500 + $4000 = $22,500$, the answer checks.

**State**.  You can expect approximately $4000 more income if you complete 4 yr of college than if you complete 1-3 yr.

13.  We locate 3100 on the DJIA scale and then move to the right until we reach the line. At that point, we move down to the Week scale and see that the DJIA closing was about 3100 in week 2.

15.  We go to the right, across the bottom scale, to the bar representing hiking. We then go to the top of the bar and, from there, back to the left to read approximately 590 calories on the Calories burned scale.

17.  **Familiarize**.  From Exercise 15 we know that approximately 590 calories are burned hiking. Following the procedure in Exercise 15 with the jogging bar, we find that approximately 650 calories are burned jogging. We let $c =$ the number by which the calories burned jogging exceed the calories burned hiking.

**Translate**.  We reword the problem and translate to an equation.

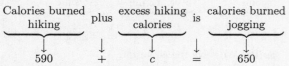

| Calories burned hiking | plus | excess hiking calories | is | calories burned jogging |
|---|---|---|---|---|
| 590 | + | $c$ | = | 650 |

**Solve**.  We solve the equation.

$$590 + c = 650$$
$$c = 60 \quad \text{Subtracting 590}$$

**Check**.  Since $590 + 60 = 650$, the answer checks.

**State**.  About 60 more calories are burned jogging than hiking.

19.  We look for the longest bar representing an activity other than jogging. This bar represents hiking, so hiking would be the most beneficial exercise if you could not jog.

21.  We locate the highest point on the graph and go down to the Year scale directly below that point. We find that estimated sales are greatest in 1998.

23.  We go across the bottom scale to 1996. Then we go up to the line and back to the left to the Estimated sales scale and read 17.0. The estimated sales in 1996 are $17.0 million.

25.  We go across the bottom scale to 2002. Then we go up to the line and back to the left to the Estimated sales scale and read about 18.9. The estimated sales in 2002 are $18.9 million.

27.  $(2,5)$ is 2 units right and 5 units up.

$(-1,3)$ is 1 unit left and 3 units up.

$(3,-2)$ is 3 units right and 2 units down.

$(-2,-4)$ is 2 units left and 4 units down.

$(0,4)$ is 0 units left or right and 4 units up.

$(0,-5)$ is 0 units left or right and 5 units down.

$(5,0)$ is 5 units right and 0 units up or down.

$(-5,0)$ is 5 units left and 0 units up or down.

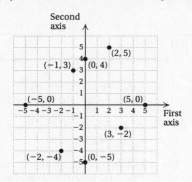

29.  Since the first coordinate is negative and the second coordinate positive, the point $(-5,3)$ is located in quadrant II.

31.  Since the first coordinate is positive and the second coordinate negative, the point $(100,-1)$ is in quadrant IV.

33.  Since both coordinates are negative, the point $(-6,-29)$ is in quadrant III.

35.  Since both coordinates are positive, the point $(3.8,9.2)$ is in quadrant I.

37.  In quadrant III, first coordinates are always <u>negative</u> and second coordinates are always <u>negative</u>.

39.

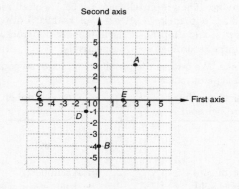

Point $A$ is 3 units right and 3 units up. The coordinates of $A$ are $(3,3)$.

Point $B$ is 0 units left or right and 4 units down. The coordinates of $B$ are $(0,-4)$.

Point $C$ is 5 units left and 0 units up or down. The coordinates of $C$ are $(-5,0)$.

Point $D$ is 1 unit left and 1 unit down. The coordinates of $D$ are $(-1,-1)$.

Point $E$ is 2 units right and 0 units up or down. The coordinates of $E$ are $(2,0)$.

41.  **Familiarize**.  Let $F =$ the amount the family making $20,500 spends on food. The ratio of amount spent on food to income for the first family is $5096/$19,600. For the second family it is $F/$20,500.

*Translate*. Assuming the two ratios are the same, we can translate to a proportion:

$$\text{Spent on food} \to \frac{\$5096}{\$19,600} = \frac{F}{\$20,500} \leftarrow \text{Spent on food}$$
$$\text{Income} \to \qquad\qquad\qquad \leftarrow \text{Income}$$

*Solve*. We solve the equation.

$$\$20,500 \cdot \frac{\$5096}{\$19,600} = \$20,500 \cdot \frac{F}{\$20,500}$$

$$\frac{\$20,500 \cdot \$5096}{\$19,600} = F$$

$$\$5330 = F$$

*Check*. $\frac{\$5096}{\$19,600} = 0.26$ and $\frac{\$5330}{\$20,500} = 0.26$.

The ratios are the same. The result checks.

*State*. A family making $20,500 would spend $5330 on food.

**43.**
$$256 \div 4 \cdot 16 - 37$$
$$= 64 \cdot 16 - 37 \qquad \text{Dividing}$$
$$= 1024 - 37 \qquad \text{Multiplying}$$
$$= 987 \qquad\qquad \text{Subtracting}$$

**45.**
$$1000 \div 100 \div 10$$
$$= 10 \div 10 \qquad \text{Doing the divisions in order}$$
$$= 1 \qquad\qquad \text{from left to right}$$

**47.**

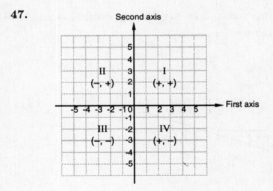

If the first coordinate is positive, then the point must be in either quadrant I or quadrant IV.

**49.** If the first and second coordinates are equal, they must either be both positive or both negative. The point must be in either quadrant I (both positive) or quadrant III (both negative).

**51.**

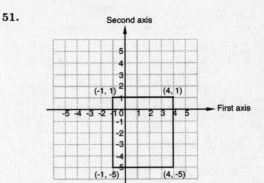

The coordinates of the fourth vertex are $(-1, -5)$.

**53.** Answers may vary.

We select eight points such that the sum of the coordinates for each point is 6.

| | |
|---|---|
| $(-1, 7)$ | $-1 + 7 = 6$ |
| $(0, 6)$ | $0 + 6 = 6$ |
| $(1, 5)$ | $1 + 5 = 6$ |
| $(2, 4)$ | $2 + 4 = 6$ |
| $(3, 3)$ | $3 + 3 = 6$ |
| $(4, 2)$ | $4 + 2 = 6$ |
| $(5, 1)$ | $5 + 1 = 6$ |
| $(6, 0)$ | $6 + 0 = 6$ |

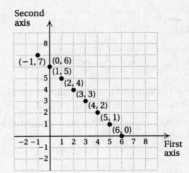

**55.**

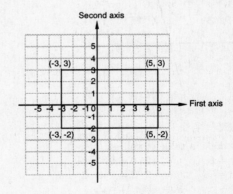

The length is 8, and the width is 5.

$$P = 2l + 2w$$
$$P = 2 \cdot 8 + 2 \cdot 5 = 16 + 10 = 26$$

## Exercise Set 6.2

**1.** $y = 3x - 1$

$$\begin{array}{c|l} 5 & 3 \cdot 2 - 1 \quad \text{Substituting 2 for } x \text{ and 5 for } y \\ & 6 - 1 \qquad \text{(alphabetical order of variables)} \\ & 5 \qquad \text{TRUE} \end{array}$$

The equation becomes true; $(2, 5)$ is a solution.

**3.** $\qquad 3x - y = 4$

$$\begin{array}{c|l} 3 \cdot 2 - (-3) & 4 \quad \text{Substituting 2 for } x \text{ and } -3 \text{ for } y \\ 6 + 3 & \\ 9 & \text{FALSE} \end{array}$$

The equation becomes false; $(2, -3)$ is not a solution.

**5.** $\qquad 4p + 2q = -9$

$$\begin{array}{c|l} 4 \cdot 0 + 2(-4) & -9 \quad \text{Substituting 0 for } p \text{ and } -4 \text{ for } q \\ 0 - 8 & \\ -8 & \text{FALSE} \end{array}$$

The equation becomes false; $(0, -4)$ is not a solution.

**7.** $y = 4x$

We first make a table of values. We choose *any* number for $x$ and then determine $y$ by substitution.

When $x = 0,$ $\quad y = 4 \cdot 0 = 0.$
When $x = -1,$ $\quad y = 4(-1) = -4.$
When $x = 1,$ $\quad y = 4 \cdot 1 = 4.$

| $x$ | $y$ |
|---|---|
| 0 | 0 |
| -1 | -4 |
| 1 | 4 |

Since two points determine a line, that is all we really need to graph a line, but you may plot a third point as a check.

Plot these points, draw the line they determine, and label the graph $y = 4x$.

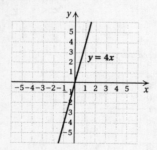

**9.** $y = -2x$

We first make a table of values.
When $x = 0,$ $\quad y = -2 \cdot 0 = 0.$
When $x = 2,$ $\quad y = -2 \cdot 2 = -4.$
When $x = -1,$ $\quad y = -2(-1) = 2.$

| $x$ | $y$ |
|---|---|
| 0 | 0 |
| 2 | -4 |
| -1 | 2 |

Plot these points, draw the line they determine, and label the graph $y = -2x$.

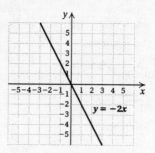

**11.** $y = \dfrac{1}{3}x$

We first make a table of values. Using multiples of 3 avoids fractions.

When $x = 0,$ $\quad y = \dfrac{1}{3} \cdot 0 = 0.$

When $x = 6,$ $\quad y = \dfrac{1}{3} \cdot 6 = 2.$

When $x = -3,$ $\quad y = \dfrac{1}{3}(-3) = -1.$

| $x$ | $y$ |
|---|---|
| 0 | 0 |
| 6 | 2 |
| -3 | -1 |

Plot these points, draw the line they determine, and label the graph $y = \dfrac{1}{3}x$.

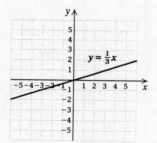

**13.** $y = -\dfrac{3}{2}x$

We first make a table of values. Using multiples of 2 avoids fractions.

When $x = 0,$ $\quad y = -\dfrac{3}{2} \cdot 0 = 0.$

When $x = 2,$ $\quad y = -\dfrac{3}{2} \cdot 2 = -3.$

When $x = -2,$ $\quad y = -\dfrac{3}{2}(-2) = 3.$

| $x$ | $y$ |
|---|---|
| 0 | 0 |
| 2 | -3 |
| -2 | 3 |

Plot these points, draw the line they determine, and label the graph $y = -\dfrac{3}{2}x$.

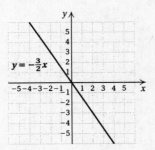

**15.** $y = x + 1$

We first make a table of values. We choose *any* number for $x$ and then determine $y$ by substitution.

When $x = 0$,    $y = 0 + 1 = 1$.
When $x = 3$,    $y = 3 + 1 = 4$.
When $x = -5$,    $y = -5 + 1 = -4$.

| $x$ | $y$ |
|---|---|
| 0 | 1 |
| 3 | 4 |
| −5 | −4 |

Plot these points, draw the line they determine, and label the graph $y = x + 1$.

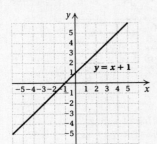

**17.** $y = 2x + 2$

We first make a table of values.

When $x = 0$,    $y = 2 \cdot 0 + 2 = 0 + 2 = 2$.
When $x = -3$,    $y = 2(-3) + 2 = -6 + 2 = -4$.
When $x = 1$,    $y = 2 \cdot 1 + 2 = 2 + 2 = 4$.

| $x$ | $y$ |
|---|---|
| 0 | 2 |
| −3 | −4 |
| 1 | 4 |

Plot these points, draw the line they determine, and label the graph $y = 2x + 2$.

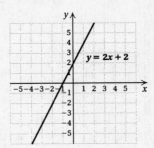

**19.** $y = \dfrac{1}{3}x - 1$

We first make a table of values. Using multiples of 3 avoids fractions.

When $x = 0$,    $y = \dfrac{1}{3} \cdot 0 - 1 = 0 - 1 = -1$.

When $x = -6$,    $y = \dfrac{1}{3}(-6) - 1 = -2 - 1 = -3$.

When $x = 3$,    $y = \dfrac{1}{3} \cdot 3 - 1 = 1 - 1 = 0$.

| $x$ | $y$ |
|---|---|
| 0 | −1 |
| −6 | −3 |
| 3 | 0 |

Plot these points, draw the line they determine, and label the graph $y = \dfrac{1}{3}x - 1$.

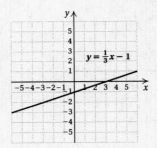

**21.** $y = -x - 3$

We first make a table of values.

When $x = 0$,    $y = -0 - 3 = -3$.
When $x = 1$,    $y = -1 - 3 = -4$.
When $x = -5$,    $y = -(-5) - 3 = 5 - 3 = 2$.

| $x$ | $y$ |
|---|---|
| 0 | −3 |
| 1 | −4 |
| −5 | 2 |

Plot these points, draw the line they determine, and label the graph.

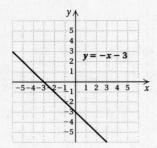

**23.** $y = \dfrac{5}{2}x + 3$

We first make a table of values. Using multiples of 2 avoids fractions.

When $x = 0,$     $y = \dfrac{5}{2} \cdot 0 + 3 = 0 + 3 = 3.$

When $x = -2,$     $y = \dfrac{5}{2}(-2) + 3 = -5 + 3 = -2.$

When $x = -4,$     $y = \dfrac{5}{2}(-4) + 3 = -10 + 3 = -7.$

| $x$ | $y$ |
|-----|-----|
| 0 | 3 |
| $-2$ | $-2$ |
| $-4$ | $-7$ |

Plot these points, draw the line they determine, and label the graph.

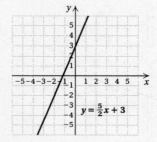

**25.** $y = -\dfrac{5}{3}x - 2$

We first make a table of values. Using multiples of 3 avoids fractions.

When $x = 0,$     $y = -\dfrac{5}{3} \cdot 0 - 2 = 0 - 2 = -2.$

When $x = -3,$     $y = -\dfrac{5}{3}(-3) - 2 = 5 - 2 = 3.$

When $x = 3,$     $y = -\dfrac{5}{3}(3) - 2 = -5 - 2 = -7.$

| $x$ | $y$ |
|-----|-----|
| 0 | $-2$ |
| $-3$ | 3 |
| 3 | $-7$ |

Plot these points, draw the line they determine, and label the graph.

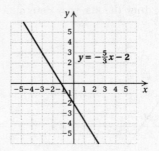

**27.** $y = x$

We first make a table of values.
When $x = -4,$   $y = -4.$
When $x = -1,$   $y = -1.$
When $x = 3,$     $y = 3.$

| $x$ | $y$ |
|-----|-----|
| $-4$ | $-4$ |
| $-1$ | $-1$ |
| 3 | 3 |

Plot these points, draw the line they determine, and label the graph.

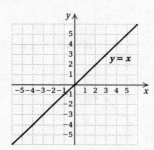

**29.** $y = 3 - 2x$

We first make a table of values.
When $x = 1,$     $y = 3 - 2 \cdot 1 = 3 - 2 = 1.$
When $x = -1,$   $y = 3 - 2 \cdot (-1) = 3 + 2 = 5.$
When $x = 3,$     $y = 3 - 2 \cdot 3 = 3 - 6 = -3.$

| $x$ | $y$ |
|-----|-----|
| 1 | 1 |
| $-1$ | 5 |
| 3 | $-3$ |

Plot these points, draw the line they determine, and label the graph.

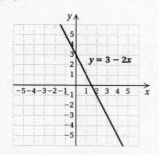

**31.** $y = \frac{4}{3} - \frac{1}{3}x$

We first make a table of values.

When $x = 1$,     $y = \frac{4}{3} - \frac{1}{3} \cdot 1 = \frac{4}{3} - \frac{1}{3} = \frac{3}{3} = 1$.

When $x = -2$,   $y = \frac{4}{3} - \frac{1}{3} \cdot (-2) = \frac{4}{3} + \frac{2}{3} = \frac{6}{3} = 2$.

When $x = 4$,     $y = \frac{4}{3} - \frac{1}{3} \cdot 4 = \frac{4}{3} - \frac{4}{3} = 0$.

When $x = 0$,     $y = \frac{4}{3} - \frac{1}{3} \cdot 0 = \frac{4}{3} - 0 = \frac{4}{3}$.

| $x$ | $y$ |
|-----|-----|
| 1   | 1   |
| -2  | 2   |
| 4   | 0   |
| 0   | $\frac{4}{3}$ |

Plot these points, draw the line they determine, and label the graph.

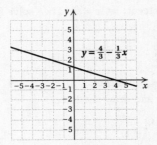

**33. Familiarize**. Let $d$ = the distance traveled and $r$ = the speed of the plane in still air. Then $r + 5$ is the speed with the tailwind and $r - 5$ is the speed against the wind. We organize the information in a table.

|  | Distance | Speed | Time |
|--|----------|-------|------|
| $d$ | $=$ | $r$ | $\cdot$ $t$ |
| With wind | $d$ | $r + 5$ | 7 |
| Against wind | $d$ | $r - 5$ | 8 |

**Translate**. We can apply the formula $d = rt$ along the rows of the table to obtain two equations:

$d = (r + 5)(7),$

$d = (r - 5)(8).$

Since the distances are the same, we get the following equation:

$(r + 5)(7) = (r - 5)(8)$

**Solve**. We solve the equation.

$(r + 5)(7) = (r - 5)(8)$

$7r + 35 = 8r - 40$

$75 = r$

**Check**. If the speed in still air is 75 km/h, then the speed with a tailwind is $75 + 5$, or 80 km/h, and in 7 hr the

plane would travel $80 \cdot 7$, or 560 km. The speed against the wind is $75 - 5$, or 70 km/h, and in 8 hr the plane would travel $70 \cdot 8$, or 560 km. Since the distances are the same, the result checks.

**State**. The speed of the plane in still air is 75 km/h.

**35.**        $25t^2 - 49 = 0$

$(5t + 7)(5t - 7) = 0$

| $5t + 7 = 0$ | or | $5t - 7 = 0$ |
|---|---|---|
| $5t = -7$ | or | $5t = 7$ |
| $t = -\frac{7}{5}$ | or | $t = \frac{7}{5}$ |

The solutions are $-\frac{7}{5}$ and $\frac{7}{5}$.

**37.**        $\frac{x^2}{x - 1} = \frac{1}{x - 1}$, LCM is $x - 1$

$\left(x - 1\right) \cdot \frac{x^2}{x - 1} = \left(x - 1\right) \cdot \frac{1}{x - 1}$

$x^2 = 1$

$x^2 - 1 = 0$

$(x + 1)(x - 1) = 0$

$x + 1 = 0$   or   $x - 1 = 0$

$x = -1$   or        $x = 1$

The numbers 1 and $-1$ are possible solutions. We look at the original equation and see that 1 makes the denominators zero and is therefore not a solution. The number $-1$ checks and is a solution.

**39.** We graph $x + y = 6$ and list the points for which both coordinates are whole numbers.

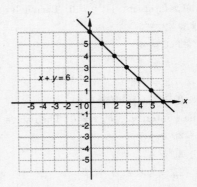

The whole-number solutions are $(0, 6)$, $(1, 5)$, $(2, 4)$, $(3, 3)$, $(4, 2)$, $(5, 1)$, and $(6, 0)$.

## Exercise Set 6.3

**1.** $x + 3y = 6$

To find the $x$-intercept, let $y = 0$. Then solve for $x$.

$x + 3y = 6$

$x + 3 \cdot 0 = 6$

$x = 6$

Thus, $(6, 0)$ is the $x$-intercept.

To find the $y$-intercept, let $x = 0$. Then solve for $y$.

$$x + 3y = 6$$
$$0 + 3y = 6$$
$$3y = 6$$
$$y = 2$$

Thus, $(0, 2)$ is the $y$-intercept.

Plot these points and draw the line.

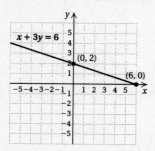

A third point should be used as a check. We substitute any value for $x$ and solve for $y$.

We let $x = 3$. Then

$$x + 3y = 6$$
$$3 + 3y = 6$$
$$3y = 3$$
$$y = 1$$

The point $(3, 1)$ is on the graph, so the graph is probably correct.

**3.** $-x + 2y = 4$

To find the $x$-intercept, let $y = 0$. Then solve for $x$.

$$-x + 2y = 4$$
$$-x + 2 \cdot 0 = 4$$
$$-x = 4$$
$$x = -4$$

Thus, $(-4, 0)$ is the $x$-intercept.

To find the $y$-intercept, let $x = 0$. Then solve for $y$.

$$-x + 2y = 4$$
$$-0 + 2y = 4$$
$$2y = 4$$
$$y = 2$$

Thus, $(0, 2)$ is the $y$-intercept.

Plot these points and draw the line.

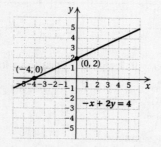

A third point should be used as a check. We substitute any value for $x$ and solve for $y$.

We let $x = 4$. Then

$$-x + 2y = 4$$
$$-4 + 2y = 4$$
$$2y = 8$$
$$y = 4$$

The point $(4, 4)$ is on the graph, so the graph is probably correct.

**5.** $3x + y = 6$

To find the $x$-intercept, let $y = 0$. Then solve for $x$.

$$3x + y = 6$$
$$3x + 0 = 6$$
$$3x = 6$$
$$x = 2$$

Thus, $(2, 0)$ is the $x$-intercept.

To find the $y$-intercept, let $x = 0$. Then solve for $y$.

$$3x + y = 6$$
$$3 \cdot 0 + y = 6$$
$$y = 6$$

Thus, $(0, 6)$ is the $y$-intercept.

Plot these points and draw the line.

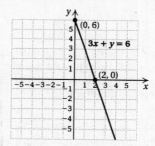

A third point should be used as a check. We substitute any value for $x$ and solve for $y$.

We let $x = 1$. Then

$$3x + y = 6$$
$$3 \cdot 1 + y = 6$$
$$3 + y = 6$$
$$y = 3$$

The point $(1, 3)$ is on the graph, so the graph is probably correct.

**7.** $2y - 2 = 6x$

To find the $x$-intercept, let $y = 0$. Then solve for $x$.

$$2y - 2 = 6x$$
$$2 \cdot 0 - 2 = 6x$$
$$-2 = 6x$$
$$-\frac{1}{3} = x$$

Thus, $\left( -\frac{1}{3}, 0 \right)$ is the $x$-intercept.

To find the $y$-intercept, let $x = 0$. Then solve for $y$.

$$2y - 2 = 6x$$
$$2y - 2 = 6 \cdot 0$$
$$2y - 2 = 0$$
$$2y = 2$$
$$y = 1$$

Thus, $(0, 1)$ is the $y$-intercept.

It is helpful to plot another point since the intercepts are so close together. This point can also serve as a check.

We let $x = 1$. Then

$$2y - 2 = 6x$$
$$2y - 2 = 6 \cdot 1$$
$$2y - 2 = 6$$
$$2y = 8$$
$$y = 4$$

Plot the point $(1, 4)$ and the intercepts and draw the line.

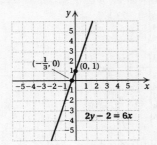

**9.** $3x - 9 = 3y$

To find the $x$-intercept, let $y = 0$. Then solve for $x$.

$$3x - 9 = 3y$$
$$3x - 9 = 3 \cdot 0$$
$$3x - 9 = 0$$
$$3x = 9$$
$$x = 3$$

Thus, $(3, 0)$ is the $x$-intercept.

To find the $y$-intercept, let $x = 0$. Then solve for $y$.

$$3x - 9 = 3y$$
$$3 \cdot 0 - 9 = 3y$$
$$-9 = 3y$$
$$-3 = y$$

Thus, $(0, -3)$ is the $y$-intercept.

Plot these points and draw the line.

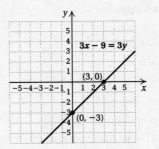

A third point should be used as a check. We substitute any value for $x$ and solve for $y$.

We let $x = 1$. Then

$$3x - 9 = 3y$$
$$3 \cdot 1 - 9 = 3y$$
$$3 - 9 = 3y$$
$$-6 = 3y$$
$$-2 = y$$

The point $(1, -2)$ is on the graph, so the graph is probably correct.

**11.** $2x - 3y = 6$

To find the $x$-intercept, let $y = 0$. Then solve for $x$.

$$2x - 3y = 6$$
$$2x - 3 \cdot 0 = 6$$
$$2x = 6$$
$$x = 3$$

Thus, $(3, 0)$ is the $x$-intercept.

To find the $y$-intercept, let $x = 0$. Then solve for $y$.

$$2x - 3y = 6$$
$$2 \cdot 0 - 3y = 6$$
$$-3y = 6$$
$$y = -2$$

Thus, $(0, -2)$ is the $y$-intercept.

Plot these points and draw the line.

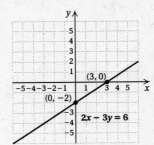

A third point should be used as a check. We substitute

any value for $x$ and solve for $y$.

We let $x = -3$.

$$2x - 3y = 6$$
$$2(-3) - 3y = 6$$
$$-6 - 3y = 6$$
$$-3y = 12$$
$$y = -4$$

The point $(-3, -4)$ is on the graph, so the graph is probably correct.

**13.** $4x + 5y = 20$

To find the $x$-intercept, let $y = 0$. Then solve for $x$.

$$4x + 5y = 20$$
$$4x + 5 \cdot 0 = 20$$
$$4x = 20$$
$$x = 5$$

Thus, $(5, 0)$ is the $x$-intercept.

To find the $y$-intercept, let $x = 0$. Then solve for $y$.

$$4x + 5y = 20$$
$$4 \cdot 0 + 5y = 20$$
$$5y = 20$$
$$y = 4$$

Thus, $(0, 4)$ is the $y$-intercept.

Plot these points and draw the graph.

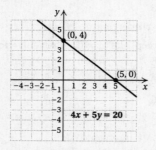

A third point should be used as a check. We substitute any value for $x$ and solve for $y$.

We let $x = 4$. Then

$$4x + 5y = 20$$
$$4 \cdot 4 + 5y = 20$$
$$16 + 5y = 20$$
$$5y = 4$$
$$y = \frac{4}{5}$$

The point $\left(4, \frac{4}{5}\right)$ is on the graph, so the graph is probably correct.

**15.** $2x + 3y = 8$

To find the $x$-intercept, let $y = 0$. Then solve for $x$.

$$2x + 3y = 8$$
$$2x + 3 \cdot 0 = 8$$
$$2x = 8$$
$$x = 4$$

Thus, $(4, 0)$ is the $x$-intercept.

To find the $y$-intercept, let $x = 0$. Then solve for $y$.

$$2x + 3y = 8$$
$$2 \cdot 0 + 3y = 8$$
$$3y = 8$$
$$y = \frac{8}{3}$$

Thus, $\left(0, \frac{8}{3}\right)$ is the $y$-intercept.

Plot these points and draw the graph.

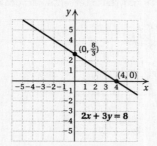

A third point should be used as a check.

We let $x = 1$. Then

$$2x + 3y = 8$$
$$2 \cdot 1 + 3y = 8$$
$$2 + 3y = 8$$
$$3y = 6$$
$$y = 2$$

The point $(1, 2)$ is on the graph, so the graph is probably correct.

**17.** $x - 3 = y$

To find the $x$-intercept, let $y = 0$. Then solve for $x$.

$$x - 3 = y$$
$$x - 3 = 0$$
$$x = 3$$

Thus, $(3, 0)$ is the $x$-intercept.

To find the $y$-intercept, let $x = 0$. Then solve for $y$.

$$x - 3 = y$$
$$0 - 3 = y$$
$$-3 = y$$

Thus, $(0, -3)$ is the $y$-intercept.

Plot these points and draw the line.

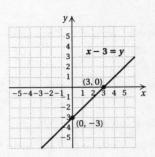

A third point should be used as a check.

We let $x = -2$. Then

$$x - 3 = y$$
$$-2 - 3 = y$$
$$-5 = y$$

The point $(-2, -5)$ is on the graph, so the graph is probably correct.

**19.** $3x - 2 = y$

To find the $x$-intercept, let $y = 0$. Then solve for $x$.

$$3x - 2 = y$$
$$3x - 2 = 0$$
$$3x = 2$$
$$x = \frac{2}{3}$$

Thus, $\left(\frac{2}{3}, 0\right)$ is the $x$-intercept.

To find the $y$-intercept, let $x = 0$. Then solve for $y$.

$$3x - 2 = y$$
$$3 \cdot 0 - 2 = y$$
$$-2 = y$$

Thus, $(0, -2)$ is the $y$-intercept.

Plot these points and draw the line.

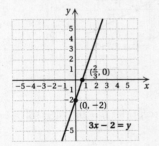

A third point should be used as a check.

We let $x = 2$. Then

$$3x - 2 = y$$
$$3 \cdot 2 - 2 = y$$
$$6 - 2 = y$$
$$4 = y$$

The point $(2, 4)$ is on the graph, so the graph is probably correct.

**21.** $6x - 2y = 18$

To find the $x$-intercept, let $y = 0$. Then solve for $x$.

$$6x - 2y = 18$$
$$6x - 2 \cdot 0 = 18$$
$$6x = 18$$
$$x = 3$$

Thus, $(3, 0)$ is the $x$-intercept.

To find the $y$-intercept, let $x = 0$. Then solve for $y$.

$$6x - 2y = 18$$
$$6 \cdot 0 - 2y = 18$$
$$-2y = 18$$
$$y = -9$$

Thus, $(0, -9)$ is the $y$-intercept.

Plot these points and draw the line.

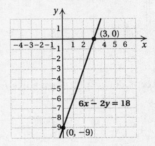

We use a third point as a check.

We let $x = 1$. Then

$$6x - 2y = 18$$
$$6 \cdot 1 - 2y = 18$$
$$6 - 2y = 18$$
$$-2y = 12$$
$$y = -6$$

The point $(1, -6)$ is on the graph, so the graph is probably correct.

**23.** $3x + 4y = 5$

To find the $x$-intercept, let $y = 0$. Then solve for $x$.

$$3x + 4y = 5$$
$$3x + 4 \cdot 0 = 5$$
$$3x = 5$$
$$x = \frac{5}{3}$$

Thus, $\left(\frac{5}{3}, 0\right)$ is the $x$-intercept.

To find the $y$-intercept, let $x = 0$. Then solve for $y$.

$$3x + 4y = 5$$
$$3 \cdot 0 + 4y = 5$$
$$4y = 5$$
$$y = \frac{5}{4}$$

Thus, $\left(0, \frac{5}{4}\right)$ is the $y$-intercept.

It is helpful to plot another point since the intercepts are so close together. This point can also serve as a check.

We let $x = 3$. Then

$$3x + 4y = 5$$
$$3 \cdot 3 + 4y = 5$$
$$9 + 4y = 5$$
$$4y = -4$$
$$y = -1$$

Plot the point $(3, -1)$ and the intercepts and draw the line.

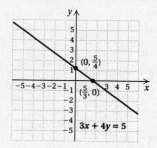

**25.** $y = -3 - 3x$

To find the $x$-intercept, let $y = 0$. Then solve for $x$.

$$y = -3 - 3x$$
$$0 = -3 - 3x$$
$$3x = -3$$
$$x = -1$$

Thus, $(-1, 0)$ is the $x$-intercept.

To find the $y$-intercept, let $x = 0$. Then solve for $y$.

$$y = -3 - 3x$$
$$y = -3 - 3 \cdot 0$$
$$y = -3$$

Thus, $(0, -3)$ is the $y$-intercept.

Plot these points and draw the graph.

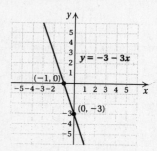

We use a third point as a check.

We let $x = -2$. Then

$$y = -3 - 3x$$
$$y = -3 - 3 \cdot (-2)$$
$$y = -3 + 6$$
$$y = 3$$

The point $(-2, 3)$ is on the graph, so the graph is probably correct.

**27.** $-4x = 8y - 5$

To find the $x$-intercept, let $y = 0$. Then solve for $x$.

$$-4x = 8 \cdot 0 - 5$$
$$-4x = -5$$
$$x = \frac{5}{4}$$

Thus, $\left(\frac{5}{4}, 0\right)$ is the $x$-intercept.

To find the $y$-intercept, let $x = 0$. Then solve for $y$.

$$-4x = 8y - 5$$
$$-4 \cdot 0 = 8y - 5$$
$$0 = 8y - 5$$
$$5 = 8y$$
$$\frac{5}{8} = y$$

Thus, $\left(0, \frac{5}{8}\right)$ is the $y$-intercept.

It is helpful to plot another point since the intercepts are so close together. This can also serve as a check.

We let $x = -5$. Then

$$-4x = 8y - 5$$
$$-4(-5) = 8y - 5$$
$$20 = 8y - 5$$
$$25 = 8y$$
$$\frac{25}{8} = y$$

Plot the point $\left(-5, \frac{25}{8}\right)$ and the intercepts and draw the line.

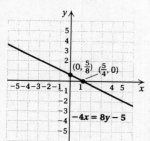

**29.** $y - 3x = 0$

To find the $x$-intercept, let $y = 0$. Then solve for $x$.

$$0 - 3x = 0$$
$$-3x = 0$$
$$x = 0$$

Thus, $(0, 0)$ is the $x$-intercept. Note that this is also the $y$-intercept.

In order to graph the line, we will find a second point.

When $x = 1$, $y - 3 \cdot 1 = 0$

$$y - 3 = 0$$
$$y = 3$$

Plot the points and draw the graph.

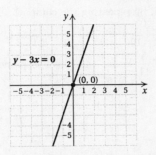

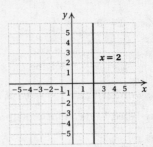

| x | y |
|---|---|
| 2 | $-1$ |
| 2 | 4 |
| 2 | 5 |

We use a third point as a check.

We let $x = -1$. Then

$$y - 3(-1) = 0$$
$$y + 3 = 0$$
$$y = -3$$

The point $(-1, -3)$ is on the graph, so the graph is probably correct.

**37.** $y = 0$

Any ordered pair $(x, 0)$ is a solution. The variable $y$ must be 0, but $x$ can be any number we choose. A few solutions are listed below. Plot these points and draw the line.

| x | y |
|---|---|
| $-5$ | 0 |
| $-1$ | 0 |
| 3 | 0 |

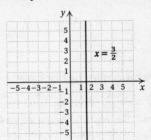

**31.** $x = -2$

Any ordered pair $(-2, y)$ is a solution. The variable $x$ must be $-2$, but $y$ can be any number we choose. A few solutions are listed below. Plot these points and draw the line.

| x | y |
|---|---|
| $-2$ | $-2$ |
| $-2$ | 0 |
| $-2$ | 4 |

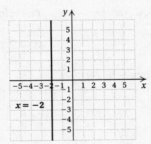

**39.** $x = \dfrac{3}{2}$

Any ordered pair $\left(\dfrac{3}{2}, y\right)$ is a solution. The variable $x$ must be $\dfrac{3}{2}$, but $y$ can be any number we choose. A few solutions are listed below. Plot these points and draw the line.

| x | y |
|---|---|
| $\dfrac{3}{2}$ | $-2$ |
| $\dfrac{3}{2}$ | 0 |
| $\dfrac{3}{2}$ | 4 |

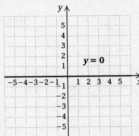

**33.** $y = 2$

Any ordered pair $(x, 2)$ is a solution. The variable $y$ must be 2, but $x$ can be any number we choose. A few solutions are listed below. Plot these points and draw the line.

| x | y |
|---|---|
| $-3$ | 2 |
| 0 | 2 |
| 2 | 2 |

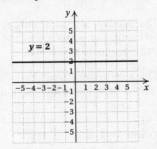

**41.** $3y = -5$

$$y = -\dfrac{5}{3} \qquad \text{Solving for } y$$

Any ordered pair $\left(x, -\dfrac{5}{3}\right)$ is a solution. A few solutions are listed below. Plot these points and draw the line.

| x | y |
|---|---|
| $-3$ | $-\dfrac{5}{3}$ |
| 0 | $-\dfrac{5}{3}$ |
| 2 | $-\dfrac{5}{3}$ |

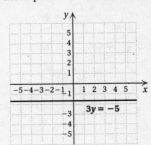

**35.** $x = 2$

Any ordered pair $(2, y)$ is a solution. The variable $x$ must be 2, but $y$ can be any number we choose. A few solutions are listed below. Plot these points and draw the line.

**43.**  $4x + 3 = 0$

$$4x = -3$$

$$x = -\frac{3}{4} \qquad \text{Solving for } x$$

Any ordered pair $\left(-\frac{3}{4}, y\right)$ is a solution. A few solutions are listed below. Plot these points and draw the line.

| $x$ | $y$ |
|---|---|
| $-\dfrac{3}{4}$ | $-2$ |
| $-\dfrac{3}{4}$ | $0$ |
| $-\dfrac{3}{4}$ | $3$ |

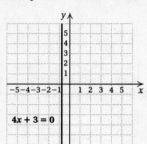

**45.**  $48 - 3y = 0$

$$-3y = -48$$

$$y = 16 \qquad \text{Solving for } y$$

Any ordered pair $(x, 16)$ is a solution. A few solutions are listed below. Plot these points and draw the line.

| $x$ | $y$ |
|---|---|
| $-4$ | $16$ |
| $0$ | $16$ |
| $2$ | $16$ |

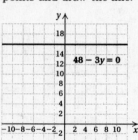

**47. Familiarize.** In 1 hr Crew $A$ does $\dfrac{1}{10}$ of the job and Crew $B$ does $\dfrac{1}{12}$ of the job, and together they do $\dfrac{1}{10} + \dfrac{1}{12}$ of the job. We want to find a number $t$ such that the fraction of the job completed by the crews together is 1.

**Translate.** From the familiarization step we see that we want some number $t$ for which

$$t\left(\frac{1}{10}\right) + t\left(\frac{1}{12}\right) = 1.$$

**Solve.** We solve the equation.

$$\frac{t}{10} + \frac{t}{12} = 1, \qquad \text{LCM} = 2 \cdot 2 \cdot 3 \cdot 5, \text{ or } 60$$

$$60\left(\frac{t}{10} + \frac{t}{12}\right) = 60 \cdot 1$$

$$60 \cdot \frac{t}{10} + 60 \cdot \frac{t}{12} = 60$$

$$6t + 5t = 60$$

$$11t = 60$$

$$t = \frac{60}{11}, \text{ or } 5\frac{5}{11}$$

**Check.** We check by repeating the computations.

$$\frac{60}{11}\left(\frac{1}{10}\right) + \frac{60}{11}\left(\frac{1}{12}\right) = \frac{6}{11} + \frac{5}{11} = \frac{11}{11} = 1$$

**State.** It takes $5\frac{5}{11}$ hr for the crews to paint the house working together.

**49.**

$$\frac{7}{1+x} - 1 = \frac{5x}{x^2 + 3x + 2},$$

$$\text{LCM} = (x+1)(x+2)$$

$$(x+1)(x+2)\left(\frac{7}{1+x} - 1\right) =$$

$$(x+1)(x+2) \cdot \frac{5x}{(x+1)(x+2)}$$

$$(x+1)(x+2) \cdot \frac{7}{1+x} + (x+1)(x+2)(-1) = 5x$$

$$7(x+2) + (x^2 + 3x + 2)(-1) = 5x$$

$$7x + 14 - x^2 - 3x - 2 = 5x$$

$$-x^2 + 4x + 12 = 5x$$

$$0 = x^2 + x - 12$$

$$0 = (x+4)(x-3)$$

$$x + 4 = 0 \quad \text{or} \quad x - 3 = 0$$

$$x = -4 \quad \text{or} \quad x = 3$$

Both numbers check. The solutions are $-4$ and $3$.

**51.**

$$\frac{1}{t} = \frac{1}{8} - \frac{1}{3}, \ \text{LCM} = 24t$$

$$24t \cdot \frac{1}{t} = 24t\left(\frac{1}{8} - \frac{1}{3}\right)$$

$$24 = 24t \cdot \frac{1}{8} - 24t \cdot \frac{1}{3}$$

$$24 = 3t - 8t$$

$$24 = -5t$$

$$-\frac{24}{5} = t$$

This number checks. The solution is $-\dfrac{24}{5}$.

**53.** The $y$-axis is a vertical line, so it is of the form $x = a$. All points on the $y$-axis are of the form $(0, y)$, so $a$ must be 0 and the equation is $x = 0$.

**55.** The $x$-coordinate must be $-3$, and the $y$-coordinate must be 6. The point of intersection is $(-3, 6)$.

## Exercise Set 6.4

**1.** We consider $(x_1, y_1)$ to be $(3, 2)$ and $(x_2, y_2)$ to be $(-1, 2)$.

$$m = \frac{2 - 2}{-1 - 3} = \frac{0}{-4} = 0$$

**3.** We consider $(x_1, y_1)$ to be $(3, 0)$ and $(x_2, y_2)$ to be $(-2, 4)$.

$$m = \frac{4 - 0}{-2 - 3} = \frac{4}{-5} = -\frac{4}{5}$$

**5.** We consider $(x_1, y_1)$ to be $(3, 0)$ and $(x_2, y_2)$ to be $(6, 2)$.
$$m = \frac{2 - 0}{6 - 3} = \frac{2}{3}$$

**7.** We consider $(x_1, y_1)$ to be $(-5, -6)$ and $(x_2, y_2)$ to be $(-3, -2)$.
$$m = \frac{-2 - (-6)}{-3 - (-5)} = \frac{4}{2} = 2$$

**9.** We consider $(x_1, y_1)$ to be $\left(-5, \frac{1}{2}\right)$ and $(x_2, y_2)$ to be $\left(-2, \frac{1}{2}\right)$.
$$m = \frac{\frac{1}{2} - \frac{1}{2}}{-2 - (-5)} = \frac{0}{3} = 0$$

**11.** We consider $(x_1, y_1)$ to be $(9, -7)$ and $(x_2, y_2)$ to be $(9, -4)$.
$$m = \frac{-4 - (-7)}{9 - 9} = \frac{-4 + 7}{9 - 9} = \frac{3}{0}$$
Since division by 0 is undefined, the slope of this line is undefined.

**13.** We solve for $y$.
$$3x + 2y = 6$$
$$2y = -3x + 6$$
$$y = \frac{1}{2}(-3x + 6)$$
$$y = -\frac{3}{2}x + 3$$
The slope is $-\frac{3}{2}$.

**15.** We solve for $y$.
$$x + 4y = 8$$
$$4y = -x + 8$$
$$y = \frac{1}{4}(-x + 8)$$
$$y = -\frac{1}{4}x + 2$$
The slope is $-\frac{1}{4}$.

**17.** We solve for $y$.
$$-2x + y = 4$$
$$y = 2x + 4$$
The slope is 2.

**19.** The line $x = -8$ is a vertical line. The slope of a vertical line is undefined.

**21.** The line $y = 2$ is a horizontal line. A horizontal line has slope 0.

**23.** The line $x = 9$ is a vertical line. The slope of a vertical line is undefined.

**25.** The line $y = -6$ is a horizontal line. A horizontal line has slope 0.

**27.** $y = -4x - 9$
The equation is already in the form $y = mx + b$. The slope is $-4$ and the $y$-intercept is $(0, -9)$.

**29.** $y = 1.8x$
We can think of $y = 1.8x$ as $y = 1.8x + 0$. The slope is 1.8 and the $y$-intercept is $(0, 0)$.

**31.** We solve for $y$.
$$-8x - 7y = 21$$
$$-7y = 8x + 21$$
$$y = -\frac{1}{7}(8x + 21)$$
$$y = -\frac{8}{7}x - 3$$
The slope is $-\frac{8}{7}$ and the $y$-intercept is $(0, -3)$.

**33.** We solve for $y$.
$$9x = 3y + 5$$
$$9x - 5 = 3y$$
$$\frac{1}{3}(9x - 5) = y$$
$$3x - \frac{5}{3} = y$$
The slope is 3 and the $y$-intercept is $\left(0, -\frac{5}{3}\right)$.

**35.** We solve for $y$.
$$5x + 4y = 12$$
$$4y = -5x + 12$$
$$y = \frac{1}{4}(-5x + 12)$$
$$y = -\frac{5}{4}x + 3$$
The slope is $-\frac{5}{4}$ and the $y$-intercept is $(0, 3)$.

**37.** $y = -17$
We can think of $y = -17$ as $y = 0x - 17$. The slope is 0 and the $y$-intercept is $(0, -17)$.

**39.** $y - y_1 = m(x - x_1)$
We substitute $-2$ for $m$, $-3$ for $x_1$, and 0 for $y_1$.
$$y - 0 = -2[x - (-3)]$$
$$y = -2(x + 3)$$
$$y = -2x - 6$$

**41.** $y - y_1 = m(x - x_1)$
We substitute $\frac{3}{4}$ for $m$, 2 for $x_1$, and 4 for $y_1$.
$$y - 4 = \frac{3}{4}(x - 2)$$
$$y - 4 = \frac{3}{4}x - \frac{3}{2}$$
$$y = \frac{3}{4}x + \frac{5}{2}$$

**43.** $y - y_1 = m(x - x_1)$

We substitute 1 for $m$, 2 for $x_1$, and $-6$ for $y_1$.

$y - (-6) = 1(x - 2)$

$y + 6 = x - 2$

$y = x - 8$

**45.** $y - y_1 = m(x - x_1)$

We substitute $-3$ for $m$, 0 for $x_1$, and 3 for $y_1$.

$y - 3 = -3(x - 0)$

$y - 3 = -3x$

$y = -3x + 3$

**47.** $(12, 16)$ and $(1, 5)$

First we find the slope.

$m = \dfrac{16 - 5}{12 - 1} = \dfrac{11}{11} = 1$

Then we use the point-slope equation.

$y - y_1 = m(x - x_1)$

We substitute 1 for $m$, 1 for $x_1$, and 5 for $y_1$.

$y - 5 = 1 \cdot (x - 1)$

$y - 5 = x - 1$

$y = x + 4$

**49.** $(0, 4)$ and $(4, 2)$

First we find the slope.

$m = \dfrac{4 - 2}{0 - 4} = \dfrac{2}{-4} = -\dfrac{1}{2}$

Then we use the point-slope equation.

$y - y_1 = m(x - x_1)$

We substitute $-\dfrac{1}{2}$ for $m$, 0 for $x_1$, and 4 for $y_1$.

$y - 4 = -\dfrac{1}{2}(x - 0)$

$y - 4 = -\dfrac{1}{2}x$

$y = -\dfrac{1}{2}x + 4$

**51.** $(3, 2)$ and $(1, 5)$

First we find the slope.

$m = \dfrac{2 - 5}{3 - 1} = \dfrac{-3}{2} = -\dfrac{3}{2}$

Then we use the point-slope equation.

$y - y_1 = m(x - x_1)$

We substitute $-\dfrac{3}{2}$ for $m$, 3 for $x_1$, and 2 for $y_1$.

$y - 2 = -\dfrac{3}{2}(x - 3)$

$y - 2 = -\dfrac{3}{2}x + \dfrac{9}{2}$

$y = -\dfrac{3}{2}x + \dfrac{13}{2}$

**53.** $(-4, 5)$ and $(-2, -3)$

First we find the slope.

$m = \dfrac{5 - (-3)}{-4 - (-2)} = \dfrac{8}{-2} = -4$

Then we use the point-slope equation.

$y - y_1 = m(x - x_1)$

We substitute $-4$ for $m$, $-4$ for $x_1$, and 5 for $y_1$.

$y - 5 = -4[x - (-4)]$

$y - 5 = -4(x + 4)$

$y - 5 = -4x - 16$

$y = -4x - 11$

**55.** We multiply and divide in order from left to right.

$11 \cdot 6 \div 3 \cdot 2 \div 7 = 66 \div 3 \cdot 2 \div 7$

$= 22 \cdot 2 \div 7$

$= 44 \div 7$

$= \dfrac{44}{7}$

**57.** $\quad [10 - 3(7 - 2)]$

$= [10 - 3 \cdot 5]$    Subtracting inside the parentheses

$= [10 - 15]$    Multiplying

$= -5$    Subtracting

**59.** $\quad \dfrac{4^2 + 2^2}{5^3 - 4^2}$

$= \dfrac{64 + 4}{125 - 16}$    Evaluating exponential expressions

$= \dfrac{68}{109}$    Adding the numerator; subtracting in the denominator

**61.** First find the slope of $3x - y + 4 = 0$.

$3x - y + 4 = 0$

$3x + 4 = y$

The slope is 3.

Then find an equation of the line containing $(2, -3)$ and having slope 3.

$y - y_1 = m(x - x_1)$

We substitute 3 for $m$, 2 for $x_1$, and $-3$ for $y_1$.

$y - (-3) = 3(x - 2)$

$y + 3 = 3x - 6$

$y = 3x - 9$

**63.** First find the slope of $3x - 2y = 8$.

$3x - 2y = 8$

$-2y = -3x + 8$

$y = \dfrac{3}{2}x - 4$

The slope is $\dfrac{3}{2}$.

Then find the $y$-intercept of $2y + 3x = -4$.

$$2y + 3x = -4$$
$$2y = -3x - 4$$
$$y = -\frac{3}{2}x - 2$$

The $y$-intercept is $(0, -2)$.

Finally, write the equation of the line with slope $\frac{3}{2}$ and $y$-intercept $(0, -2)$.

$$y = mx + b$$
$$y = \frac{3}{2}x + (-2)$$
$$y = \frac{3}{2}x - 2$$

## Exercise Set 6.5

**1.** 1. The first equation is already solved for $y$:
$$y = x + 4$$
  2.   We solve the second equation for y:
$$y - x = -3$$
$$y = x - 3$$
The slope of each line is 1. The $y$-intercepts, $(0, 4)$ and $(0, -3)$, are different. The lines are parallel.

**3.** We solve each equation for $y$:
1. $y + 3 = 6x$       2. $-6x - y = 2$
$\quad y = 6x - 3$         $-y = 6x + 2$
$\qquad\qquad\qquad\qquad\qquad y = -6x - 2$

The slope of the first line is 6 and of the second is $-6$. Since the slopes are different, the lines are not parallel.

**5.** We solve each equation for $y$:
1. $10y + 32x = 16.4$     2. $y + 3.5 = 0.3125x$
$\quad 10y = -32x + 16.4$      $y = 0.3125x - 3.5$
$\qquad y = -3.2x + 1.64$

The slope of the first line is $-3.2$ and of the second is $0.3125$. Since the slopes are different, the lines are not parallel.

**7.** 1. The first equation is already solved for $y$:
$$y = 2x + 7$$
  2.   We solve the second equation for y:
$$5y + 10x = 20$$
$$5y = -10x + 20$$
$$y = -2x + 4$$
The slope of the first line is 2 and of the second is $-2$. Since the slopes are different, the lines are not parallel.

**9.** We solve each equation for $y$:
1. $3x - y = -9$     2. $2y - 6x = -2$
$\quad 3x + 9 = y$          $2y = 6x - 2$
$\qquad\qquad\qquad\qquad\qquad y = 3x - 1$

The slope of each line is 3. The $y$-intercepts, $(0, 9)$ and $(0, -1)$ are different. The lines are parallel.

**11.** $x = 3$,
$\quad x = 4$

These are vertical lines with equations of the form $x = p$ and $x = q$, where $p \neq q$. Thus, they are parallel.

**13.** 1. The first equation is already solved for $y$:
$$y = -4x + 3$$
  2.   We solve the second equation for y:
$$4y + x = -1$$
$$4y = -x - 1$$
$$y = -\frac{1}{4}x - \frac{1}{4}$$
The slopes are $-4$ and $-\frac{1}{4}$. Their product is
$-4\left(-\frac{1}{4}\right) = 1$. Since the product of the slopes is not $-1$, the lines are not perpendicular.

**15.** We solve each equation for $y$:
1. $x + y = 6$      2. $4y - 4x = 12$
$\quad y = -x + 6$        $4y = 4x + 12$
$\qquad\qquad\qquad\qquad\qquad y = x + 3$

The slopes are $-1$ and 1. Their product is $-1 \cdot 1 = -1$. The lines are perpendicular.

**17.** 1. The first equation is already solved for $y$:
$$y = -0.3125x + 11$$
  2.   We solve the second equation for y:
$$y - 3.2x = -14$$
$$y = 3.2x - 14$$
The slopes are $-0.3125$ and $3.2$. Their product is $-0.3125(3.2) = -1$. The lines are perpendicular.

**19.** 1. The first equation is already solved for $y$:
$$y = -x + 8$$
  2.   We solve the second equation for y:
$$x - y = -1$$
$$x + 1 = y$$
The slopes are $-1$ and 1. Their product is $-1 \cdot 1 = -1$. The lines are perpendicular.

**21.** We solve each equation for $y$:

1.
$$\frac{3}{8}x - \frac{y}{2} = 1$$
$$8\left(\frac{3}{8}x - \frac{y}{2}\right) = 8 \cdot 1$$
$$8 \cdot \frac{3}{8}x - 8 \cdot \frac{y}{2} = 8$$
$$3x - 4y = 8$$
$$-4y = -3x + 8$$
$$y = \frac{3}{4}x - 2$$

2. $\frac{4}{3}x - y + 1 = 0$
$$\frac{4}{3}x + 1 = y$$

The slopes are $\frac{3}{4}$ and $\frac{4}{3}$. Their product is $\frac{3}{4}\left(\frac{4}{3}\right) = 1$. Since the product of the slopes is not $-1$, the lines are not perpendicular.

**23. Familiarize.** We let $t =$ the time the second train travels before it overtakes the first train. Then the first train travels for $t + 2$ hr before it is overtaken. The trains travel the same distance. We organize the information in a table.

|             | Distance | Speed | Time  |
|-------------|----------|-------|-------|
| First train | $d$      | 70    | $t+2$ |
| Second train| $d$      | 90    | $t$   |

**Translate.** From the rows of the table we get two equations:
$$d = 70(t + 2)$$
$$d = 90t$$

Since the right sides of both equations are equal to $d$, we set them equal to each other.
$$70(t + 2) = 90t$$

**Solve.** We solve the equation.
$$70(t + 2) = 90t$$
$$70t + 140 = 90t$$
$$140 = 20t$$
$$7 = t$$

**Check.** In 7 hr the second train travels $90 \cdot 7$, or 630 mi. In $7 + 2$, or 9 hr, the first train travels $70 \cdot 9$, or 630 mi. Since the distances are the same, the result checks.

**State.** The second train will overtake the first train 7 hr after the second train leaves the station (or 9 hr after the first train leaves the station).

**25.**
$$x^2 - 10x + 25 = 0$$
$$(x - 5)(x - 5) = 0$$
$$x - 5 = 0 \quad \text{or} \quad x - 5 = 0$$
$$x = 5 \quad \text{or} \qquad x = 5$$

The solution is 5.

**27.**
$$\frac{2}{3} - \frac{5}{6} = \frac{1}{x}, \quad \text{LCM is } 6x$$
$$6x\left(\frac{2}{3} - \frac{5}{6}\right) = 6x \cdot \frac{1}{x}$$
$$6x \cdot \frac{2}{3} - 6x \cdot \frac{5}{6} = 6$$
$$4x - 5x = 6$$
$$-x = 6$$
$$x = -6$$

The number $-6$ checks and is the solution.

**29.** First we find the slope of the given line:
$$y - 3x = 4$$
$$y = 3x + 4$$
The slope is 3.

Then we use the slope-intercept equation to write the equation of a line with slope 3 and $y$-intercept $(0, 6)$:
$$y = mx + b$$
$$y = 3x + 6 \qquad \text{Substituting 3 for } m \text{ and 6 for } b$$

**31.** First we find the slope of the given line:
$$3y - x = 0$$
$$3y = x$$
$$y = \frac{1}{3}x$$

The slope is $\frac{1}{3}$.

We can find the slope of the line perpendicular to the given line by taking the reciprocal of $\frac{1}{3}$ and changing the sign. We get $-3$.

Then we use the slope-intercept equation to write the equation of a line with slope $-3$ and $y$-intercept $(0, 2)$:
$$y = mx + b$$
$$y = -3x + 2 \quad \text{Substituting } -3 \text{ for } m \text{ and 2 for } b$$

**33.** First we find the slope of the given line:
$$4x - 8y = 12$$
$$-8y = -4x + 12$$
$$y = \frac{1}{2}x - \frac{3}{2}$$

The slope is $\frac{1}{2}$.

Then we use the point-slope equation to find the equation of a line with slope $\frac{1}{2}$ containing the point $(-2, 0)$:

$$y - y_1 = m(x - x_1)$$
$$y - 0 = \frac{1}{2}[x - (-2)]$$
$$y = \frac{1}{2}(x + 2)$$
$$y = \frac{1}{2}x + 1$$

**35.** We find the slope of each line:

1. $4y = kx - 6$         2. $5x + 20y = 12$

$$y = \frac{k}{4}x - \frac{3}{2}$$      $$20y = -5x + 12$$
$$y = -\frac{1}{4}x + \frac{3}{5}$$

The slopes are $\frac{k}{4}$ and $-\frac{1}{4}$. If the lines are perpendicular, the product of their slopes is $-1$.

$$\frac{k}{4}\left(-\frac{1}{4}\right) = -1$$
$$-\frac{k}{16} = -1$$
$$k = 16$$

**37.** First we find the equation of $A$, a line containing the points $(1, -1)$ and $(4, 3)$:

The slope is $\dfrac{3 - (-1)}{4 - 1} = \dfrac{4}{3}$.

Use the point-slope equation:

$$y - y_1 = m(x - x_1)$$
$$y - 3 = \frac{4}{3}(x - 4)$$
$$y - 3 = \frac{4}{3}x - \frac{16}{3}$$
$$y = \frac{4}{3}x - \frac{7}{3}$$

The slope of $A$ is $\frac{4}{3}$. Since $A$ and $B$ are perpendicular we find the slope of $B$ by taking the reciprocal of $\frac{4}{3}$ and changing the sign. We get $-\frac{3}{4}$. Then we use the point-slope equation to find the equation of $B$, a line with slope $-\frac{3}{4}$ and containing the point $(1, -1)$:

$$y - y_1 = m(x - x_1)$$
$$y - (-1) = -\frac{3}{4}(x - 1)$$
$$y + 1 = -\frac{3}{4}x + \frac{3}{4}$$
$$y = -\frac{3}{4}x - \frac{1}{4}$$

## Exercise Set 6.6

**1.** We use alphabetical order to replace $x$ by $-3$ and $y$ by $-5$.

$$\begin{array}{c|c} \multicolumn{2}{c}{-x - 3y < 18} \\ \hline -(-3) - 3(-5) & 18 \\ 3 + 15 & \\ 18 & \text{FALSE} \end{array}$$

Since $18 < 18$ is false, $(-3, -5)$ is not a solution.

**3.** We use alphabetical order to replace $x$ by $\frac{1}{2}$ and $y$ by $-\frac{1}{4}$.

$$\begin{array}{c|c} \multicolumn{2}{c}{7y - 9x \le -3} \\ \hline 7\left(-\frac{1}{4}\right) - 9 \cdot \frac{1}{2} & -3 \\ -\frac{7}{4} - \frac{9}{2} & \\ -\frac{7}{4} - \frac{18}{4} & \\ -\frac{25}{4} & \\ -6\frac{1}{4} & \text{TRUE} \end{array}$$

Since $-6\frac{1}{4} \le -3$ is true, $\left(\frac{1}{2}, -\frac{1}{4}\right)$ is a solution.

**5.** Graph $x > 2y$.

First graph the line $x = 2y$, or $y = \frac{1}{2}x$. Two points on the line are $(0, 0)$ and $(4, 2)$. We draw a dashed line since the inequality symbol is $>$. Then we pick a test point that is not on the line. We try $(-2, 1)$.

$$\begin{array}{c|c} \multicolumn{2}{c}{x > 2y} \\ \hline -2 & 2 \cdot 1 \\ & 2 \quad \text{FALSE} \end{array}$$

We see that $(-2, 1)$ is not a solution of the inequality, so we shade the points in the region that does not contain $(-2, 1)$.

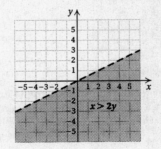

**7.** Graph $y \leq x - 3$.

First graph the line $y = x - 3$. The intercepts are $(0, -3)$ and $(3, 0)$. We draw a solid line since the inequality symbol is $\leq$. Then we pick a test point that is not on the line. We try $(0, 0)$.

$$\frac{y \leq x - 3}{0 \;\bigm|\; 0 - 3}$$
$$\phantom{0\;\bigm|\;} -3 \qquad \text{FALSE}$$

We see that $(0, 0)$ is not a solution of the inequality, so we shade the region that does not contain $(0, 0)$.

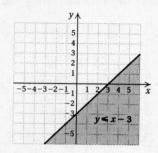

**9.** Graph $y < x + 1$.

First graph the line $y = x + 1$. The intercepts are $(0, 1)$ and $(-1, 0)$. We draw a dashed line since the inequality symbol is $<$. Then we pick a test point that is not on the line. We try $(0, 0)$.

$$\frac{y < x + 1}{0 \;\bigm|\; 0 + 1}$$
$$\phantom{0\;\bigm|\;} 1 \qquad \text{TRUE}$$

Since $(0, 0)$ is a solution of the inequality, we shade the region that contains $(0, 0)$.

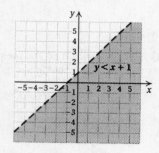

**11.** Graph $y \geq x - 2$.

First graph the line $y = x - 2$. The intercepts are $(0, -2)$ and $(2, 0)$. We draw a solid line since the inequality symbol is $\geq$. Then we test the point $(0, 0)$.

$$\frac{y \geq x - 2}{0 \;\bigm|\; 0 - 2}$$
$$\phantom{0\;\bigm|\;} -2 \qquad \text{TRUE}$$

Since $(0, 0)$ is a solution of the inequality, we shade the region containing $(0, 0)$.

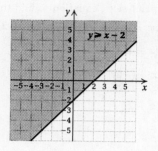

**13.** Graph $y \leq 2x - 1$.

First graph the line $y = 2x - 1$. The intercepts are $(0, -1)$ and $\left(\frac{1}{2}, 0\right)$. We draw a solid line since the inequality symbol is $\leq$. Then we test the point $(0, 0)$.

$$\frac{y \leq 2x - 1}{0 \;\bigm|\; 2 \cdot 0 - 1}$$
$$\phantom{0\;\bigm|\;} -1 \qquad \text{FALSE}$$

Since $(0, 0)$ is not a solution of the inequality, we shade the region that does not contain $(0, 0)$.

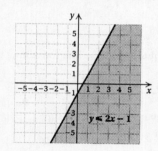

**15.** Graph $x + y \leq 3$.

First graph the line $x + y = 3$. The intercepts are $(0, 3)$ and $(3, 0)$. We draw a solid line since the inequality symbol is $\leq$. Then we test the point $(0, 0)$.

$$\frac{x + y \leq 3}{0 + 0 \;\bigm|\; 3}$$
$$\phantom{0+0\;\bigm|\;} 0 \qquad \text{TRUE}$$

Since $(0, 0)$ is a solution of the inequality, we shade the region that contains $(0, 0)$.

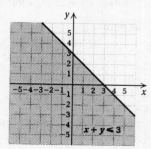

**17.** Graph $x - y > 7$.

First graph the line $x - y = 7$. The intercepts are $(0, -7)$ and $(7, 0)$. We draw a dashed line since the inequality symbol is $>$. Then we test the point $(0, 0)$.

$$\frac{x - y > 7}{\begin{array}{c|c} 0 - 0 & 7 \\ \hline 0 & \text{FALSE} \end{array}}$$

Since $(0, 0)$ is not a solution of the inequality, we shade the region that does not contain $(0, 0)$.

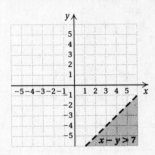

**19.** Graph $2x + 3y \leq 12$.

First graph the line $2x + 3y = 12$. The intercepts are $(0, 4)$ and $(6, 0)$. We draw a solid line since the inequality symbol is $\leq$. Then we test the point $(0, 0)$.

$$\frac{2x + 3y \leq 12}{\begin{array}{c|c} 2 \cdot 0 + 3 \cdot 0 & 12 \\ \hline 0 & \text{TRUE} \end{array}}$$

Since $(0, 0)$ is a solution of the inequality, we shade the region containing $(0, 0)$.

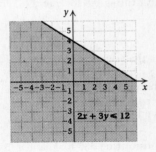

**21.** Graph $y \geq 1 - 2x$.

First graph the line $y = 1 - 2x$. The intercepts are $(0, 1)$ and $\left(\frac{1}{2}, 0\right)$. We draw a solid line since the inequality symbol is $\geq$. Then we test the point $(0, 0)$.

$$\frac{y \geq 1 - 2x}{\begin{array}{c|c} 0 & 1 - 2 \cdot 0 \\ \hline & 1 \quad \text{FALSE} \end{array}}$$

Since $(0, 0)$ is not a solution of the inequality, we shade the region that does not contain $(0, 0)$.

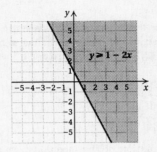

**23.** Graph $y + 4x > 0$.

First graph the line $y + 4x = 0$, or $y = -4x$. Two points on the line are $(0, 0)$ and $(1, -4)$. We draw a dashed line since the inequality symbol is $>$. Then we test the point $(2, -3)$, which is not a point on the line.

$$\frac{y + 4x > 0}{\begin{array}{c|c} -3 + 4 \cdot 2 & 0 \\ \hline -3 + 8 & \\ 5 & \text{TRUE} \end{array}}$$

Since $(2, -3)$ is a solution of the inequality, we shade the region containing $(2, -3)$.

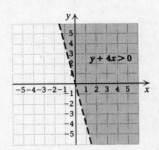

**25.** Graph $y \leq 3$.

First graph the line $y = 3$ using a solid line since the inequality symbol is $\leq$. Then pick a test point that is not on the line. We choose $(1, -2)$. We can write the inequality as $0x + y \leq 3$.

$$\frac{0x + y \leq 3}{\begin{array}{c|c} 0 \cdot 1 + (-2) & 3 \\ \hline -2 & \text{TRUE} \end{array}}$$

Since $(1, -2)$ is a solution of the inequality, we shade the region containing $(1, -2)$.

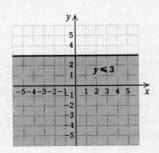

**27.** Graph $x \geq -1$.

Graph the line $x = 1$ using a solid line since the inequality symbol is $\geq$. Then pick a test point that is not on the line. We choose $(2, 3)$. We can write the inequality as $x + 0y \geq -1$.

$$\begin{array}{c|c} x + 0y \geq -1 \\ \hline 2 + 0 \cdot 3 & -1 \\ \hline 2 & \text{TRUE} \end{array}$$

Since $(2, 3)$ is a solution of the inequality, we shade the region containing $(2, 3)$.

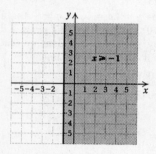

**29.** We do the divisions in order from left to right.

$$3^7 \div 3^4 \div 3^3 \div 3 = 3^3 \div 3^3 \div 3$$
$$= 1 \div 3$$
$$= \frac{1}{3}$$

**31.**
$$\frac{12}{x} = \frac{48}{x + 9}, \quad \text{LCM is } x(x + 9)$$

$$x(x + 9) \cdot \frac{12}{x} = x(x + 9) \cdot \frac{48}{x + 9}$$

$$12(x + 9) = 48x$$

$$12x + 108 = 48x$$

$$108 = 36x$$

$$3 = x$$

The number 3 checks and is the solution.

**33.**
$$x^2 + 16 = 8x$$
$$x^2 - 8x + 16 = 0$$
$$(x - 4)(x - 4) = 0$$
$$x - 4 = 0 \quad \text{or} \quad x - 4 = 0$$
$$x = 4 \quad \text{or} \quad x = 4$$

The solution is 4.

**35.** The $c$ children weigh $35c$ kg, and the $a$ adults weigh $75a$ kg. Together, the children and adults weigh $35c + 75a$ kg. When this total is more than 1000 kg the elevator is overloaded, so we have $35c + 75a > 1000$. (Of course, $c$ and $a$ would also have to be nonnegative, but we will not deal with nonnegativity constraints here.)

To graph $35c + 75a > 1000$, we first graph $35c + 75a = 1000$ using a dashed line. Two points on the line are $(4, 20)$

and $(11, 5)$. (We are using alphabetical order of variables.) Then we test the point $(0, 0)$.

$$\begin{array}{c|c} 35c + 75a > 1000 \\ \hline 35 \cdot 0 + 75 \cdot 0 & 1000 \\ \hline 0 & \text{FALSE} \end{array}$$

Since $(0, 0)$ is not a solution of the inequality, we shade the region that does not contain $(0, 0)$.

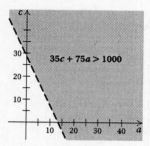

## Exercise Set 6.7

**1.** We substitute to find $k$.

$$y = kx$$
$$36 = k \cdot 9 \quad \text{Substituting 36 for } y \text{ and 9 for } x$$
$$\frac{36}{9} = k$$
$$4 = k \qquad k \text{ is the variation constant.}$$

The equation of the variation is $y = 4x$.

**3.** We substitute to find $k$.

$$y = kx$$
$$0.8 = k \cdot 0.5 \quad \text{Substituting 0.8 for } y \text{ and 0.5 for } x$$
$$\frac{0.8}{0.5} = k$$
$$\frac{8}{5} = k \qquad k \text{ is the variation constant.}$$

The equation of the variation is $y = \frac{8}{5}x$.

**5.** We substitute to find $k$.

$$y = kx$$
$$630 = k \cdot 175 \quad \text{Substituting 630 for } y \text{ and 175 for } x$$
$$\frac{630}{175} = k$$
$$3.6 = k \qquad k \text{ is the variation constant.}$$

The equation of the variation is $y = 3.6x$.

**7.** We substitute to find $k$.

$$y = kx$$
$$500 = k \cdot 60 \quad \text{Substituting 500 for } y \text{ and 60 for } x$$
$$\frac{500}{60} = k$$
$$\frac{25}{3} = k \qquad k \text{ is the variation constant.}$$

The equation of the variation is $y = \frac{25}{3}x$.

**9. *Familiarize and Translate*.** The problem states that we have direct variation between the variables $P$ and $H$. Thus, an equation $P = kH$, $k > 0$, applies. As the number of hours increases, the paycheck increases.

***Solve*.**

a) First find an equation of variation.

$$P = kH$$

$78.75 = k \cdot 15$   Substituting 78.75 for $P$ and 15 for $H$

$$\frac{78.75}{15} = k$$

$$5.25 = k$$

The equation of variation is $P = 5.25H$.

b) Use the equation to find the pay for 35 hours work.

$$P = 5.25H$$

$P = 5.25(35)$   Substituting 35 for $H$

$$P = 183.75$$

***Check*.** This check might be done by repeating the computations. We might also do some reasoning about the answer. The paycheck increased from \$78.75 to \$183.75. Similarly, the hours increased from 15 to 35.

***State*.** For 35 hours work, the paycheck is \$183.75.

**11. *Familiarize and Translate*.** This problem states that we have direct variation between the variables $S$ and $W$. Thus, an equation $S = kW$, $k > 0$, applies. As the weight increases, the number of servings increases.

***Solve*.**

a) First find an equation of variation.

$$S = kW$$

$40 = k \cdot 14$   Substituting 40 for $S$ and 14 for $W$

$$\frac{40}{14} = k$$

$$\frac{20}{7} = k$$

The equation of variation is $S = \frac{20}{7}W$.

b) Use the equation to find the number of servings from an 8-kg turkey.

$$S = \frac{20}{7}W$$

$S = \frac{20}{7} \cdot 8$   Substituting 8 for $W$

$$S = \frac{160}{7}, \text{ or } 22\frac{6}{7}$$

***Check*.** A check can always be done by repeating the computations. We can also do some reasoning about the answer. The number of servings decreased from 40 to $22\frac{6}{7}$. Similarly, the weight decreased from 14 kg to 8 kg.

***State*.** $22\frac{6}{7}$ servings can be obtained from an 8-kg turkey.

**13. *Familiarize and Translate*.** The problem states that we have direct variation between the variables $M$ and $E$. Thus, an equation $M = kE$, $k > 0$, applies. As the weight on earth increases, the weight on the moon increases.

***Solve*.**

a) First find an equation of variation.

$$M = kE$$

$28.6 = k \cdot 171.6$   Substituting 28.6 for $M$ and 171.6 for $E$

$286 = 1716k$   Clearing decimals

$$\frac{286}{1716} = k$$

$$\frac{1}{6} = k$$

The equation of variation is $M = \frac{1}{6}E$.

b) Use the equation to find how much a 220-lb person would weigh on the moon.

$$M = \frac{1}{6}E$$

$M = \frac{1}{6} \cdot 220$   Substituting 220 for $E$

$$M = \frac{220}{6}, \text{ or } 36.\overline{6}$$

***Check*.** In addition to repeating the computations we can do some reasoning. The weight on the earth increased from 171.6 lb to 220 lb. Similarly, the weight on the moon increased from 28.6 lb to $36.\overline{6}$ lb.

***State*.** A 220-lb person would weigh $36.\overline{6}$ lb on the moon.

**15. *Familiarize and Translate*.** The problem states that we have direct variation between the variables $N$ and $S$. Thus, an equation $N = kS$, $k > 0$, applies. As the speed of the internal processor increases, the number of instructions increases.

***Solve*.**

a) First find an equation of variation.

$$N = kS$$

$2,000,000 = k \cdot 25$   Substituting 2,000,000 for $N$ and 25 for $S$

$$\frac{2,000,000}{25} = k$$

$$80,000 = k$$

The equation of variation is $N = 80,000S$.

b) Use the equation to find how many instructions the processor will perform at a speed of 40 megahertz.

$$N = 80,000S$$

$N = 80,000 \cdot 40$   Substituting 40 for $S$

$$N = 3,200,000$$

***Check*.** In addition to repeating the computations we can do some reasoning. The speed of the processor increased from 25 to 40 megahertz. Similarly, the number of instructions increased from 2,000,000 to 3,200,000.

*State*.  The processor will perform 3,200,000 instructions running at a speed of 40 megahertz.

**17.** We substitute to find $k$.

$$y = \frac{k}{x}$$

$$3 = \frac{k}{25} \quad \text{Substituting 3 for } y \text{ and 25 for } x$$

$$25 \cdot 3 = k$$

$$75 = k$$

The equation of variation is $y = \dfrac{75}{x}$.

**19.** We substitute to find $k$.

$$y = \frac{k}{x}$$

$$10 = \frac{k}{8} \quad \text{Substituting 10 for } y \text{ and 8 for } x$$

$$8 \cdot 10 = k$$

$$80 = k$$

The equation of variation is $y = \dfrac{80}{x}$.

**21.** We substitute to find $k$.

$$y = \frac{k}{x}$$

$$6.25 = \frac{k}{0.16} \quad \begin{array}{l}\text{Substituting 6.25 for } y \text{ and 0.16} \\ \text{for } x\end{array}$$

$$0.16(6.25) = k$$

$$1 = k$$

The equation of variation is $y = \dfrac{1}{x}$.

**23.** We substitute to find $k$.

$$y = \frac{k}{x}$$

$$50 = \frac{k}{42} \quad \text{Substituting 50 for } y \text{ and 42 for } x$$

$$42 \cdot 50 = k$$

$$2100 = k$$

The equation of variation is $y = \dfrac{2100}{x}$.

**25.** We substitute to find $k$.

$$y = \frac{k}{x}$$

$$0.2 = \frac{k}{0.3} \quad \text{Substituting 0.2 for } y \text{ and 0.3 for } x$$

$$0.06 = k$$

The equation of variation is $y = \dfrac{0.06}{x}$.

**27.**  a)  It seems reasonable that, as the number of hours of production increases, the number of compact-disc players produced will increase, so direct variation might apply.

b)  *Familiarize*.  Let $H$ = the number of hours the production line is working, and let $P$ = the number of compact-disc players produced.  An equation $P = kH$, $k > 0$, applies.  (See part (a)).

*Translate*.  We write an equation of variation.

Number of players produced varies directly as hours of production.  This translates to $P = kH$.

*Solve*.

a)  First we find an equation of variation.

$$P = kH$$

$$15 = k \cdot 8 \quad \text{Substituting 8 for } H \text{ and 15 for } P$$

$$\frac{15}{8} = k$$

The equation of variation is $P = \dfrac{15}{8}H$.

b)  Use the equation to find the number of players produced in 37 hr.

$$P = \frac{15}{8}H$$

$$P = \frac{15}{8} \cdot 37 \qquad \text{Substituting 37 for } H$$

$$P = \frac{555}{8} = 69\frac{3}{8}$$

*Check*.  In addition to repeating the computations, we can do some reasoning.  The number of hours increased from 8 to 37.  Similarly, the number of compact disc players produced increased from 15 to $69\frac{3}{8}$.

*State*.  About $69\frac{3}{8}$ compact-disc players can be produced in 37 hr.

**29.**  a)  It seems reasonable that, as the number of workers increases, the number of hours required to do the job decreases, so inverse variation might apply.

b)  *Familiarize*.  Let $T$ = the time required to cook the meal and $N$ = the number of cooks.  An equation $T = k/N$, $k > 0$, applies.  (See part (a)).

*Translate*.  We write an equation of variation.  Time varies inversely as the number of cooks.  This translates to $T = \dfrac{k}{N}$.

*Solve.*

a)  First find the equation of variation.

$$T = \frac{k}{N}$$

$$4 = \frac{k}{9} \quad \text{Substituting 4 for } T \text{ and 9 for } N$$

$$36 = k$$

The equation of variation is $T = \frac{36}{N}$.

b)  Use the equation to find the amount of time it takes 8 cooks to prepare the dinner.

$$T = \frac{36}{N}$$

$$T = \frac{36}{8} \quad \text{Substituting 8 for } N$$

$$T = 4.5$$

*Check.* The check might be done by repeating the computation. We might also analyze the results. The number of cooks decreased from 9 to 8, and the time increased from 4 hr to 4.5 hr. This is what we would expect with inverse variation.

*State.* It will take 8 cooks 4.5 hr to prepare the dinner.

**31.** *Familiarize.* The problem states that we have inverse variation between the variables $I$ and $R$. Thus, an equation $I = k/R$, $k > 0$, applies. As the resistance decreases, the current increases.

*Translate.* We write an equation of variation. Current varies inversely as resistance. This translates to $I = \frac{k}{R}$.

*Solve.*

a) First find an equation of variation.

$$I = \frac{k}{R}$$

$$2 = \frac{k}{960} \quad \text{Substituting 2 for } I \text{ and 960 for } R$$

$$1920 = k$$

The equation of variation is $I = \frac{1920}{R}$.

b)  Use the equation to find the current when the resistance is 540 ohms.

$$I = \frac{1920}{R}$$

$$I = \frac{1920}{540} \quad \text{Substituting 540 for } R$$

$$I = \frac{32}{9}, \text{ or } 3\frac{5}{9}$$

*Check.* The check might be done by repeating the computations. We might also analyze the results. The resistance decreased from 960 ohms to 540 ohms, and the current increased from 2 amperes to $3\frac{5}{9}$ amperes. This is what we would expect with inverse variation.

*State.* The current is $3\frac{5}{9}$ amperes when the resistance is 540 ohms.

**33.** *Familiarize.* Let $S =$ the size of the files. The problem states that we have inverse variation between the variables $N$ and $S$. Thus, an equation $N = k/S$, $k > 0$, applies. As the size of the files increases, the number of files that can be held decreases.

*Translate.* We write an equation of variation. Number of files varies inversely as the size of the files. This translates to $N = \frac{k}{S}$.

*Solve.*

a) First find an equation of variation.

$$N = \frac{k}{S}$$

$$1600 = \frac{k}{50,000} \quad \text{Substituting 1600 for } N \text{ and 50,000 for } S$$

$$80,000,000 = k$$

The equation of variation is $N = \frac{80,000,000}{S}$.

b) Use the equation to find the number of files the disk will hold if each is 125,000 bytes.

$$N = \frac{80,000,000}{S}$$

$$N = \frac{80,000,000}{125,000} \quad \text{Substituting 125,000 for } S$$

$$N = 640$$

*Check.* The check might be done by repeating the computations. We might also analyze the results. The size of each file increased from 50,000 to 125,000 bytes and the number of files decreased from 1600 to 640. This is what we would expect with inverse variation.

*State.* The disk will hold 640 files when each is 125,000 bytes.

**35.** *Familiarize.* The problem states that we have inverse variation between the variables $A$ and $d$. Thus, an equation $A = k/d$, $k > 0$, applies. As the distance increases, the apparent size decreases.

*Translate.* We write an equation of variation. Apparent size varies inversely as the distance. This translates to $A = \frac{k}{d}$.

*Solve.*

a) First find an equation of variation.

$$A = \frac{k}{d}$$

$$27.5 = \frac{k}{30} \quad \text{Substituting 27.5 for } A \text{ and 30 for } d$$

$$825 = k$$

The equation of variation is $A = \frac{825}{d}$.

b) Use the equation to find the apparent size
when the distance is 100 ft.

$$A = \frac{825}{d}$$

$$A = \frac{825}{100} \quad \text{Substituting 100 for } d$$

$$A = 8.25$$

**Check**. The check might be done by repeating the computations. We might also analyze the results. The distance increased from 30 ft to 100 ft, and the apparent size decreased from 27.5 ft to 8.25 ft. This is what we would expect with inverse variation.

**State**. The flagpole will appear to be 8.25 ft tall when it is 100 ft from the observer.

**37.**
$$\frac{x+2}{x+5} = \frac{x-4}{x-6}, \text{ LCM is } (x+5)(x-6)$$

$$(x+5)(x-6) \cdot \frac{x+2}{x+5} = (x+5)(x-6) \cdot \frac{x-4}{x-6}$$

$$(x-6)(x+2) = (x+5)(x-4)$$

$$x^2 - 4x - 12 = x^2 + x - 20$$

$$-4x - 12 = x - 20 \quad \text{Subtracting } x^2$$

$$-5x = -8 \quad \begin{array}{l}\text{Subtracting } x \text{ and} \\ \text{adding 12}\end{array}$$

$$x = \frac{8}{5}$$

The number $\frac{8}{5}$ checks and is the solution.

**39.**  $x^2 - 25x + 144 = 0$

$(x-9)(x-16) = 0$

$x - 9 = 0 \quad \text{or} \quad x - 16 = 0$

$x = 9 \quad \text{or} \quad x = 16$

The solutions are 9 and 16.

**41.**
$$35x^2 + 8 = 34x$$

$$35x^2 - 34x + 8 = 0$$

$$(7x - 4)(5x - 2) = 0$$

$$7x - 4 = 0 \quad \text{or} \quad 5x - 2 = 0$$

$$7x = 4 \quad \text{or} \quad 5x = 2$$

$$x = \frac{4}{7} \quad \text{or} \quad x = \frac{2}{5}$$

The solutions are $\frac{4}{7}$ and $\frac{2}{5}$.

**43.** $P = ks$; $k = n$, where $n$ is the number of sides of the polygon

**45.** $C = kA$

**47.** $P^2 = kt$

**49.** $P = kV^3$

**51.** $N = \frac{k}{C}$

# Chapter 7

# Systems of Equations

## Exercise Set 7.1

**1.** We check by substituting alphabetically 1 for $x$ and 5 for $y$.

$$\frac{5x - 2y = -5}{\begin{array}{c|c} 5 \cdot 1 - 2 \cdot 5 & -5 \\ 5 - 10 & \\ -5 & \text{TRUE} \end{array}}$$

$$\frac{3x - 7y = -32}{\begin{array}{c|c} 3 \cdot 1 - 7 \cdot 5 & -32 \\ 3 - 35 & \\ -32 & \text{TRUE} \end{array}}$$

The ordered pair $(1, 5)$ is a solution of both equations, so it is a solution of the system of equations.

**3.** We check by substituting alphabetically 4 for $a$ and 2 for $b$.

$$\frac{3b - 2a = -2}{\begin{array}{c|c} 3 \cdot 2 - 2 \cdot 4 & -2 \\ 6 - 8 & \\ -2 & \text{TRUE} \end{array}}$$

$$\frac{b + 2a = 8}{\begin{array}{c|c} 2 + 2 \cdot 4 & 8 \\ 2 + 8 & \\ 10 & \text{FALSE} \end{array}}$$

The ordered pair $(4, 2)$ is not a solution of $b + 2a = 8$, so it is not a solution of the system of equations.

**5.** We check by substituting alphabetically 15 for $x$ and 20 for $y$.

$$\frac{3x - 2y = 5}{\begin{array}{c|c} 3 \cdot 15 - 2 \cdot 20 & 5 \\ 45 - 40 & \\ 5 & \text{TRUE} \end{array}}$$

$$\frac{6x - 5y = -10}{\begin{array}{c|c} 6 \cdot 15 - 5 \cdot 20 & -10 \\ 90 - 100 & \\ -10 & \text{TRUE} \end{array}}$$

The ordered pair $(15, 20)$ is a solution of both equations, so it is a solution of the system of equations.

**7.** We check by substituting alphabetically $-1$ for $x$ and 1 for $y$.

$$\frac{x = -1}{\begin{array}{c|c} -1 & -1 \end{array} \text{ TRUE}}$$

$$\frac{x - y = -2}{\begin{array}{c|c} -1 - 1 & -2 \\ -2 & \text{TRUE} \end{array}}$$

The ordered pair $(-1, 1)$ is a solution of both equations, so it is a solution of the system of equations.

**9.** We check by substituting alphabetically 18 for $x$ and 3 for $y$.

$$y = \frac{1}{6}x$$
$$\begin{array}{c|c} 3 & \frac{1}{6} \cdot 18 \\ & 3 \quad \text{TRUE} \end{array}$$

$$\frac{2x - y = 33}{\begin{array}{c|c} 2 \cdot 18 & 33 \\ 36 - 3 & \\ 33 & \text{TRUE} \end{array}}$$

The ordered pair $(18, 3)$ is a solution of both equations, so it is a solution of the system of equations.

**11.** We graph the equations.

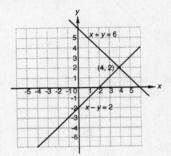

The point of intersection looks as if it has coordinates $(4, 2)$.

Check:

$$\frac{x - y = 2}{\begin{array}{c|c} 4 - 2 & 2 \\ 2 & \text{TRUE} \end{array}}$$

$$\frac{x + y = 6}{\begin{array}{c|c} 4 + 2 & 6 \\ 6 & \text{TRUE} \end{array}}$$

The solution is $(4, 2)$.

**13.** We graph the equations.

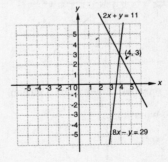

The point of intersection looks as if it has coordinates $(4, 3)$.

Check:

$$\frac{8x - y = 29}{\begin{array}{c|c} 8 \cdot 4 - 3 & 29 \\ \hline 32 - 3 & \\ 29 & \text{TRUE} \end{array}}$$

$$\frac{2x + y = 11}{\begin{array}{c|c} 2 \cdot 4 + 3 & 11 \\ \hline 8 + 3 & \\ 11 & \text{TRUE} \end{array}}$$

The solution is $(4, 3)$.

**15.** We graph the equations.

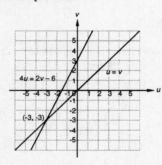

The point of intersection looks as if it has coordinates $(-3, -3)$.

Check:

$$\frac{u = v}{\begin{array}{c|c} -3 & -3 \end{array}} \text{ TRUE}$$

$$\frac{4u = 2v - 6}{\begin{array}{c|c} 4(-3) & 2(-3) - 6 \\ \hline -12 & -6 - 6 \\ & -12 \quad \text{TRUE} \end{array}}$$

The solution is $(-3, -3)$.

**17.** We graph the equations.

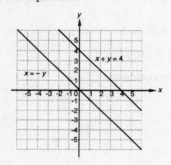

The lines are parallel. There is no solution.

**19.** We graph the equations.

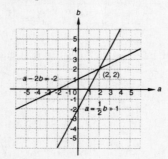

The point of intersection looks as if it has coordinates $(2, 2)$.

Check:

$$\frac{a = \frac{1}{2}b + 1}{\begin{array}{c|c} 2 & \frac{1}{2} \cdot 2 + 1 \\ \hline & 1 + 1 \\ & 2 \quad \text{TRUE} \end{array}}$$

$$\frac{a - 2b = -2}{\begin{array}{c|c} 2 - 2 \cdot 2 & -2 \\ \hline 2 - 4 & \\ -2 & \text{TRUE} \end{array}}$$

The solution is $(2, 2)$.

**21.** We graph the equations.

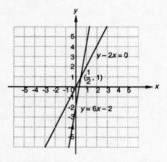

The point of intersection looks as if it has coordinates $\left(\frac{1}{2}, 1\right)$.

Check:

$$\frac{y - 2x = 0}{\begin{array}{c|c} 1 - 2 \cdot \frac{1}{2} & 0 \\ \hline 1 - 1 & \\ 0 & \text{TRUE} \end{array}}$$

$$\frac{y = 6x - 2}{\begin{array}{c|c} 1 & 6 \cdot \frac{1}{2} - 2 \\ \hline & 3 - 2 \\ & 1 \quad \text{TRUE} \end{array}}$$

The solution is $\left(\frac{1}{2}, 1\right)$.

**23.** We graph the equations.

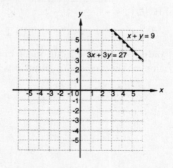

The lines coincide. The system has an infinite number of solutions.

**25.** We graph the equations.

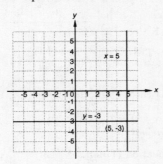

The point of intersection looks as if it has coordinates $(5, -3)$.

Check:

$$\frac{x = 5}{5 \mid 5} \text{ TRUE} \qquad \frac{y = -3}{-3 \mid -3} \text{ TRUE}$$

The solution is $(5, -3)$.

**27.** $(9x^{-5})(12x^{-8}) = 9 \cdot 12 \cdot x^{-5} \cdot x^{-8}$

$= 108x^{-5+(-8)}$

$= 108x^{-13}$

$= \dfrac{108}{x^{13}}$

**29.** $\dfrac{1}{x} - \dfrac{1}{x^2} + \dfrac{1}{x+1}$, LCM is $x^2(x+1)$

$= \dfrac{1}{x} \cdot \dfrac{x(x+1)}{x(x+1)} - \dfrac{1}{x^2} \cdot \dfrac{x+1}{x+1} + \dfrac{1}{x+1} \cdot \dfrac{x^2}{x^2}$

$= \dfrac{x(x+1) - (x+1) + x^2}{x^2(x+1)}$

$= \dfrac{x^2 + x - x - 1 + x^2}{x^2(x+1)}$

$= \dfrac{2x^2 - 1}{x^2(x+1)}$

**31.** $\dfrac{x+2}{x-4} - \dfrac{x+1}{x+4}$, LCM is $(x-4)(x+4)$

$= \dfrac{x+2}{x-4} \cdot \dfrac{x+4}{x+4} - \dfrac{x+1}{x+4} \cdot \dfrac{x-4}{x-4}$

$= \dfrac{(x+2)(x+4) - (x+1)(x-4)}{(x-4)(x+4)}$

$= \dfrac{x^2 + 6x + 8 - (x^2 - 3x - 4)}{(x-4)(x+4)}$

$= \dfrac{x^2 + 6x + 8 - x^2 + 3x + 4}{(x-4)(x+4)}$

$= \dfrac{9x + 12}{(x-4)(x+4)}$

**33.** $(2, -3)$ is a solution of $Ax - 3y = 13$. Substitute 2 for $x$

and $-3$ for $y$ and solve for $A$.

$$Ax - 3y = 13$$
$$A \cdot 2 - 3(-3) = 13$$
$$2A + 9 = 13$$
$$2A = 4$$
$$A = 2$$

$(2, -3)$ is a solution of $x - By = 8$. Substitute 2 for $x$ and $-3$ for $y$ and solve for $B$.

$$x - By = 8$$
$$2 - B(-3) = 8$$
$$2 + 3B = 8$$
$$3B = 6$$
$$B = 2$$

**35.** Answers may vary. Any two equations with a solution of $(6, -2)$ will do. One possibility is

$$x + y = 4,$$
$$x - y = 8.$$

---

## Exercise Set 7.2

**1.** $x + y = 10,$ (1)

$y = x + 8$ (2)

We substitute $x + 8$ for $y$ in Equation (1) and solve for $x$.

$$x + y = 10 \qquad \text{Equation (1)}$$
$$x + (x + 8) = 10 \qquad \text{Substituting}$$
$$2x + 8 = 10 \qquad \text{Collecting like terms}$$
$$2x = 2 \qquad \text{Subtracting 8}$$
$$x = 1 \qquad \text{Dividing by 2}$$

Next we substitute 1 for $x$ in either equation of the original system and solve for $y$. We choose Equation (2) since it has $y$ alone on one side.

$$y = x + 8 \qquad \text{Equation (2)}$$
$$y = 1 + 8 \qquad \text{Substituting}$$
$$y = 9$$

We check the ordered pair $(1, 9)$.

$$\frac{x + y = 10}{1 + 9 \mid 10} \qquad \frac{y = x + 8}{9 \mid 1 + 8}$$
$$\quad 10 \qquad \text{TRUE} \qquad \quad 9 \qquad \text{TRUE}$$

Since $(1, 9)$ checks in both equations, it is the solution.

**3.** $y = x - 6,$ (1)

$x + y = -2$ (2)

We substitute $x - 6$ for $y$ in Equation (2) and solve for $x$.

$$x + y = -2 \qquad \text{Equation (2)}$$
$$x + (x - 6) = -2 \qquad \text{Substituting}$$
$$2x - 6 = -2 \qquad \text{Collecting like terms}$$
$$2x = 4 \qquad \text{Adding 6}$$
$$x = 2 \qquad \text{Dividing by 2}$$

Next we substitute 2 for $x$ in either equation of the original system and solve for $y$. We choose Equation (1) since it has $y$ alone on one side.

$$y = x - 6 \quad \text{Equation (1)}$$
$$y = 2 - 6 \quad \text{Substituting}$$
$$y = -4$$

We check the ordered pair $(2, -4)$.

| $y = x - 6$ | | $x + y = -2$ | |
|---|---|---|---|
| $-4$ | $2 - 6$ | $2 + (-4)$ | $-2$ |
| | $-4$ \quad TRUE | | $-2$ \quad TRUE |

Since $(2, -4)$ checks in both equations, it is the solution.

**5.** $y = 2x - 5, \quad (1)$
$3y - x = 5 \quad (2)$

We substitute $2x - 5$ for $y$ in Equation (2) and solve for $x$.

$$3y - x = 5 \qquad \text{Equation (2)}$$
$$3(2x - 5) - x = 5 \qquad \text{Substituting}$$
$$6x - 15 - x = 5 \qquad \text{Removing parentheses}$$
$$5x - 15 = 5 \qquad \text{Collecting like terms}$$
$$5x = 20 \qquad \text{Adding 15}$$
$$x = 4 \qquad \text{Dividing by 5}$$

Next we substitute 4 for $x$ in either equation of the original system and solve for $y$.

$$y = 2x - 5 \qquad \text{Equation (1)}$$
$$y = 2 \cdot 4 - 5 \qquad \text{Substituting}$$
$$y = 8 - 5$$
$$y = 3$$

We check the ordered pair $(4, 3)$.

| $y = 2x - 5$ | | $3y - x = 5$ | |
|---|---|---|---|
| $3$ | $2 \cdot 4 - 5$ | $3 \cdot 3 - 4$ | $5$ |
| | $8 - 5$ | $9 - 4$ | |
| | $3$ \quad TRUE | $5$ | TRUE |

Since $(4, 3)$ checks in both equations, it is the solution.

**7.** $x = -2y, \quad (1)$
$x + 4y = 2 \quad (2)$

We substitute $-2y$ for $x$ in Equation (2) and solve for $y$.

$$x + 4y = 2 \qquad \text{Equation (2)}$$
$$-2y + 4y = 2 \qquad \text{Substituting}$$
$$2y = 2 \qquad \text{Collecting like terms}$$
$$y = 1 \qquad \text{Dividing by 2}$$

Next we substitute 1 for $y$ in either equation of the original system and solve for $x$.

$$x = -2y \qquad \text{Equation (1)}$$
$$x = -2 \cdot 1$$
$$x = -2$$

We check the ordered pair $(-2, 1)$.

| $x = -2y$ | | $3y - x = 5$ | |
|---|---|---|---|
| $-2$ | $-2 \cdot 1$ | $3 \cdot 1 - (-2)$ | $5$ |
| | $-2$ \quad TRUE | $3 + 2$ | |
| | | $5$ | TRUE |

Since $(-2, 1)$ checks in both equations, it is the solution.

**9.** $x - y = 6, \quad (1)$
$x + y = -2 \quad (2)$

We solve Equation (1) for $x$.

$$x - y = 6 \qquad \text{Equation (1)}$$
$$x = y + 6 \qquad \text{Adding } y \qquad (3)$$

We substitute $y + 6$ for $x$ in Equation (2) and solve for $y$.

$$x + y = -2 \qquad \text{Equation (2)}$$
$$(y + 6) + y = -2 \qquad \text{Substituting}$$
$$2y + 6 = -2 \qquad \text{Collecting like terms}$$
$$2y = -8 \qquad \text{Subtracting 6}$$
$$y = -4 \qquad \text{Dividing by 2}$$

Now we substitute $-4$ for $y$ in Equation (3) and compute $x$.

$$x = y + 6 = -4 + 6 = 2$$

The ordered pair $(2, -4)$ checks in both equations. It is the solution.

**11.** $y - 2x = -6, \quad (1)$
$2y - x = 5 \quad (2)$

We solve Equation (1) for $y$.

$$y - 2x = -6 \qquad \text{Equation (1)}$$
$$y = 2x - 6 \qquad \qquad (3)$$

We substitute $2x - 6$ for $y$ in Equation (2) and solve for $x$.

$$2y - x = 5 \qquad \text{Equation (2)}$$
$$2(2x - 6) - x = 5 \qquad \text{Substituting}$$
$$4x - 12 - x = 5 \qquad \text{Removing parentheses}$$
$$3x - 12 = 5 \qquad \text{Collecting like terms}$$
$$3x = 17 \qquad \text{Adding 12}$$
$$x = \frac{17}{3} \qquad \text{Dividing by 3}$$

We substitute $\frac{17}{3}$ for $x$ in Equation (3) and compute $y$.

$$y = 2x - 6 = 2\left(\frac{17}{3}\right) - 6 = \frac{34}{3} - \frac{18}{3} = \frac{16}{3}$$

The ordered pair $\left(\frac{17}{3}, \frac{16}{3}\right)$ checks in both equations. It is the solution.

**13.** $2x + 3y = -2, \quad (1)$
$2x - y = 9 \quad (2)$

We solve Equation (2) for $y$.

$$2x - y = 9 \qquad \text{Equation (2)}$$
$$2x = 9 + y \qquad \text{Adding } y$$
$$2x - 9 = y \qquad \text{Subtracting 9} \qquad (3)$$

We substitute $2x - 9$ for $y$ in Equation (1) and solve for $x$.

$$2x + 3y = -2 \qquad \text{Equation (1)}$$
$$2x + 3(2x - 9) = -2 \qquad \text{Substituting}$$
$$2x + 6x - 27 = -2 \qquad \text{Removing parentheses}$$
$$8x - 27 = -2 \qquad \text{Collecting like terms}$$
$$8x = 25 \qquad \text{Adding 27}$$
$$x = \frac{25}{8} \qquad \text{Dividing by 8}$$

Now we substitute $\dfrac{25}{8}$ for $x$ in Equation (3) and compute $y$.

$$y = 2x - 9 = 2\left(\frac{25}{8}\right) - 9 = \frac{25}{4} - \frac{36}{4} = -\frac{11}{4}$$

The ordered pair $\left(\dfrac{25}{8}, -\dfrac{11}{4}\right)$ checks in both equations. It is the solution.

**15.**   $x - y = -3, \qquad (1)$

     $2x + 3y = -6 \quad (2)$

We solve Equation (1) for $x$.

$$x - y = -3 \qquad \text{Equation (1)}$$
$$x = y - 3 \qquad\qquad (3)$$

We substitute $y - 3$ for $x$ in Equation (2) and solve for $y$.

$$2x + 3y = -6 \qquad \text{Equation (2)}$$
$$2(y - 3) + 3y = -6 \qquad \text{Substituting}$$
$$2y - 6 + 3y = -6 \qquad \text{Removing parentheses}$$
$$5y - 6 = -6 \qquad \text{Collecting like terms}$$
$$5y = 0 \qquad \text{Adding 6}$$
$$y = 0 \qquad \text{Dividing by 5}$$

Now we substitute 0 for $y$ in Equation (3) and compute $x$.

$$x = y - 3 = 0 - 3 = -3$$

The ordered pair $(-3, 0)$ checks in both equations. It is the solution.

**17.**   $r - 2s = 0, \qquad (1)$

     $4r - 3s = 15 \quad (2)$

We solve Equation (1) for $r$.

$$r - 2s = 0 \qquad \text{Equation (1)}$$
$$r = 2s \qquad\qquad (3)$$

We substitute $2s$ for $r$ in Equation (2) and solve for $s$.

$$4r - 3s = 15 \qquad \text{Equation (2)}$$
$$4(2s) - 3s = 15 \qquad \text{Substituting}$$
$$8s - 3s = 15 \qquad \text{Removing parentheses}$$
$$5s = 15 \qquad \text{Collecting like terms}$$
$$s = 3 \qquad \text{Dividing by 5}$$

Now we substitute 3 for $s$ in Equation (3) and compute $r$.

$$r = 2s = 2 \cdot 3 = 6$$

The ordered pair $(6, 3)$ checks in both equations. It is the solution.

**19.** ***Familiarize.*** We let $x =$ the larger number and $y =$ the smaller number.

***Translate.*** We translate the first statement.

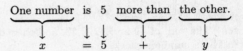

Now we translate the second statement.

$$\underbrace{\text{One number}}_{x} \ \underset{=}{\text{is}} \ \underset{5}{5} \ \underbrace{\text{more than}}_{+} \ \underbrace{\text{the other.}}_{y}$$

The resulting system is

$$x + y = 37, \quad (1)$$
$$x = 5 + y. \quad (2)$$

***Solve.*** We solve the system of equations. We substitute $5 + y$ for $x$ in Equation (1) and solve for $y$.

$$x + y = 37 \qquad \text{Equation (1)}$$
$$(5 + y) + y = 37 \qquad \text{Substituting}$$
$$5 + 2y = 37 \qquad \text{Collecting like terms}$$
$$2y = 32 \qquad \text{Subtracting 5}$$
$$y = 16 \qquad \text{Dividing by 2}$$

We go back to the original equations and substitute 16 for $y$. We use Equation (2).

$$x = 5 + y \qquad \text{Equation (2)}$$
$$x = 5 + 16 \qquad \text{Substituting}$$
$$x = 21$$

***Check.*** The sum of 21 and 16 is 37. The number 21 is 5 more than the number 16. These numbers check.

***State.*** The numbers are 21 and 16.

**21.** ***Familiarize.*** Let $x =$ one number and $y =$ the other.

***Translate.*** We reword and translate.

$$\underbrace{\text{The sum of two numbers}}_{x + y} \ \underset{=}{\text{is}} \ \underset{52}{52}.$$

$$\underbrace{\text{The difference of two numbers}}_{x - y} \ \underset{=}{\text{is}} \ \underset{28}{28}.$$

(The second statement could also be translated as $y - x = 28$.)

The resulting system is

$$x + y = 52, \quad (1)$$
$$x - y = 28. \quad (2)$$

***Solve.*** We solve the system. First we solve Equation (2) for $x$.

$$x - y = 28 \qquad \text{Equation (2)}$$
$$x = y + 28 \qquad \text{Adding } y \qquad (3)$$

We substitute $y + 28$ for $x$ in Equation (1) and solve for $y$.

$$x + y = 52 \qquad \text{Equation (1)}$$
$$(y + 28) + y = 52 \qquad \text{Substituting}$$
$$2y + 28 = 52 \qquad \text{Collecting like terms}$$
$$2y = 24 \qquad \text{Subtracting 28}$$
$$y = 12 \qquad \text{Dividing by 2}$$

Now we substitute 12 for $y$ in Equation (3) and compute $x$.

$$x = y + 28 = 12 + 28 = 40$$

**Check**. The sum of 40 and 12 is 52, and their difference is 28. These numbers check.

**State**. The numbers are 40 and 12.

**23. Familiarize**. We let $x$ = the larger number and $y$ = the smaller number.

**Translate**. We translate the first statement.

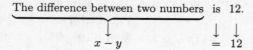

Now we translate the second statement.

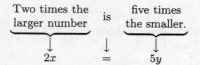

The resulting system is

$$x - y = 12, \quad (1)$$
$$2x = 5y. \qquad (2)$$

**Solve**. We solve the system. First we solve Equation (1) for $x$.

$$x - y = 12 \qquad \text{Equation (1)}$$
$$x = y + 12 \qquad \text{Adding } y \qquad (3)$$

We substitute $y + 12$ for $x$ in Equation (2) and solve for $y$.

$$2x = 5y \qquad \text{Equation (2)}$$
$$2(y + 12) = 5y \qquad \text{Substituting}$$
$$2y + 24 = 5y \qquad \text{Removing parentheses}$$
$$24 = 3y \qquad \text{Subtracting } 2y$$
$$8 = y \qquad \text{Dividing by 3}$$

Now we substitute 8 for $y$ in Equation (3) and compute $x$.

$$x = y + 12 = 8 + 12 = 20$$

**Check**. The difference between 20 and 8 is 12. Two times 20, or 40, is five times 8. These numbers check.

**State**. The numbers are 20 and 8.

**25. Familiarize**. From the drawing in the text we see that we have a rectangle with length $l$ and width $w$.

**Translate**. The perimeter of a rectangle is $2l + 2w$. We translate the first statement.

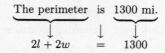

Then we reword and translate the second statement.

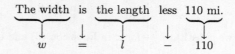

The resulting system is

$$2l + 2w = 1300, \quad (1)$$
$$w = l - 110. \qquad (2)$$

**Solve**. We solve the system. We substitute $l - 110$ for $w$ in Equation (1) and solve for $l$.

$$2l + 2w = 1300 \qquad \text{Equation (1)}$$
$$2l + 2(l - 110) = 1300 \qquad \text{Substituting}$$
$$2l + 2l - 220 = 1300 \qquad \text{Removing parentheses}$$
$$4l - 220 = 1300 \qquad \text{Collecting like terms}$$
$$4l = 1520 \qquad \text{Adding 220}$$
$$l = 380 \qquad \text{Dividing by 4}$$

Now we substitute 380 for $l$ in Equation (2) and solve for $w$.

$$w = l - 110 \qquad \text{Equation (2)}$$
$$w = 380 - 110 \qquad \text{Substituting}$$
$$w = 270$$

**Check**. A possible solution is a length of 380 mi and a width of 270 mi. The perimeter would be $2 \cdot 380 + 2 \cdot 270$, or $760 + 540$, or 1300. Also, the width is 110 mi less than the length. These numbers check.

**State**. The length is 380 mi, and the width is 270 mi.

**27. Familiarize**. We make a drawing. We let $l$ = the length and $w$ = the width.

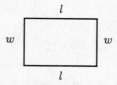

**Translate**. The perimeter is $2l + 2w$. We translate the first statement.

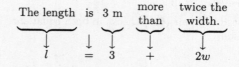

We translate the second statement.

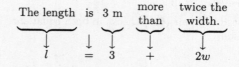

The resulting system is

$$2l + 2w = 400, \quad (1)$$
$$l = 3 + 2w. \quad (2)$$

**Solve**. We solve the system. We substitute $3 + 2w$ for $l$ in Equation (1) and solve for $w$.

$$\begin{array}{ll} 2l + 2w = 400 & \text{Equation (1)} \\ 2(3 + 2w) + 2w = 400 & \text{Substituting} \\ 6 + 4w + 2w = 400 & \text{Removing parentheses} \\ 6 + 6w = 400 & \text{Collecting like terms} \\ 6w = 394 & \text{Subtracting 6} \\ w = \dfrac{394}{6} & \text{Dividing by 6} \\ w = \dfrac{197}{3}, \text{ or } 65\dfrac{2}{3} & \end{array}$$

Now we substitute $\dfrac{197}{3}$ for $w$ in Equation (2) and solve for $l$.

$$\begin{array}{ll} l = 3 + 2w & \text{Equation (2)} \\ l = 3 + 2\left(\dfrac{197}{3}\right) & \text{Substituting} \\ l = \dfrac{9}{3} + \dfrac{394}{3} & \\ l = \dfrac{403}{3}, \text{ or } 134\dfrac{1}{3} & \end{array}$$

**Check**. A possible solution is a length of $134\dfrac{1}{3}$ m and a width of $65\dfrac{2}{3}$ m. The perimeter would be $2\left(134\dfrac{1}{3}\right) + 2\left(65\dfrac{2}{3}\right)$, or $2\left(\dfrac{403}{3}\right) + 2\left(\dfrac{197}{3}\right)$, or $\dfrac{806}{3} + \dfrac{394}{3}$, or $\dfrac{1200}{3}$, or 400. Also, 3 more than twice the width is $3 + 2\left(65\dfrac{2}{3}\right)$, or $3 + 2\left(\dfrac{197}{3}\right)$, or $\dfrac{9}{3} + \dfrac{394}{3}$, or $\dfrac{403}{3}$, or $134\dfrac{1}{3}$, which is the length. These numbers check.

**State**. The length is $134\dfrac{1}{3}$ m, and the width is $65\dfrac{2}{3}$ m.

**29.** Graph: $2x - 3y = 6$

To find the $x$-intercept, let $y = 0$. Then solve for $x$.

$$\begin{array}{l} 2x - 3 \cdot 0 = 6 \\ 2x = 6 \\ x = 3 \, . \end{array}$$

The $x$-intercept is $(3, 0)$.

To find the $y$-intercept, let $x = 0$. Then solve for $y$.

$$\begin{array}{l} 2 \cdot 0 - 3y = 6 \\ -3y = 6 \\ y = -2 \end{array}$$

The $y$-intercept is $(0, -2)$.

We plot these points and draw the line.

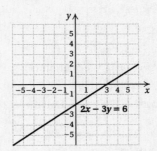

A third point should be used as a check. We let $x = -3$:

$$\begin{array}{l} 2(-3) - 3y = 6 \\ -6 - 3y = 6 \\ -3y = 12 \\ y = -4 \end{array}$$

The point $(-3, -4)$ is on the graph, so our graph is probably correct.

**31.**
$$\begin{array}{l} 2x - 3 = 0 \\ 2x = 3 \\ x = \dfrac{3}{2} \end{array}$$

Any ordered pair $\left(\dfrac{3}{2}, y\right)$ is a solution, so the line is parallel to the $y$-axis with $x$-intercept $\left(\dfrac{3}{2}, 0\right)$.

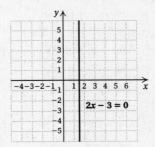

**33.** Graph: $x = -3$

Any ordered pair $(-3, y)$ is a solution, so the line is parallel to the $y$-axis with $x$-intercept $(-3, 0)$.

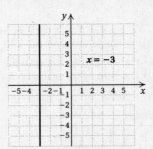

**35.**  $y - 2.35x = -5.97$,    (1)

$2.14y - x = 4.88$        (2)

Solve Equation (1) for $y$.

$y - 2.35x = -5.97$    Equation (1)

$y = 2.35x - 5.97$            (3)

Substitute $2.35x - 5.97$ for $y$ in Equation (2) and solve for $x$.

$2.14(2.35x - 5.97) - x = 4.88$

$5.029x - 12.7758 - x = 4.88$

$4.029x = 17.6558$

$x \approx 4.3821792$

Substitute 4.382 for $x$ in Equation (3) and solve for $y$.

$y = 2.35x - 5.97 = 2.35(4.382) - 5.97 \approx 4.3281211$

The ordered pair $(4.3821792, 4.3281211)$ checks in both equations. It is the solution.

**37.**  $\dfrac{x}{2} + \dfrac{3y}{2} = 2$,    (1)

$\dfrac{x}{5} - \dfrac{y}{2} = 3$    (2)

Multiply Equation (1) by 2 and Equation (2) by 10 to clear fractions.

$x + 3y = 4$,    (1a)

$2x - 5y = 30$    (2a)

Solve Equation (1a) for $x$.

$x + 3y = 4$    Equation (1a)

$x = -3y + 4$            (3)

Substitute $-3y + 4$ for $x$ in Equation (2a) and solve for $y$.

$2(-3y + 4) - 5y = 30$

$-6y + 8 - 5y = 30$

$-11y + 8 = 30$

$-11y = 22$

$y = -2$

Substitute $-2$ for $y$ in Equation (3) and solve for $x$.

$x = -3y + 4 = -3(-2) + 4 = 6 + 4 = 10$

The ordered pair $(10, -2)$ checks in both equations. It is the solution.

**39.** *Familiarize*.  We let $s$ = the length of each of the two equal sides of the triangle and $b$ = the length of the base.

*Translate*.  We translate the first statement.  Recall that the perimeter is the sum of the lengths of the sides of the triangle.

The perimeter  is  324 cm.

$s + s + b$ $=$ $324$, or

$2s + b$ $=$ $324$

We translate the second statement.

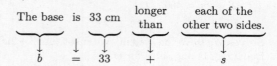

The base is 33 cm longer than each of the other two sides.

$b = 33 + s$

The resulting system is

$2s + b = 324$,    (1)

$b = 33 + s$.    (2)

*Solve*.  We solve the system.  We substitute $33 + s$ for $b$ in Equation (1) and solve for $s$.

$2s + b = 324$    Equation (1)

$2s + (33 + s) = 324$

$3s + 33 = 324$

$3s = 291$

$s = 97$

Now we substitute 97 for $s$ in Equation (2) and solve for $b$.

$b = 33 + s$    Equation (2)

$b = 33 + 97$

$b = 130$

*Check*.  A possible solution is two sides of 97 cm and a base of 130 cm.  The perimeter would be $97 + 97 + 130$, or 324.  Also, the base is 33 cm longer than each of the other two sides.  These numbers check.

*State*.  The lengths of the sides of the triangle are 97 cm, 97 cm, and 130 cm.

**41.**  $y = x + 5$,        (1)

$-3x + 3y = 15$    (2)

Substitute $x + 5$ for $y$ in Equation (2) and solve for $x$.

$-3x + 3(x + 5) = 15$

$-3x + 3x + 15 = 15$

$15 = 15$

We obtain an equation that is true for all values of $x$ and $y$, so the system has infinitely many solutions.  The equations have the same graph.

---

## Exercise Set 7.3

**1.**  $x - y = 7$    (1)

$\underline{x + y = 5}$    (2)

$2x \quad = 12$    Adding

$x = 6$    Dividing by 2

Substitute 6 for $x$ in either of the original equations and solve for $y$.

$x + y = 5$    Equation (2)

$6 + y = 5$    Substituting

$y = -1$    Subtracting 6

Check:

$$\begin{array}{c|c} x - y = 7 \\ \hline 6 - (-1) & 7 \\ 6 + 1 & \\ 7 & \text{TRUE} \end{array}$$

$$\begin{array}{c|c} x + y = 5 \\ \hline 6 + (-1) & 5 \\ & 5 \quad \text{TRUE} \end{array}$$

Since $(6, -1)$ checks, it is the solution.

3.　$x + y = 8$　(1)

　$-x + 2y = 7$　(2)

　$\quad\quad 3y = 15$　Adding

　$\quad\quad\ y = 5$　Dividing by 3

Substitute 5 for $y$ in either of the original equations and solve for $x$.

$x + y = 8$　Equation (1)

$x + 5 = 8$　Substituting

$\quad\ x = 3$

Check:

$$\begin{array}{c|c} x + y = 8 \\ \hline 3 + 5 & 8 \\ 8 & \text{TRUE} \end{array}$$

$$\begin{array}{c|c} -x + 2y = 7 \\ \hline -3 + 2 \cdot 5 & 7 \\ -3 + 10 & \\ & 7 \quad \text{TRUE} \end{array}$$

Since $(3, 5)$ checks, it is the solution.

5.　$5x - y = 5$　(1)

　$3x + y = 11$　(2)

　$8x \quad\ = 16$　Adding

　$\quad x = 2$　Dividing by 8

Substitute 2 for $x$ in either of the original equations and solve for $y$.

$3x + y = 11$　Equation (2)

$3 \cdot 2 + y = 11$　Substituting

$6 + y = 11$

$\quad\quad y = 5$

Check:

$$\begin{array}{c|c} 5x - y = 5 \\ \hline 5 \cdot 2 - 5 & 5 \\ 10 - 5 & \\ 5 & \text{TRUE} \end{array}$$

$$\begin{array}{c|c} 3x + y = 11 \\ \hline 3 \cdot 2 + 5 & 11 \\ 6 + 5 & \\ 11 & \text{TRUE} \end{array}$$

Since $(2, 5)$ checks, it is the solution.

7.　$4a + 3b = 7$　(1)

　$-4a + b = 5$　(2)

　$\quad\quad 4b = 12$　Adding

　$\quad\quad\ b = 3$

Substitute 3 for $b$ in either of the original equations and solve for $a$.

$4a + 3b = 7$　Equation (1)

$4a + 3 \cdot 3 = 7$　Substituting

$4a + 9 = 7$

$\quad 4a = -2$

$\quad\ a = -\dfrac{1}{2}$

Check:

$$\begin{array}{c|c} 4a + 3b = 7 \\ \hline 4\left(-\dfrac{1}{2}\right) + 3 \cdot 3 & 7 \\ -2 + 9 & \\ 7 & \text{TRUE} \end{array}$$

$$\begin{array}{c|c} -4a + b = 5 \\ \hline -4\left(-\dfrac{1}{2}\right) + 3 & 5 \\ 2 + 3 & \\ 5 & \text{TRUE} \end{array}$$

Since $\left(-\dfrac{1}{2}, 3\right)$ checks, it is the solution.

9.　$8x - 5y = -9$　(1)

　$3x + 5y = -2$　(2)

　$11x \quad\ = -11$　Adding

　$\quad x = -1$

Substitute $-1$ for $x$ in either of the original equations and solve for $y$.

$3x + 5y = -2$　Equation (2)

$3(-1) + 5y = -2$　Substituting

$-3 + 5y = -2$

$\quad\quad 5y = 1$

$\quad\quad\ y = \dfrac{1}{5}$

Check:

$$\begin{array}{c|c} 8x - 5y = -9 \\ \hline 8(-1) - 5\left(\dfrac{1}{5}\right) & -9 \\ -8 - 1 & \\ -9 & \text{TRUE} \end{array}$$

$$\begin{array}{c|c} 3x + 5y = -2 \\ \hline 3(-1) + 5\left(\dfrac{1}{5}\right) & -2 \\ -3 + 1 & \\ -2 & \text{TRUE} \end{array}$$

Since $\left(-1, \dfrac{1}{5}\right)$ checks, it is the solution.

11.　$4x - 5y = 7$

　$-4x + 5y = 7$

　$\quad\quad 0 = 14$　Adding

We obtain a false equation, $0 = 14$, so there is no solution.

13.　$x + y = -7,$　(1)

　$3x + y = -9$　(2)

We multiply on both sides of Equation (1) by $-1$ and then add.

$-x - y = 7$　Multiplying by $-1$

$3x + y = -9$　Equation (2)

$2x \quad\ = -2$　Adding

$\quad x = -1$

Substitute $-1$ for $x$ in one of the original equations and solve for $y$.

$$x + y = -7 \quad \text{Equation (1)}$$
$$-1 + y = -7 \quad \text{Substituting}$$
$$y = -6$$

Check:

| $x + y = -7$ | |
|---|---|
| $-1 + (-6)$ | $-7$ |
| $-7$ | TRUE |

| $3x + y = -9$ | |
|---|---|
| $3(-1) + (-6)$ | $-9$ |
| $-3 - 6$ | |
| $-9$ | TRUE |

Since $(-1, -6)$ checks, it is the solution.

**15.**  $3x - y = 8, \quad (1)$
$\quad\;\; x + 2y = 5 \quad (2)$

We multiply on both sides of Equation (1) by 2 and then add.

$$6x - 2y = 16 \quad \text{Multiplying by 2}$$
$$\underline{x + 2y = 5} \quad \text{Equation (2)}$$
$$7x \quad\;\; = 21 \quad \text{Adding}$$
$$x = 3$$

Substitute 3 for $x$ in one of the original equations and solve for $y$.

$$x + 2y = 5 \quad \text{Equation (2)}$$
$$3 + 2y = 5 \quad \text{Substituting}$$
$$2y = 2$$
$$y = 1$$

Check:

| $3x - y = 8$ | |
|---|---|
| $3 \cdot 3 - 1$ | $8$ |
| $9 - 1$ | |
| $8$ | TRUE |

| $x + 2y = 5$ | |
|---|---|
| $3 + 2 \cdot 1$ | $5$ |
| $3 + 2$ | |
| $5$ | TRUE |

Since $(3, 1)$ checks, it is the solution.

**17.**  $x - y = 5, \quad (1)$
$\quad\; 4x - 5y = 17 \quad (2)$

We multiply on both sides of Equation (1) by $-4$ and then add.

$$-4x + 4y = -20 \quad \text{Multiplying by } -4$$
$$\underline{4x - 5y = 17} \quad \text{Equation (2)}$$
$$-y = -3 \quad \text{Adding}$$
$$y = 3$$

Substitute 3 for $y$ in one of the original equations and solve for $x$.

$$x - y = 5 \quad \text{Equation (1)}$$
$$x - 3 = 5 \quad \text{Substituting}$$
$$x = 8$$

Check:

| $x - y = 5$ | |
|---|---|
| $8 - 3$ | $5$ |
| $5$ | TRUE |

| $4x - 5y = 17$ | |
|---|---|
| $4 \cdot 8 - 5 \cdot 3$ | $17$ |
| $32 - 15$ | |
| $17$ | TRUE |

Since $(8, 3)$ checks, it is the solution.

**19.**  $2w - 3z = -1, \quad (1)$
$\quad\; 3w + 4z = 24 \quad (2)$

We use the multiplication principle with both equations and then add.

$$8w - 12z = -4 \quad \text{Multiplying (1) by 4}$$
$$\underline{9w + 12z = 72} \quad \text{Multiplying (2) by 3}$$
$$17w \quad\quad\; = 68 \quad \text{Adding}$$
$$w = 4$$

Substitute 4 for $w$ in one of the original equations and solve for $z$.

$$3w + 4z = 24 \quad \text{Equation (2)}$$
$$3 \cdot 4 + 4z = 24 \quad \text{Substituting}$$
$$12 + 4z = 24$$
$$4z = 12$$
$$z = 3$$

Check:

| $2w - 3z = -1$ | |
|---|---|
| $2 \cdot 4 - 3 \cdot 3$ | $-1$ |
| $8 - 9$ | |
| $-1$ | TRUE |

| $3w + 4z = 24$ | |
|---|---|
| $3 \cdot 4 + 4 \cdot 3$ | $24$ |
| $12 + 12$ | |
| $24$ | TRUE |

Since $(4, 3)$ checks, it is the solution.

**21.**  $2a + 3b = -1, \quad (1)$
$\quad\; 3a + 5b = -2 \quad (2)$

We use the multiplication principle with both equations and then add.

$$-10a - 15b = 5 \quad \text{Multiplying (1) by } -5$$
$$\underline{9a + 15b = -6} \quad \text{Multiplying (2) by 3}$$
$$-a \quad\quad\;\; = -1 \quad \text{Adding}$$
$$a = 1$$

Substitute 1 for $a$ in one of the original equations and solve for $b$.

$$2a + 3b = -1 \quad \text{Equation (1)}$$
$$2 \cdot 1 + 3b = -1 \quad \text{Substituting}$$
$$2 + 3b = -1$$
$$3b = -3$$
$$b = -1$$

Check:

| $2a + 3b = -1$ | | | $3a + 5b = -2$ | |
|---|---|---|---|---|
| $2 \cdot 1 + 3(-1)$ | $-1$ | | $3 \cdot 1 + 5(-1)$ | $-2$ |
| $2 - 3$ | | | $3 - 5$ | |
| $-1$ | TRUE | | $-2$ | TRUE |

Since $(1, -1)$ checks, it is the solution.

**23.**    $x = 3y,$    (1)

$5x + 14 = y$    (2)

We first get each equation in the form $Ax + By = C$.

$x - 3y = 0,$    (1a)    Adding $-3y$

$5x - y = -14$    (2a)    Adding $-y - 14$

We multiply by $-5$ on both sides of Equation (1a) and add.

$-5x + 15y = \phantom{-}0$    Multiplying by $-5$

$\underline{5x - \phantom{1}y = -14}$

$14y = -14$    Adding

$y = \phantom{-}-1$

Substitute $-1$ for $y$ in Equation (1) and solve for $x$.

$x - 3y = 0$

$x - 3(-1) = 0\cdot$    Substituting

$x + 3 = 0$

$x = -3$

Check:

| $x - 3y = 0$ | | | $5x - y = -14$ | |
|---|---|---|---|---|
| $-3 - 3(-1)$ | $0$ | | $5(-3) - (-1)$ | $-14$ |
| $-3 + 3$ | | | $-15 + 1$ | |
| $0$ | TRUE | | $-14$ | TRUE |

Since $(-3, -1)$ checks, it is the solution.

**25.**  $2x + 5y = 16,$    (1)

$3x - 2y = \phantom{1}5$    (2)

We use the multiplication principle with both equations and then add.

$4x + 10y = 32$    Multiplying (1) by 2

$\underline{15x - 10y = 25}$    Multiplying (2) by 5

$19x \phantom{- 10y} = 57$

$x = \phantom{-}3$

Substitute 3 for $x$ in one of the original equations and solve for $y$.

$2x + 5y = 16$    Equation (1)

$2 \cdot 3 + 5y = 16$    Substituting

$6 + 5y = 16$

$5y = 10$

$y = 2$

Check:

| $2x + 5y = 16$ | | | $3x - 2y = 5$ | |
|---|---|---|---|---|
| $2 \cdot 3 + 5 \cdot 2$ | $16$ | | $3 \cdot 3 - 2 \cdot 2$ | $5$ |
| $6 + 10$ | | | $9 - 4$ | |
| $16$ | TRUE | | $5$ | TRUE |

Since $(3, 2)$ checks, it is the solution.

**27.**    $p = 32 + q,$    (1)

$3p = 8q + 6$    (2)

First we write each equation in the form $Ap + Bq = C$.

$p - q = 32,$    (1a)    Subtracting $q$

$3p - 8q = 6$    (2a)    Subtracting $8q$

Now we multiply both sides of Equation (1a) by $-3$ and then add.

$-3p + \phantom{1}3q = -96$    Multiplying by $-3$

$\underline{3p - \phantom{1}8q = \phantom{-}6}$    Equation (2a)

$-5q = -90$    Adding

$q = \phantom{-}18$

Substitute 18 for $q$ in Equation (1) and solve for $p$.

$p = 32 + q$

$p = 32 + 18$    Substituting

$p = 50$

Check:

| $p - q = 32$ | | | $3p - 8q = 6$ | |
|---|---|---|---|---|
| $50 - 18$ | $32$ | | $3 \cdot 50 - 8 \cdot 18$ | $6$ |
| $32$ | TRUE | | $150 - 144$ | |
| | | | $6$ | TRUE |

Since $(50, 18)$ checks, it is the solution.

**29.**  $3x - 2y = \phantom{-}10,$    (1)

$-6x + 4y = -20$    (2)

We multiply by 2 on both sides of Equation (1) and add.

$6x - 4y = \phantom{-}20$

$\underline{-6x + 4y = -20}$

$0 = \phantom{-}0$

We get an obviously true equation, so the system has an infinite number of solutions.

**31.**  $0.06x + 0.05y = 0.07,$

$0.04x - 0.03y = 0.11$

We first multiply each equation by 100 to clear the decimals.

$6x + 5y = 7,$    (1)

$4x - 3y = 11$    (2)

We use the multiplication principle with both equations of the resulting system.

$18x + 15y = 21$    Multiplying (1) by 3

$\underline{20x - 15y = 55}$    Multiplying (2) by 5

$38x \phantom{- 15y} = 76$    Adding

$x = \phantom{-}2$

Substitute 2 for $x$ in Equation (1) and solve for $y$.

$$6x + 5y = 7$$
$$6 \cdot 2 + 5y = 7$$
$$12 + 5y = 7$$
$$5y = -5$$
$$y = -1$$

Check:

| $0.06x + 0.05y = 0.07$ | |
|---|---|
| $0.06(2) + 0.05(-1)$ | $0.07$ |
| $0.12 - 0.05$ | |
| $0.07$ | TRUE |

| $0.04x - 0.03y = 0.11$ | |
|---|---|
| $0.04(2) - 0.03(-1)$ | $0.11$ |
| $0.08 + 0.03$ | |
| $0.11$ | TRUE |

Since $(2, -1)$ checks, it is the solution.

**33.** $\dfrac{1}{3}x + \dfrac{3}{2}y = \dfrac{5}{4}$,

$\dfrac{3}{4}x - \dfrac{5}{6}y = \dfrac{3}{8}$

First we clear the fractions. We multiply on both sides of the first equation by 12 and on both sides of the second equation by 24.

$$12\left(\frac{1}{3}x + \frac{3}{2}y\right) = 12 \cdot \frac{5}{4}$$
$$12 \cdot \frac{1}{3}x + 12 \cdot \frac{3}{2}y = 15$$
$$4x + 18y = 15$$
$$24\left(\frac{3}{4}x - \frac{5}{6}y\right) = 24 \cdot \frac{3}{8}$$
$$24 \cdot \frac{3}{4}x - 24 \cdot \frac{5}{6}y = 9$$
$$18x - 20y = 9$$

The resulting system is

$$4x + 18y = 15, \quad (1)$$
$$18x - 20y = 9. \quad (2)$$

We use the multiplication principle with both equations.

$\begin{array}{ll} 72x + 324y = 270 & \text{Multiplying (1) by 18} \\ -72x + 80y = -36 & \text{Multiplying (2) by } -4 \\ \hline 404y = 234 & \end{array}$

$$y = \frac{234}{404}, \text{ or } \frac{117}{202}$$

Substitute $\dfrac{117}{202}$ for $y$ in Equation (1) and solve for $x$.

$$4x + 18\left(\frac{117}{202}\right) = 15$$
$$4x + \frac{1053}{101} = 15$$
$$4x = \frac{462}{101}$$
$$x = \frac{1}{4} \cdot \frac{462}{101}$$
$$x = \frac{231}{202}$$

The ordered pair $\left(\dfrac{231}{202}, \dfrac{117}{202}\right)$ checks in both equations. It is the solution.

**35. Familiarize.** We let $M$ = the number of miles driven and $C$ = the total cost of the rental.

**Translate.** We reword and translate the first statement, using \$0.39 for 39 cents.

$\begin{array}{ccccccc} \text{\$19.95 plus} & \text{39 cents times} & \text{the number of miles driven} & \text{is} & \text{the cost.} \\ \downarrow & \downarrow & \downarrow & \downarrow & \downarrow & \downarrow & \downarrow \\ \$19.95 & + & \$0.39 & \cdot & M & = & C \end{array}$

We reword and translate the second statement using \$0.29 for 29 cents.

$\begin{array}{ccccccc} \text{\$39.95 plus} & \text{29 cents times} & \text{the number of miles driven} & \text{is} & \text{the cost.} \\ \downarrow & \downarrow & \downarrow & \downarrow & \downarrow & \downarrow & \downarrow \\ \$39.95 & + & \$0.29 & \cdot & M & = & C \end{array}$

The resulting system is

$$19.95 + 0.39M = C,$$
$$39.95 + 0.29M = C.$$

**Solve.** We solve the system of equations. We clear the decimals by multiplying on both sides of each equation by 100.

$$1995 + 39M = 100C, \quad (1)$$
$$3995 + 29M = 100C \quad (2)$$

We multiply Equation (2) by $-1$ and add.

$\begin{array}{rcl} 1995 + 39M &=& 100C \\ -3995 - 29M &=& -100C \\ \hline -2000 + 10M &=& 0 \\ 10M &=& 2000 \\ M &=& 200 \end{array}$

**Check.** For 200 mi, the cost of the Quick-Haul van is $\$19.95 + \$0.39(200)$, or $\$19.95 + \$78 = \$97.95$. The cost of the other company's van is $\$39.95 + \$0.29(200)$, or $\$39.95 + \$58$, or $\$97.95$. Thus the costs are the same when the mileage is 200.

**State.** When the moving vans are driven 200 miles, the costs will be the same.

**37. Familiarize.** Let $x$ = the smaller angle and $y$ = the larger angle.

***Translate.*** We reword the problem.

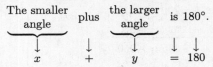

The smaller angle plus the larger angle is 180°.
$$x + y = 180$$

The larger angle is 30° more than 2 times the smaller angle.
$$y = 30 + 2 \cdot x$$

The resulting system is

$$x + y = 180,$$
$$y = 30 + 2x.$$

***Solve.*** We solve the system. We will use the elimination method although we could also easily use the substitution method. First we get the second equation in the form $Ax + By = C$.

$$x + y = 180 \quad (1)$$
$$-2x + y = 30 \quad (2) \text{ Adding } -2x$$

Now we multiply Equation (2) by $-1$ and add.

$$\begin{array}{r} x + y = 180 \\ \underline{2x - y = -30} \\ 3x \phantom{ - y} = 150 \\ x = 50 \end{array}$$

Then we substitute 50 for $x$ in Equation (1) and solve for $y$.

$$x + y = 180 \quad \text{Equation (1)}$$
$$50 + y = 180 \quad \text{Substituting}$$
$$y = 130$$

***Check.*** The sum of the angles is $50° + 130°$, or $180°$, so the angles are supplementary. Also, $30°$ more than two times the $50°$ angle is $30° + 2 \cdot 50°$, or $30° + 100°$, or $130°$, the other angle. These numbers check.

***State.*** The angles are $50°$and $130°$.

**39.** ***Familiarize.*** We let $x = $ the larger angle and $y = $ the smaller angle.

***Translate.*** We reword and translate the first statement.

The sum of two angles is 90°.
$$x + y = 90$$

We reword and translate the second statement.

The difference of two angles is 34°.
$$x - y = 34$$

The resulting system is

$$x + y = 90,$$
$$x - y = 34.$$

***Solve.*** We solve the system.

$$\begin{array}{r} x + y = 90, \quad (1) \\ \underline{x - y = 34} \quad (2) \\ 2x \phantom{ - y} = 124 \quad \text{Adding} \\ x = 62 \end{array}$$

Now we substitute 62 for $x$ in Equation (1) and solve for $y$.

$$x + y = 90 \quad \text{Equation (1)}$$
$$62 + y = 90 \quad \text{Substituting}$$
$$y = 28$$

***Check.*** The sum of the angles is $62° + 28°$, or $90°$, so the angles are complementary. The difference of the angles is $62° - 28°$, or $34°$. These numbers check.

***State.*** The angles are $62°$and $28°$.

**41.** ***Familiarize.*** We let $x = $ the number of hectares of hay that should be planted and $y = $ the number of hectares of oats that should be planted.

***Translate.*** We reword and translate the first statement.

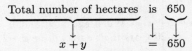

Total number of hectares is 650
$$x + y = 650$$

Now we reword and translate the second statement.

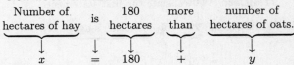

Number of hectares of hay is 180 hectares more than number of hectares of oats.
$$x = 180 + y$$

The resulting system is

$$x + y = 650,$$
$$x = 180 + y$$

***Solve.*** We solve the system. We will use the elimination method, although we could also easily use the substitution method. First we get the second equation in the form $Ax + By = C$. Then we add the equations.

$$\begin{array}{r} x + y = 650 \quad (1) \\ \underline{x - y = 180} \quad (2) \quad \text{Subtracting } y \\ 2x \phantom{ - y} = 830 \quad \text{Adding} \\ x = 415 \end{array}$$

Now we substitute 415 for $x$ in Equation (1) and solve for $y$.

$$x + y = 650 \quad \text{Equation (1)}$$
$$415 + y = 650 \quad \text{Substituting}$$
$$y = 235$$

***Check.*** The total number of hectares is $415 + 235$, or $650$. Also, the number of hectares of hay is 180 more than the number of hectares of oats. These numbers check.

***State.*** The owners should plant 415 hectares of hay and 235 hectares of oats.

**43.** $(a^2 b^{-3})(a^5 b^{-6}) = a^{2+5} b^{-3+(-6)} = a^7 b^{-9} = \dfrac{a^7}{b^9}$

**45.** $\dfrac{x^2 - 5x + 6}{x^2 - 4} = \dfrac{(x-3)(x-2)}{(x+2)(x-2)}$

$\qquad\qquad\qquad = \dfrac{(x-3)\cancel{(x-2)}}{(x+2)\cancel{(x-2)}}$

$\qquad\qquad\qquad = \dfrac{x-3}{x+2}$

**47.** $\dfrac{x-2}{x+3} - \dfrac{2x-5}{x-4}$    LCD is $(x+3)(x-4)$

$= \dfrac{x-2}{x+3} \cdot \dfrac{x-4}{x-4} - \dfrac{2x-5}{x-4} \cdot \dfrac{x+3}{x+3}$

$= \dfrac{(x-2)(x-4)}{(x+3)(x-4)} - \dfrac{(2x-5)(x+3)}{(x-4)(x+3)}$

$= \dfrac{x^2-6x+8}{(x+3)(x-4)} - \dfrac{2x^2+x-15}{(x-4)(x+3)}$

$= \dfrac{x^2-6x+8-(2x^2+x-15)}{(x+3)(x-4)}$

$= \dfrac{x^2-6x+8-2x^2-x+15}{(x+3)(x-4)}$

$= \dfrac{-x^2-7x+23}{(x+3)(x-4)}$

**49.** *Familiarize*. We let $x$ = Will's age now and $y$ = his father's age now. In 20 years their ages will be $x + 20$ and $y + 20$.

*Translate*. We translate the first statement.

Will's age  is   20%  of  his father's age.

$\qquad x \quad = \quad 20\% \quad \cdot \qquad y,\text{ or}$

$\qquad x \quad = \quad 0.2 \quad \cdot \qquad y$

We reword and translate the second statement.

Will's age    will    52%  of   his father's age
in 20 years   be                in 20 years.

$\quad x + 20 \quad = \quad 52\% \quad \cdot \quad (y+20),\text{ or}$

$\quad x + 20 \quad = \quad 0.52 \quad \cdot \quad (y+20)$

The resulting system is

$\qquad x = 0.2y,$

$\qquad x + 20 = 0.52(y+20).$

*Solve*. We solve the system of equations. We will use the elimination method, although we could also easily use the substitution method. First we clear decimals by multiplying on both sides of the first equation by 10 and on both sides of the second equation by 100.

$\qquad 10x = 2y,$

$\qquad 100x + 2000 = 52(y+20)$

Now we write each equation in the form $Ax + By = C$.

$\qquad 10x - 2y = 0, \qquad (1)$

$\qquad 100x - 52y = -960 \qquad (2)$

We multiply Equation (1) by $-10$ and then add.

$\qquad -100x + \;\; 20y = \qquad 0$

$\qquad \underline{\;\;\;100x - \;\; 52y = -960\;\;\;}$

$\qquad\qquad\qquad -32y = -960$

$\qquad\qquad\qquad\quad\; y = \quad\; 30$

Substitute 30 for $y$ in Equation (1) and solve for $x$.

$\qquad 10x - 2y = 0$

$\qquad 10x - 2 \cdot 30 = 0$

$\qquad 10x - 60 = 0$

$\qquad\quad 10x = 60$

$\qquad\qquad x = 6$

*Check*. Will's age now, 6, is 20% of 30, his father's age now. Will's age 20 years from now, 26, is 52% of 50, his father's age 20 years from now. These numbers check.

*State*. Will is 6 years old now, and his father is 30 years old.

**51.** *Familiarize*. Let $b$ and $h$ represent the original base and height of the triangle. Then $b + 2 =$ the base increased by 2 ft, and $h - 1 =$ the height decreased by 1 ft.

*Translate*. We reword and translate.

The decreased    is  one-third  of   the increased
height                                    base.

$\qquad h - 1 \qquad = \qquad \dfrac{1}{3} \qquad \cdot \qquad (b+2)$

The new area    is    24 ft².

$\qquad \dfrac{1}{2}(b+2)(h-1) \;\; = \qquad 24$

The resulting system is

$\qquad h - 1 = \dfrac{1}{3}(b+2), \qquad (1)$

$\qquad \dfrac{1}{2}(b+2)(h-1) = 24. \quad (2)$

*Solve*. We solve the system using the substitution method. We substitute $\dfrac{1}{3}(b+2)$ for $h - 1$ in Equation (2).

$\quad \dfrac{1}{2}(b+2)(h-1) = 24 \qquad$ Equation (2)

$\quad \dfrac{1}{2}(b+2) \cdot \dfrac{1}{3}(b+2) = 24 \qquad$ Substituting

$\qquad \dfrac{1}{6}(b+2)^2 = 24 \qquad$ Multiplying on the left

$\qquad\;\; (b+2)^2 = 144 \qquad$ Multiplying by 6

$\qquad b^2 + 4b + 4 = 144$

$\qquad b^2 + 4b - 140 = 0$

$\qquad (b+14)(b-10) = 0$

$\quad b + 14 = 0 \qquad \text{or} \quad b - 10 = 0$

$\qquad b = -14 \quad \text{or} \qquad\quad b = 10$

Since the dimensions cannot be negative, we consider only 10. We substitute 10 for $b$ in Equation (1) and solve for $h$.

$$h - 1 = \frac{1}{3}(b + 2) \qquad \text{Equation (1)}$$

$$h - 1 = \frac{1}{3}(10 + 2) \qquad \text{Substituting}$$

$$h - 1 = \frac{1}{3} \cdot 12$$

$$h - 1 = 4$$

$$h = 5$$

**Check**. If the original base and height are 10 ft and 5 ft, respectively, then the increased base is $10 + 2$, or 12 ft, and the decreased height is $5 - 1$, or 4 ft. The new height is one-third of the new base, and the new area is $\frac{1}{2} \cdot 12 \cdot 4$, or 24 ft$^2$. These numbers check.

**State**. The original base of the triangle is 10 ft, and original height is 5 ft.

**53.** $3(x - y) = 9$,

$x + y = 7$

First we remove parentheses in the first equation.

$$3x - 3y = 9, \quad (1)$$
$$x + y = 7 \quad (2)$$

Then we multiply Equation (2) by 3 and add.

$$3x - 3y = 9$$
$$\underline{3x + 3y = 21}$$
$$6x \qquad = 30$$
$$x = 5$$

Now we substitute 5 for $x$ in Equation (2) and solve for $y$.

$$x + y = 7$$
$$5 + y = 7$$
$$y = 2$$

The ordered pair $(5, 2)$ checks and is the solution.

**55.** $2(5a - 5b) = 10$,

$-5(6a + 2b) = 10$

First we remove parentheses.

$$10a - 10b = 10, \quad (1)$$
$$-30a - 10b = 10 \quad (2)$$

Then we multiply Equation (2) by $-1$ and add.

$$10a - 10b = 10$$
$$\underline{30a + 10b = -10}$$
$$40a \qquad = 0$$
$$a = 0$$

Substitute 0 for $a$ in Equation (1) and solve for $b$.

$$10 \cdot 0 - 10b = 10$$
$$-10b = 10$$
$$b = -1$$

The ordered pair $(0, -1)$ checks and is the solution.

## Exercise Set 7.4

**1. Familiarize**. We let $h$ = the number of bags of trash the Huxtables produce each month and $s$ = the number of bags of trash the Simpsons produce each month.

**Translate**. We reword and translate.

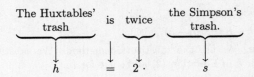

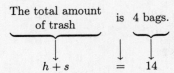

The resulting system is

$$h = 2s, \qquad (1)$$
$$h + s = 14. \qquad (2)$$

**Solve**. We use the substitution method. We substitute $2s$ for $h$ in Equation (2) and solve for $s$.

$$h + s = 14$$
$$2s + s = 14$$
$$3s = 14$$
$$s = \frac{14}{3}$$

We find $h$ by substituting $\frac{14}{3}$ for $s$ in Equation (1).

$$h = 2s$$
$$h = 2 \cdot \frac{14}{3}$$
$$h = \frac{28}{3}$$

**Check**. The amount of trash generated by the Huxtables each month, $\frac{28}{3}$ bags, is twice as much as the amount generated by the Simpsons, $\frac{14}{3}$ bags. The total amount of trash is $\frac{28}{3} + \frac{14}{3}$, or $\frac{42}{3}$, or 14 bags. These numbers check.

**State**. The Huxtables produce $\frac{28}{3}$ bags of trash, and the Simpsons produce $\frac{14}{3}$ bags.

**3. Familiarize**. Let $k$ = the age of the Kuyatt's house now and $m$ = the age of the Marconi's house now. Eight years ago the houses' ages were $k - 8$ and $m - 8$.

**Translate**. We reword and translate.

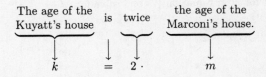

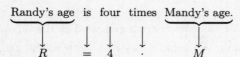

$$k - 8 \quad = \quad 3 \quad \cdot \quad m - 8$$

The resulting system is

$$k = 2m, \qquad (1)$$
$$k - 8 = 3(m - 8). \qquad (2)$$

**Solve.** We use the substitution method. We substitute $2m$ for $k$ in Equation (2) and solve for $m$.

$$k - 8 = 3(m - 8)$$
$$2m - 8 = 3(m - 8)$$
$$2m - 8 = 3m - 24$$
$$-8 = m - 24$$
$$16 = m$$

We find $k$ by substituting 16 for $m$ in Equation (1).

$$k = 2m$$
$$k = 2 \cdot 16$$
$$k = 32$$

**Check.** The age of the Kuyatt's house, 32 years, is twice the age of the Marconi's house, 16 years. Eight years ago, when the Kuyatt's house was 24 years old and the Marconi's house was 8 years old, the Kuyatt's house was three times as old as the Marconi's house. These numbers check.

**State.** The Kuyatt's house is 32 years old, and the Marconi's house is 16 years old.

**5. Familiarize.** Let $R = $ Randy's age now and $M = $ Mandy's age now. In twelve years their ages will be $R + 12$ and $M + 12$.

**Translate.** We reword and translate.

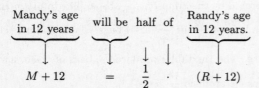

$$R \quad = \quad 4 \quad \cdot \quad M$$

$$M + 12 \quad = \quad \frac{1}{2} \quad \cdot \quad (R + 12)$$

The resulting system is

$$R = 4M, \qquad (1)$$
$$M + 12 = \frac{1}{2}(R + 12). \qquad (2)$$

**Solve.** We use the substitution method. We substitute $4M$ for $R$ in Equation (2) and solve for $M$.

$$M + 12 = \frac{1}{2}(R + 12)$$
$$M + 12 = \frac{1}{2}(4M + 12)$$
$$M + 12 = 2M + 6$$
$$12 = M + 6$$
$$6 = M$$

We find $R$ by substituting 6 for $M$ in Equation (1).

$$R = 4M$$
$$R = 4 \cdot 6$$
$$R = 24$$

**Check.** Randy's age now, 24, is 4 times 6, Mandy's age. In 12 yr, when Randy will be 36 and Mandy 18, Mandy's age will be half of Randy's age. These numbers check.

**State.** Randy is 24 years old now, and Mandy is 6.

**7. Familiarize.** Let $d$ represent the number of dimes and $q$ the number of quarters. Then, $10d$ represents the value of the dimes in cents, and $25q$ represents the value of the quarters in cents. The total value is $15.25, or 1525¢. The total number of coins is 103.

**Translate.**

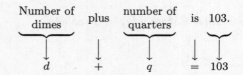

$$d \quad + \quad q \quad = \quad 103$$

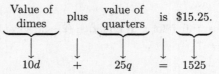

$$10d \quad + \quad 25q \quad = \quad 1525$$

The resulting system is

$$d + q = 103, \qquad (1)$$
$$10d + 25q = 1525. \qquad (2)$$

**Solve.** We use the addition method. We multiply Equation (1) by $-10$ and then add.

$$
\begin{array}{rl}
-10d - 10q = -1030 & \text{Multiplying by } -10 \\
\underline{10d + 25q = \phantom{-}1525} & \\
15q = \phantom{-}495 & \text{Adding} \\
q = \phantom{-}33 &
\end{array}
$$

Now we substitute 33 for $q$ in one of the original equations and solve for $d$.

$$d + q = 103 \quad (1)$$
$$d + 33 = 103 \quad \text{Substituting}$$
$$d = 70$$

**Check.** The number of dimes plus the number of quarters is $70 + 33$, or 103. The total value in cents is $10 \cdot 70 + 25 \cdot 33$, or $700 + 825$, or 1525. This is equal to $15.25. This checks.

**State.** There are 70 dimes and 33 quarters.

**9. Familiarize**. Let $p =$ the cost of one slice of pizza and $s =$ the cost of one soda.

**Translate**.

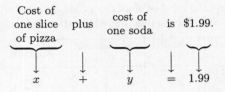

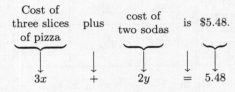

We multiply both equations by 100 to clear decimals. The resulting system is

$$100x + 100y = 199, \quad (1)$$
$$300x + 200y = 548. \quad (2)$$

**Solve**. We use the elimination method. We multiply Equation (1) by $-2$ and then add.

$$
\begin{array}{ll}
-200x - 200y = -398 & \text{Multiplying by } -2 \\
\underline{\phantom{-}300x + 200y = \phantom{-}548} & \\
\phantom{-}100x \phantom{+200y} = \phantom{-}150 & \\
\phantom{-100}x \phantom{+200y} = \phantom{-}1.5 &
\end{array}
$$

We go back to Equation (1) and substitute 1.5 for $x$.

$$100x + 100y = 199$$
$$100(1.5) + 100y = 199$$
$$150 + 100y = 199$$
$$100y = 49$$
$$y = 0.49$$

**Check**. If one slice of pizza costs $1.50 and one soda costs $0.49, then they cost $1.50 + $0.49, or $1.99 together. Also, three slices of pizza and two sodas cost $3(\$1.50) + 2(\$0.49)$, or $4.50 + $0.98, or $5.48. These numbers check.

**State**. One slice of pizza costs $1.50, and one soda costs $0.49.

**11. Familiarize**. Let $x =$ the number of cardholders tickets that were sold and $y =$ the number of non-cardholders tickets. We arrange the information in a table.

|  | Card-holders | Non-card-holders | Total |
|---|---|---|---|
| Price | $1.25 | $2 |  |
| Number sold | $x$ | $y$ | 203 |
| Money taken in | $1.25x$ | $2y$ | $310 |

**Translate**. The last two rows of the table give us two equations. The total number of tickets sold was 203, so we

have

$$x + y = 203.$$

The total amount of money collected was $310, so we have

$$1.25x + 2y = 310.$$

We can multiply the second equation on both sides by 100 to clear decimals. The resulting system is

$$x + y = 203, \quad (1)$$
$$125x + 200y = 31,000. \quad (2)$$

**Solve**. We use the elimination method. We multiply on both sides of Equation (1) by $-125$ and then add.

$$
\begin{array}{ll}
-125x - 125y = -25,375 & \text{Multiplying by } -125 \\
\underline{\phantom{-}125x + 200y = \phantom{-}31,000} & \\
\phantom{-125x +}75y = \phantom{-}5625 & \\
\phantom{-125x +}y = \phantom{-}75 &
\end{array}
$$

We go back to Equation (1) and substitute 75 for $y$.

$$x + y = 203$$
$$x + 75 = 203$$
$$x = 128$$

**Check**. The number of tickets sold was $128 + 75$, or 203. The money collected was $\$1.25(128) + \$2(75)$, or $160 + $150, or $310. These numbers check.

**State**. 128 cardholders tickets and 75 non-cardholders tickets were sold.

**13. Familiarize**. Let $b =$ the number of Upper Box tickets and $r =$ the number of Lower Reserved tickets that were bought. We arrange the information in a table.

|  | Upper Box | Lower Reserved | Total |
|---|---|---|---|
| Price | $10 | $9.50 |  |
| Number bought | $b$ | $r$ | 29 |
| Cost | $10b$ | $9.5r$ | $284 |

**Translate**. The last two rows of the table give us two equations. The total number of tickets bought was 29, so we have

$$b + r = 29.$$

The total cost of the tickets was $284, so we have

$$10b + 9.5r = 284.$$

We multiply the second equation on both sides by 10 to clear decimals. The resulting system is

$$b + r = 29, \quad (1)$$
$$100b + 95r = 2840. \quad (2)$$

**Solve**. We use the elimination method. We multiply on both sides of Equation (1) by $-95$ and then add.

$$
\begin{array}{ll}
-95b - 95r = -2755 & \text{Multiplying by } -95 \\
\underline{\phantom{-}100b + 95r = \phantom{-}2840} & \\
\phantom{-100b}5b \phantom{+ 95r} = \phantom{-}85 & \\
\phantom{-100b}b \phantom{+ 95r} = \phantom{-}17 &
\end{array}
$$

We go back to Equation (1) and substitute 17 for $b$.

$$b + r = 29$$
$$17 + r = 29$$
$$r = 12$$

**Check**. The number of tickets sold was $17 + 12$, or 29. The total cost was $\$10(17) + \$9.5(12)$, or $\$170 + \$114$, or $\$284$. These numbers check.

**State**. They bought 17 Upper Box and 12 Lower Reserved tickets.

15. **Familiarize**. We complete the table in the text. Note that $x$ represents the number of liters of solution $A$ to be used and $y$ represents the number of liters of solution $B$.

| Type of solution | $A$ | $B$ | Mixture |
|---|---|---|---|
| Amount of solution | $x$ | $y$ | 100 L |
| Percent of acid | 50% | 80% | 68% |
| Amount of acid in solution | 50%$x$ | 80%$y$ | 68% × 100, or 68 L |

Equation from first row:  $x + y = 100$

Equation from second row:  $50\%x + 80\%y = 68$

**Translate**. The first and third rows of the table give us two equations. Since the total amount of solution is 100 liters, we have

$$x + y = 100.$$

The amount of acid in the mixture is to be 68% of 100, or 68 liters. The amounts of acid from the two solutions are $50\%x$ and $80\%y$. Thus

$$50\%x + 80\%y = 68,$$
$$\text{or} \quad 0.5x + 0.8y = 68,$$
$$\text{or} \quad 5x + 8y = 680 \quad \text{Clearing decimals}$$

The resulting system is

$$x + y = 100, \quad (1)$$
$$5x + 8y = 680. \quad (2)$$

**Solve**. We use the elimination method. We multiply on both sides of Equation (1) by $-5$ and then add.

$$-5x - 5y = -500 \quad \text{Multiplying by } -5$$
$$\underline{5x + 8y = \phantom{-}680}$$
$$3y = \phantom{-}180$$
$$y = \phantom{-}60$$

We go back to Equation (1) and substitute 60 for $y$.

$$x + y = 100$$
$$x + 60 = 100$$
$$x = 40$$

**Check**. We consider $x = 40$ and $y = 60$. The sum is 100. Now 50% of 40 is 20 and 80% of 60 is 48. These add up to 68. The numbers check.

**State**. 40 liters of solution $A$ and 60 liters of solution $B$ should be used.

17. **Familiarize**. We let $x =$ the number of gallons of Skylite Pink and $y =$ the number of gallons of MacIntosh Red that should be used. We arrange the information in a table.

| Type of paint | Skylite Pink | MacIntosh Red | Mixture |
|---|---|---|---|
| Amount of paint | $x$ | $y$ | 1 gal |
| Percent of red pigment | 12.5% | 20% | 17% |
| Amount of red pigment in paint | 12.5%$x$ | 20%$y$ | 17% × 1, or 0.17 gal |

**Translate**. The first and last rows of the table give us two equations. The total amount of paint is 1 gal, so we have

$$x + y = 1.$$

The amount of red pigment in the mixture is to be 17% of 1 gal, or 0.17 gal. The amounts of red pigment from the two paints are $12.5\%x$ and $20\%y$. Thus

$$12.5\%x + 20\%y = 0.17, \quad \text{or}$$
$$0.125x + 0.2y = 0.17, \quad \text{or}$$
$$125x + 200y = 170 \quad \text{Clearing decimals}$$

The resulting system is

$$x + y = 1, \quad (1)$$
$$125x + 200y = 170. \quad (2)$$

**Solve**. We use the elimination method. We multiply on both sides of Equation (1) by $-125$ and then add.

$$-125x - 125y = -125 \quad \text{Multiplying by } -125$$
$$\underline{125x + 200y = \phantom{-}170}$$
$$75y = \phantom{-}45$$
$$y = \phantom{-}0.6$$

We go back to Equation (1) and substitute 0.6 for $y$.

$$x + y = 1$$
$$x + 0.6 = 1$$
$$x = 0.4$$

**Check**. The sum of 0.4 and 0.6 is 1. Also, 12.5% of 0.4 is 0.05 and 20% of 0.6 is 0.12, and $0.05 + 0.12 = 0.17$. These numbers check.

**State**. Gayle should pick up 0.4 gal of Skylite Pink paint and 0.6 gal of MacIntosh Red paint.

19. **Familiarize**. We complete the table in the text. Note that $x$ represents the number of pounds of Brazilian coffee to be used and $y$ represents the number of pounds of Turkish coffee.

| Type of coffee | Brazilian | Turkish | Mixture |
|---|---|---|---|
| Cost of coffee | $5 | $8 | $7 |
| Amount (in pounds) | $x$ | $y$ | 300 |
| Mixture | $5x$ | $8y$ | $7(300)$, or $2100 |

Equation from second row:  $x + y = 300$

Equation from third row:   $5x + 8y = 2100$

**Translate.** The second and third rows of the table give us two equations. Since the total amount of the mixture is 300 lb, we have

$x + y = 300.$

The value of the Brazilian coffee is $5x$ ($x$ lb at $5 per pound), the value of the Turkish coffee is $8y$ ($y$ lb at $8 per pound), and the value of the mixture is $7(300) or $2100. Thus we have

$5x + 8y = 2100.$

The resulting system is

$x + y = 300,$   (1)

$5x + 8y = 2100.$   (2)

**Solve.** We use the elimination method. We multiply on both sides of Equation (1) by $-5$ and then add.

$$-5x - 5y = -1500 \quad \text{Multiplying by } -5$$
$$\underline{5x + 8y = \phantom{-}2100}$$
$$3y = \phantom{-}600$$
$$y = \phantom{-}200$$

We go back to Equation (1) and substitute 200 for $y$.

$$x + y = 300$$
$$x + 200 = 300$$
$$x = 100$$

**Check.** The sum of 100 and 200 is 300. The value of the mixture is $5(100) + $8(200), or $500 + $1600, or $2100. These values check.

**State.** 100 lb of Brazilian coffee and 200 lb of Turkish coffee should be used.

21. **Familiarize.** We arrange the information in a table. Let $x =$ the amount of seed $A$ and $y =$ the amount of seed $B$ to be used.

| Type of seed | $A$ | $B$ | Mixture |
|---|---|---|---|
| Cost of seed | $2.50 | $1.75 | $2.14 |
| Amount (in pounds) | $x$ | $y$ | 75 |
| Mixture | $2.50x$ | $1.75y$ | $2.14(75)$, or $160.50 |

**Translate.** The last two rows of the table give us two equations.

Since the total amount of grass seed is 75 lb, we have

$x + y = 75.$

The value of seed $A$ is $2.50x$ ($x$ lb at $2.50 per pound), and the value of seed $B$ is $1.75y$ ($y$ lb at $1.75 per pound). The value of the mixture is $2.14(75), or $160.50, so we have

$2.50x + 1.75y = 160.50,$  or

$250x + 175y = 16,050$   Clearing decimals

The resulting system is

$x + y = 75,$   (1)

$250x + 175y = 16,050.$   (2)

**Solve.** We use the elimination method. We multiply Equation (1) by $-250$ and then add.

$$-250x - \phantom{0}250y = -18,750$$
$$\underline{250x + \phantom{0}175y = \phantom{-0}16,050}$$
$$-75y = \phantom{-0}-2700$$
$$y = \phantom{-00000}36$$

Next we substitute 36 for $y$ in one of the original equations and solve for $x$.

$$x + y = 75 \quad (1)$$
$$x + 36 = 75$$
$$x = 39$$

**Check.** We consider $x = 39$ lb and $y = 36$ lb. The sum is 75 lb. The value of the mixture is $2.50(39) + $1.75(36), or $97.50 + $63.00, or $160.50. These values check.

**State.** 39 lb of seed $A$ and 36 lb of seed $B$ should be used.

23. **Familiarize.** We arrange the information in a table. Let $a =$ the number of type $A$ questions and $b =$ the number of type $B$ questions.

| Type of question | $A$ | $B$ | Mixture (Test) |
|---|---|---|---|
| Number | $a$ | $b$ | 16 |
| Time | 3 min | 6 min | |
| Value | 10 points | 15 points | |
| Mixture (Test) | $3a$ min, $10a$ points | $6b$ min, $15b$ points | 60 min, 180 points |

**Translate.** The table actually gives us three equations. Since the total number of questions is 16, we have

$a + b = 16.$

The total time is 60 min, so we have

$3a + 6b = 60.$

The total number of points is 180, so we have

$10a + 15b = 180.$

The resulting system is

$a + b = 16,$   (1)

$3a + 6b = 60,$   (2)

$10a + 15b = 180.$   (3)

*Solve*. We will solve the system composed of Equations (1) and (2) and then check to see that this solution also satisfies Equation (3). We multiply equation (1) by $-3$ and add.

$$-3a - 3b = -48$$
$$\underline{3a + 6b = \phantom{0}60}$$
$$3b = \phantom{0}12$$
$$b = \phantom{00}4$$

Now we substitute 4 for $b$ in Equation (1) and solve for $a$.

$$a + b = 16$$
$$a + 4 = 16$$
$$a = 12$$

*Check*. We consider $a = 12$ questions and $b = 4$ questions. The total number of questions is 16. The time required is $3 \cdot 12 + 6 \cdot 4$, or $36 + 24$, or 60 min. The total points are $10 \cdot 12 + 15 \cdot 4$, or $120 + 60$, or 180. These values check.

*State*. 12 questions of type $A$ and 4 questions of type $B$ were answered correctly.

**25.** *Familiarize*. We let $x =$ the number of pages in large type and $y =$ the number of pages in small type. We arrange the information in a table.

| Size of type | Large | Small | Mixture (Book) |
|---|---|---|---|
| Words per page | 1300 | 1850 | |
| Number of pages | $x$ | $y$ | 12 |
| Number of words | $1300x$ | $1850y$ | $18,526$ |

*Translate*. The last two rows of the table give us two equations. The total number of pages in the document is 12, so we have

$$x + y = 12.$$

The number of words on the pages with large type is $1300x$ ($x$ pages with 1300 words per page), and the number of words on the pages with small type is $1850y$ ($y$ pages with 1850 words per page). The total number of words is 18,526, so we have

$$1300x + 1850y = 18,526.$$

The resulting system is

$$x + y = 12, \tag{1}$$
$$1300x + 1850y = 18,526. \tag{2}$$

*Solve*. We use the elimination method. We multiply on both sides of Equation (1) by $-1300$ and then add.

$$-1300x - 1300y = -15,600 \quad \text{Multiplying by } -1300$$
$$\underline{1300x + 1850y = \phantom{0}18,526}$$
$$550y = \phantom{0000}2926$$
$$y = \phantom{0000}5.32$$

We go back to Equation (1) and substitute 5.32 for $y$.

$$x + y = 12$$
$$x + 5.32 = 12$$
$$x = 6.68$$

*Check*. The sum of 6.68 and 5.32 is 12. The number of words in large type is $1300(6.68)$, or 8684, and the number of words in small type is $1850(5.32)$, or 9842. Then the total number of words is $8684 + 9842$, or 18,526. These numbers check.

*State*. There were 6.68, or $6\frac{17}{25}$ pages, in large type and 5.32, or $5\frac{8}{25}$ pages, in small type.

**27.** *Familiarize*. In a table we arrange the information regarding the solution <u>after</u> some of the 30% solution is drained and replaced with pure antifreeze. We let $x$ represent the amount of the original (30%) solution remaining, and we let $y$ represent the amount of the 30% mixture that is drained and replaced with pure antifreeze.

| Type of solution | Original (30%) | Pure antifreeze | Mixture |
|---|---|---|---|
| Amount of solution | $x$ | $y$ | 16 |
| Percent of antifreeze | 30% | 100% | 50% |
| Amount of antifreeze in solution | $0.3x$ | $1 \cdot y$, or $y$ | $0.5(16)$, or 8 |

*Translate*. The table gives us two equations.

Amount of solution: $x + y = 16$

Amount of antifreeze in solution: $0.3x + y = 8$, or $3x + 10y = 80$

The resulting system is

$$x + y = 16, \tag{1}$$
$$3x + 10y = 80. \tag{2}$$

*Solve*. We multiply Equation (1) by $-3$ and then add.

$$-3x - \phantom{0}3y = -48$$
$$\underline{3x + 10y = \phantom{0}80}$$
$$7y = \phantom{0}32$$
$$y = \phantom{0}\frac{32}{7}, \text{ or } 4\frac{4}{7}$$

Then we substitute $4\frac{4}{7}$ for $y$ in Equation (1) and solve for $x$.

$$x + y = 16$$
$$x + 4\frac{4}{7} = 16$$
$$x = 11\frac{3}{7}$$

*Check*. When $x = 11\frac{3}{7}$ L and $y = 4\frac{4}{7}$ L, the total is 16 L. The amount of antifreeze in the mixture is $0.3\left(11\frac{3}{7}\right) + 4\frac{4}{7}$, or $\frac{3}{10} \cdot \frac{80}{7} + \frac{32}{7}$, or $\frac{24}{7} + \frac{32}{7} = \frac{56}{7}$, or 8 L. This is 50% of

16 L, so the numbers check.

*State*. $4\frac{4}{7}$ of the original mixture should be drained and replaced with pure antifreeze.

29. *Familiarize*. Let $x$ = the number of people who paid full price and $y$ = the number of people who used a coupon. The admission price with a 20% off coupon is 80% of $8.50, or 0.8($8.50), or $6.80. We arrange the information in a table.

| | Full price | Coupon | Total |
|---|---|---|---|
| Admission | $8.50 | $6.80 | |
| Number admitted | $x$ | $y$ | 1315 |
| Receipts | $8.50x$ | $6.80y$ | $10,242.50 |

*Translate*. The table gives us two equations.

Number admitted: $x + y = 1315$

Receipts: $8.50x + 6.80y = 10,242.50$

We multiply the second equation by 10 to clear decimals. The resulting system is

$$x + y = 1315, \qquad (1)$$
$$85x + 68y = 102,425. \quad (2)$$

*Solve*. We multiply Equation (1) by $-68$ and then add.

$$-68x - 68y = -89,420$$
$$\underline{85x + 68y = 102,425}$$
$$17x \qquad = 13,005$$
$$x = 765$$

We go back to Equation (1) and substitute 765 for $x$.

$$x + y = 1315$$
$$765 + y = 1315$$
$$y = 550$$

*Check*. The total number of admissions was $765 + 550$, or 1315. The receipts from the full-price admissions were $8.50(765), or $6502.50, and the receipts from the coupon admissions were $6.80(550), or $3740. The total receipts were $6502.50 + $3740, or $10,242.50. These numbers check.

*State*. 550 people used a coupon.

31. *Familiarize*. We arrange the information in a table. Let $x$ = the number of liters of skim milk and $y$ = the number of liters of 3.2% milk.

| Type of milk | 4.6% | Skim | 3.2% (Mixture) |
|---|---|---|---|
| Amount of milk | 100 L | $x$ | $y$ |
| Percent of butterfat | 4.6% | 0% | 3.2% |
| Amount of butterfat in milk | 4.6% × 100, or 4.6 L | 0% · $x$, or 0 L | 3.2%$y$ |

*Translate*. The first and third rows of the table give us two equations.

Amount of milk: $100 + x = y$

Amount of butterfat: $4.6 + 0 = 3.2\%y$, or $4.6 = 0.032y$.

The resulting system is

$$100 + x = y,$$
$$4.6 = 0.032y.$$

*Solve*. We solve the second equation for $y$.

$$4.6 = 0.032y$$
$$\frac{4.6}{0.032} = y$$
$$143.75 = y$$

We substitute 143.75 for $y$ in the first equation and solve for $x$.

$$100 + x = y$$
$$100 + x = 143.75$$
$$x = 43.75$$

*Check*. We consider $x = 43.75$ L and $y = 143.75$ L. The difference between 143.75 L and 43.75 L is 100 L. There is no butterfat in the skim milk. There are 4.6 liters of butterfat in the 100 liters of the 4.6% milk. Thus there are 4.6 liters of butterfat in the mixture. This checks because 3.2% of 143.75 is 4.6.

*State*. 43.75 L of skim milk should be used.

33. *Familiarize*. Let $x$ represent the part invested at 12% and $y$ represent the part invested at 13%. The interest earned from the 12% investment is $12\% \cdot x$. The interest earned from the 13% investment is $13\% \cdot y$. The total investment is $27,000, and the total interest earned is $3385.

*Translate*.

$$x + y = 27,000,$$
$$12\%x + 13\%y = \$3385, \text{ or } 12x + 13y = 338,500$$

*Solve*. Multiply the first equation by $-12$ and add.

$$-12x - 12y = -324,000$$
$$\underline{12x + 13y = 338,500}$$
$$y = 14,500$$

Substitute 14,500 for $y$ in the first equation and solve for $x$.

$$x + y = 27,000$$
$$x + 14,500 = 27,000$$
$$x = 12,500$$

*Check*. We consider $12,500 invested at 12% and $14,500 invested at 13%. The sum of the investments is $27,000. The interest earned is $12\% \cdot 12,500 + 13\% \cdot 14,500$, or $1500 + 1885$, or $3385. These numbers check.

*State*. $12,500 was invested at 12%, and $14,500 was invested at 13%.

35. *Familiarize*. We let $x$ = the number of ounces of flavoring that should be used and $y$ = the number of ounces of sugar.

We arrange the information in a table.

|  | Flavoring | Sugar | Mixture |
|---|---|---|---|
| Number of ounces | $x$ | $y$ | 20 |
| Price per ounce | $1.45 | $0.05 | $0.106 |
| Cost | $1.45x$ | $0.05y$ | 20($0.106), or $2.12 |

**Translate**. The table gives us two equations.

Number of ounces: $x + y = 20$

Cost: $1.45x + 0.05y = 2.12$.

We multiply the second equation by 100 to clear decimals. The resulting system is

$$x + y = 20, \quad (1)$$
$$145x + 5y = 212. \quad (2)$$

**Solve**. We multiply Equation (1) by $-5$ and then add.

$$-5x - 5y = -100$$
$$\underline{145x + 5y = \phantom{-}212}$$
$$140x \phantom{+ 5y} = \phantom{-}112$$
$$x = \phantom{-}0.8$$

We go back to Equation (1) and substitute 0.8 for $x$.

$$x + y = 20$$
$$0.8 + y = 20$$
$$y = 19.2$$

**Check**. The total weight is $0.8 + 19.2$, or 20 oz. The cost of 0.8 oz of flavoring is $145(0.8)$, or $1.16, and the cost of 19.2 oz of sugar is $0.05(19.2)$, or $0.96. The total cost of the mixture is $1.16 + $0.96, or $2.12. These numbers check.

**State**. 0.8 oz of flavoring and 19.2 oz of sugar should be used.

**37. Familiarize**. We let $x = $ the cost of the glove and $y = $ the cost of the ball. Then $x + 40 = $ the cost of the bat.

**Translate**. We reword and translate.

Cost of bat + cost of ball + cost of glove is $204.85.

$$(x + 40) + y + x = 204.85$$

Cost of glove is $75 more than cost of ball.

$$x = 75 + y$$

We multiply the first equation by 100 to clear decimals and put both equations in the form $Ax + By = C$. The resulting system is

$$200x + 100y = 16,485, \quad (1)$$
$$x - y = 75. \quad (2)$$

**Solve**. We multiply Equation (2) by 100 and then add.

$$200x + 100y = 16,485$$
$$\underline{100x - 100y = \phantom{16,}7500}$$
$$300x \phantom{+ 100y} = 23,985$$
$$x = \phantom{2}79.95$$

We go back to Equation (2) and substitute 79.95 for $x$.

$$x - y = 75$$
$$79.95 - y = 75$$
$$-y = -4.95$$
$$y = 4.95$$

**Check**. If the cost of the glove is $79.95, then the cost of the bat is $79.95 + $40, or $119.95. The total cost of the three items is $79.95 + $4.95 + $119.95, or $204.85. The cost of the glove, $79.95, is $75 more than the cost of the ball, $4.95. The cost of the bat, $119.95, is $40 more than the cost of the glove, $79.95.

**State**. The bat costs $119.95, the ball costs $4.95, and the glove costs $79.95.

## Exercise Set 7.5

**1. Familiarize**. We first make a drawing.

| | 30 mph | |
|---|---|---|
| Slow car | $t$ hours | $d$ miles |

| | 46 mph | |
|---|---|---|
| Fast car | $t$ hours | $d + 72$ miles |

We let $d = $ the distance the slow car travels. Then $d + 72 = $ the distance the fast car travels. We call the time $t$. We complete the table in the text, filling in the distances as well as the other information.

$$d = r \cdot t$$

| | Distance | Speed | Time |
|---|---|---|---|
| Slow car | $d$ | 30 | $t$ |
| Fast car | $d + 72$ | 46 | $t$ |

**Translate**. We get an equation $d = rt$ from each row of the table. Thus we have

$$d = 30t, \quad (1)$$
$$d + 72 = 46t. \quad (2)$$

**Solve**. We use the substitution method. We substitute $30t$ for $d$ in Equation (2).

$$d + 72 = 46t$$
$$30t + 72 = 46t \quad \text{Substituting}$$
$$72 = 16t \quad \text{Subtracting } 30t$$
$$4.5 = t \quad \text{Dividing by 16}$$

**Check**. In 4.5 hr the slow car travels 30(4.5), or 135 mi, and the fast car travels 46(4.5), or 207 mi. Since 207 is 72 more than 135, our result checks.

**State**. The trains will be 72 mi apart in 4.5 hr.

**3. Familiarize**. First make a drawing.

| Station | 72 mph | |
|---|---|---|
| Slow train | $t + 3$ hours | $d$ miles |

| Station | 120 mph | |
|---|---|---|
| Fast train | $t$ hours | $d$ miles |

Trains meet here.

From the drawing we see that the distances are the same. Let's call the distance $d$. Let $t$ represent the time for the faster train and $t+3$ represent the time for the slower train. We complete the table in the text.

$$d = r \cdot t$$

| | Distance | Speed | Time |
|---|---|---|---|
| Slow train | $d$ | 72 | $t + 3$ |
| Fast train | $d$ | 120 | $t$ |

Equation from first row: $d = 72(t + 3)$

Equation from second row: $d = 120t$

**Translate**. Using $d = rt$ in each row of the table, we get the following system of equations:

$$d = 72(t + 3), \quad (1)$$
$$d = 120t. \quad (2)$$

**Solve**. Substitute $120t$ for $d$ in Equation (1) and solve for $t$.

$$d = 72(t + 3)$$
$$120t = 72(t + 3) \quad \text{Substituting}$$
$$120t = 72t + 216$$
$$48t = 216$$
$$t = \frac{216}{48}$$
$$t = 4.5$$

**Check**. When $t = 4.5$ hours, the faster train will travel $120(4.5)$, or 540 mi, and the slower train will travel $72(7.5)$, or 540 mi. In both cases we get the distance 540 mi.

**State**. In 4.5 hours after the second train leaves, the second train will overtake the first train. We can also state the answer as 7.5 hours after the first train leaves.

**5. Familiarize**. We first make a drawing.

| With the current | $r + 6$ | |
|---|---|---|
| 4 hours | | $d$ kilometers |

| Against the current | $r - 6$ | |
|---|---|---|
| 10 hours | | $d$ kilometers |

From the drawing we see that the distances are the same. Let $d$ represent the distance. Let $r$ represent the speed of the canoe in still water. Then, when the canoe is traveling with the current, its speed is $r + 6$. When it is traveling against the current, its speed is $r - 6$. We complete the table in the text.

$$d = r \cdot t$$

| | Distance | Speed | Time |
|---|---|---|---|
| With current | $d$ | $r + 6$ | 4 |
| Against current | $d$ | $r - 6$ | 10 |

Equation from first row: $d = (r + 6)4$

Equation from second row: $d = (r - 6)10$

**Translate**. Using $d = rt$ in each row of the table, we get the following system of equations:

$$d = (r + 6)4, \quad (1)$$
$$d = (r - 6)10 \quad (2)$$

**Solve**. Substitute $(r + 6)4$ for $d$ in Equation (2) and solve for $r$.

$$d = (r - 6)10$$
$$(r + 6)4 = (r - 6)10 \quad \text{Substituting}$$
$$4r + 24 = 10r - 60$$
$$84 = 6r$$
$$14 = r$$

**Check**. When $r = 14$, $r + 6 = 20$ and $20 \cdot 4 = 80$, the distance. When $r = 14$, $r - 6 = 8$ and $8 \cdot 10 = 80$. In both cases, we get the same distance.

**State**. The speed of the canoe in still water is 14 km/h.

**7. Familiarize**. First make a drawing.

| Passenger | 96 km/h | |
|---|---|---|
| $t - 2$ hours | | $d$ kilometers |

| Freight | 64 km/h | |
|---|---|---|
| $t$ hours | | $d$ kilometers |

Central City                         Clear Creek

From the drawing we see that the distances are the same. Let $d$ represent the distance. Let $t$ represent the time for the freight train. Then the time for the passenger train is $t - 2$. We organize the information in a table.

$$d = r \cdot t$$

| | Distance | Speed | Time |
|---|---|---|---|
| Passenger | $d$ | 96 | $t - 2$ |
| Freight | $d$ | 64 | $t$ |

**Translate**. From each row of the table we get an equation.

$$d = 96(t - 2), \quad (1)$$
$$d = 64t \quad (2)$$

*Solve*. Substitute $64t$ for $d$ in Equation (1) and solve for $t$.

$$d = 96(t - 2)$$
$$64t = 96(t - 2) \quad \text{Substituting}$$
$$64t = 96t - 192$$
$$192 = 32t$$
$$6 = t$$

Next we substitute 6 for $t$ in one of the original equations and solve for $d$.

$$d = 64t \quad \text{Equation (2)}$$
$$d = 64 \cdot 6 \quad \text{Substituting}$$
$$d = 384$$

*Check*. If the time is 6 hr, then the distance the passenger train travels is $96(6 - 2)$, or 384 km. The freight train travels $64(6)$, or 384 km. The distances are the same.

*State*. It is 384 km from Central City to Clear Creek.

**9.** *Familiarize*. We first make a drawing.

Against the wind        $r - w$

2 hours               600 miles

With the wind          $r + w$

$1\frac{2}{3}$ hours          600 miles

We let $r$ represent the speed of the airplane in still air and $w$ represent the speed of the wind. Then when flying against a head wind, the rate is $r - w$, and when flying with the wind, the rate is $r + w$. We organize the information in a chart.

|         | Distance | Speed | Time |
|---------|----------|-------|------|
| Against | 600      | $r - w$ | 2 |
| With    | 600      | $r + w$ | $1\frac{2}{3}$, or $\frac{5}{3}$ |

$$d = r \cdot t$$

*Translate*. Using $d = rt$ in each row of the table, we get the following system of equations:

$$600 = (r - w)2 \quad \text{or} \quad 300 = r - w \quad \text{Multiplying by } \frac{1}{2}$$
$$600 = (r + w)\frac{5}{3} \quad \text{or} \quad 360 = r + w \quad \text{Multiplying by } \frac{3}{5}$$

*Solve*. We use the elimination method with the resulting system.

$$300 = r - w$$
$$\underline{360 = r + w}$$
$$660 = 2r$$
$$330 = r$$

Next we substitute 330 for $r$ in the second equation and solve for $w$.

$$360 = r + w$$
$$360 = 300 + w \quad \text{Substituting}$$
$$30 = w$$

*Check*. If $r = 330$ and $w = 30$, then $r - w = 300$, and $r + w = 360$. If the plane flies 2 hours against the wind, it travels $300 \cdot 2$ or 600 mi. If the plane flies $1\frac{2}{3}$ hours with the wind, it travels $360 \cdot \frac{5}{3}$, or 600 mi. All values check.

*State*. The speed of the plane in still air is 330 mph.

**11.** *Familiarize*. We first make a drawing.

230 ft/min

Toddler        $t + 1$ min        $d$ ft

660 ft/min

Mother        $t$ min        $d$ ft

They meet here.

From the drawing we see that the distances are the same. Let's call the distance $d$. Let $t =$ the time the mother runs. Then $t + 1 =$ the time the toddler runs. We arrange the information in a table.

$$d = r \cdot t$$

|         | Distance | Speed | Time |
|---------|----------|-------|------|
| Toddler | $d$      | 230   | $t + 1$ |
| Mother  | $d$      | 660   | $t$ |

*Translate*. Using $d = rt$ in each row of the table we get two equations.

$$d = 230(t + 1), \quad (1)$$
$$d = 660t \quad (2)$$

*Solve*. Substitute $660t$ for $d$ in Equation (1) and solve for $t$.

$$d = 230(t + 1)$$
$$660t = 230(t + 1) \quad \text{Substituting}$$
$$660t = 230t + 230$$
$$430t = 230$$
$$t = \frac{230}{430}, \text{ or } \frac{23}{43}$$

*Check*. When $t = \frac{23}{43}$ the toddler will travel $230\left(1\frac{23}{43}\right)$, or $230 \cdot \frac{66}{43}$, or $\frac{15,180}{43}$ ft and the mother will travel $660 \cdot \frac{23}{43}$, or $\frac{15,180}{43}$ ft. Since the distances are the same, our result checks.

*State*. The mother will overtake the toddler $\frac{23}{43}$ min after she starts running. We can also state the answer as $1\frac{23}{43}$ min after the toddler starts running.

**13.** *Familiarize*. First make a drawing.

Home   $t$ hr   45 mph | $(2 - t)$ hr  6 mph  Work
Motorcycle distance   |   Walking distance

$\longleftarrow$ 25 miles $\longrightarrow$

Let $t$ represent the time the motorcycle was driven. Then $2 - t$ represents the time the rider walked. We organize the information in a table.

| | Distance | Speed | Time |
|---|---|---|---|
| Motorcycling | Motorcycle distance | 45 | $t$ |
| Walking | Walking distance | 6 | $2 - t$ |
| Total | 25 | | |

$$d = r \cdot t$$

**Translate.** From the drawing we see that

Motorcycle distance + Walking distance = 25

Then using $d = rt$ in each row of the table we get

$$45t + 6(2 - t) = 25$$

**Solve.** We solve this equation for $t$.

$$45t + 12 - 6t = 25$$
$$39t + 12 = 25$$
$$39t = 13$$
$$t = \frac{13}{39}$$
$$t = \frac{1}{3}$$

**Check.** The problem asks us to find how far the motorcycle went before it broke down. If $t = \frac{1}{3}$, then $45t$ (the distance the motorcycle traveled) $= 45 \cdot \frac{1}{3}$, or 15 and $6(2 - t)$ (the distance walked) $= 6\left(2 - \frac{1}{3}\right) = 6 \cdot \frac{5}{3}$, or 10. The total of these distances is 25, so $\frac{1}{3}$ checks.

**State.** The motorcycle went 15 miles before it broke down.

**15.** $\dfrac{x^2 - 3x - 10}{x^2 - 2x - 15} = \dfrac{(x-5)(x+2)}{(x-5)(x+3)}$

$\qquad = \dfrac{(x-5)(x+2)}{(x-5)(x+3)}$

$\qquad = \dfrac{x+2}{x+3}$

**17.** $\dfrac{6x^2 + 15x - 36}{2x^2 - 5x + 3} = \dfrac{3(2x^2 + 5x - 12)}{(2x-3)(x-1)}$

$\qquad = \dfrac{3(2x-3)(x+4)}{(2x-3)(x-1)}$

$\qquad = \dfrac{3(2x-3)(x+4)}{(2x-3)(x-1)}$

$\qquad = \dfrac{3(x+4)}{x-1}$

**19.** $\dfrac{5x^8 y^4}{10x^3 y} = \dfrac{5 \cdot x^3 \cdot x^5 \cdot y \cdot y^3}{5 \cdot 2 \cdot x^3 \cdot y}$

$\qquad = \dfrac{5 \cdot x^3 \cdot x^5 \cdot y \cdot y^3}{5 \cdot 2 \cdot x^3 \cdot y}$

$\qquad = \dfrac{x^5 y^3}{2}$

**21. Familiarize.** We arrange the information in a table. Let $d$ = the length of the route and $t$ = Lindbergh's time. Note that 16 hr and 57 min = $16\frac{57}{60}$ hr = 16.95 hr.

$$d = r \cdot t$$

| | Distance | Speed | Time |
|---|---|---|---|
| Lindbergh | $d$ | 107.4 | $t$ |
| Hughes | $d$ | 217.1 | $t - 16.95$ |

**Translate.** From the rows of the table we get two equations.

$$d = 107.4t, \qquad (1)$$
$$d = 217.1(t - 16.95) \quad (2)$$

**Solve.** We substitute $107.4t$ for $d$ in Equation (2) and solve for $t$.

$$d = 217.1(t - 16.95)$$
$$107.4t = 217.1(t - 16.95)$$
$$107.4t = 217.1t - 3679.845$$
$$-109.7t = -3679.845$$
$$t \approx 33.54$$

Now we go back to Equation (1) and substitute 33.54 for $t$.

$$d = 107.4t$$
$$d = 107.4(33.54)$$
$$d \approx 3602$$

**Check.** When $t \approx 33.54$, Lindbergh traveled $107.4(33.54) \approx 3602$ mi, and Hughes traveled $217.1(16.59) \approx 3602$ mi. Since the distances are the same, our result checks.

**State.** The route was 3602 mi long. (Answers may vary slightly due to rounding differences.)

**23. Familiarize.** We arrange the information in a table. Let's call the distance $d$. When the riverboat is traveling upstream its speed is $12 - 4$, or 8 mph. Its speed traveling downstream is $12 + 4$, or 16 mph.

$$d = r \cdot t$$

| | Distance | Speed | Time |
|---|---|---|---|
| Upstream | $d$ | 8 | Time upstream |
| Downstream | $d$ | 16 | Time downstream |
| Total | | | 1 |

**Translate.** From the table we see that (Time upstream) + (Time downstream) = 1. Then using $d = rt$, in the form $\dfrac{d}{r} = t$, in each row of the table we get

$$\frac{d}{8} + \frac{d}{16} = 1.$$

*Solve*. We solve the equation. The LCM is 16.

$$\frac{d}{8} + \frac{d}{16} = 1$$

$$16\left(\frac{d}{8} + \frac{d}{16}\right) = 16 \cdot 1$$

$$16 \cdot \frac{d}{8} + 16 \cdot \frac{d}{16} = 16$$

$$2d + d = 16$$

$$3d = 16$$

$$d = \frac{16}{3}, \text{ or } 5\frac{1}{3}$$

*Check*. When $d = \frac{16}{3}$,

(Time upstream) + (Time downstream)

$$= \frac{\frac{16}{3}}{8} + \frac{\frac{16}{3}}{16}$$

$$= \frac{16}{3} \cdot \frac{1}{8} + \frac{16}{3} \cdot \frac{1}{16}$$

$$= \frac{2}{3} + \frac{1}{3}$$

$$= 1 \text{ hr}$$

Thus the distance of $\frac{16}{3}$ mi, or $5\frac{1}{3}$ mi checks.

*State*. The pilot should travel $5\frac{1}{3}$ mi upstream before turning around.

# Chapter 8

# Radical Expressions and Equations

## Exercise Set 8.1

**1.** The square roots of 4 are 2 and $-2$, because $2^2 = 4$ and $(-2)^2 = 4$.

**3.** The square roots of 9 are 3 and $-3$, because $3^2 = 9$ and $(-3)^2 = 9$.

**5.** The square roots of 100 are 10 and $-10$, because $10^2 = 100$ and $(-10)^2 = 100$.

**7.** The square roots of 169 are 13 and $-13$, because $13^2 = 169$ and $(-13)^2 = 169$.

**9.** $\sqrt{4} = 2$, taking the principal square root.

**11.** $\sqrt{9} = 3$, so $-\sqrt{9} = -3$.

**13.** $\sqrt{36} = 6$, so $-\sqrt{36} = -6$.

**15.** $\sqrt{225} = 15$, so $-\sqrt{225} = -15$.

**17.** $\sqrt{361} = 19$, taking the principal square root.

**19.** 2.236

**21.** 4.123

**23.** 9.539

**25.** a) We substitute 25 into the formula:
$$N = 2.5\sqrt{25} = 2.5(5) = 12.5 \approx 13$$

   b) We substitute 89 into the formula and use Table 2 or a calculator to find an approximation.
$$N = 2.5\sqrt{89} \approx 2.5(9.434) = 23.585 \approx 24$$

**27.** The radicand is the expression under the radical, $a - 4$.

**29.** The radicand is the expression under the radical, $t^2 + 1$.

**31.** The radicand is the expression under the radical, $\dfrac{3}{x+2}$.

**33.** No, because the radicand is negative

**35.** Yes, because the radicand is nonnegative

**37.** $\sqrt{c^2} = c$    Since $c$ is assumed to be nonnegative

**39.** $\sqrt{9x^2} = \sqrt{(3x)^2} = 3x$    Since $3x$ is assumed to be nonnegative

**41.** $\sqrt{(ab)^2} = ab$

**43.** $\sqrt{(34d)^2} = 34d$

**45.** $\sqrt{(x+3)^2} = x + 3$

**47.** $\sqrt{a^2 - 10a + 25} = \sqrt{(a-5)^2} = a - 5$

**49.** $\sqrt{4a^2 - 20a + 25} = \sqrt{(2a-5)^2} = 2a - 5$

**51.** *Familiarize*. This problem states that we have direct variation between $F$ and $I$. Thus, an equation $F = kI$, $k > 0$, applies. As the income increases, the amount spent on food increases.

*Translate*. We write an equation of variation.

Amount spent on food varies directly as the income.

This translates to $F = kI$.

*Solve*.

a) First find an equation of variation.
$$F = kI$$
$$5096 = k \cdot 19{,}600 \quad \text{Substituting 5096 for } F$$
$$\text{and 19,600 for } I$$
$$\frac{5096}{19{,}600} = k$$
$$0.26 = k$$

   The equation of variation is $F = 0.26I$.

b) We use the equation to find how much a family spends on food when their income is \$20,500.
$$F = 0.26I$$
$$F = 0.26(\$20{,}500) \quad \text{Substituting \$20,500 for } I$$
$$F = \$5330$$

*Check*. Let us do some reasoning about the answer. The income increased from \$19,600 to \$20,500. Similarly, the amount spend on food increased from \$5096 to \$5330. This is what we would expect with direct variation.

*State*. The amount spent on food is \$5330.

**53.**
$$\frac{x^2 - x - 2}{x - 1} \div \frac{x - 2}{x^2 - 1} = \frac{x^2 - x - 2}{x - 1} \cdot \frac{x^2 - 1}{x - 2}$$
$$= \frac{(x^2 - x - 2)(x^2 - 1)}{(x - 1)(x - 2)}$$
$$= \frac{(x - 2)(x + 1)(x + 1)(x - 1)}{(x - 1)(x - 2)}$$
$$= \frac{(x - 2)(x + 1)(x + 1)(x - 1)}{(x - 1)(x - 2)(1)}$$
$$= (x + 1)(x + 1), \text{ or } (x + 1)^2$$

**55.** $\dfrac{x^4-16}{x^4-1} \div \dfrac{x^2+4}{x^2+1} = \dfrac{x^4-16}{x^4-1} \cdot \dfrac{x^2+1}{x^2+4}$

$= \dfrac{(x^4-16)(x^2+1)}{(x^4-1)(x^2+4)}$

$= \dfrac{(x^2+4)(x^2-4)(x^2+1)}{(x^2+1)(x^2-1)(x^2+4)}$

$= \dfrac{(x^2+4)(x^2+1)}{(x^2+4)(x^2+1)} \cdot \dfrac{x^2-4}{x^2-1}$

$= \dfrac{x^2-4}{x^2-1}$, or $\dfrac{(x+2)(x-2)}{(x+1)(x-1)}$

**57.** $\sqrt{3^2+4^2} = \sqrt{9+16} = \sqrt{25} = 5$

**59.** 3.578

**61.** 32.309

**63.** $\sqrt{y^2} = -7$ has no solution, because the principal square root is nonnegative.

**65.** *Familiarize.* Let $s$ represent the length of a side of the square. Then the area is $s \cdot s$, or $s^2$.

*Translate.*

$\underbrace{\text{The area of a square}}_{s^2} \ \underset{=}{\text{is}} \ \underset{3}{3.}$

*Solve.* We solve the equation.

$s^2 = 3$

$s = \sqrt{3} \quad \text{or} \quad s = -\sqrt{3}$

(The solutions $s$ of the equation are the square roots of 3, since $s^2 = 3$.)

*Check.* Since length cannot be negative, we only need to check $\sqrt{3}$. If the length of a side of a square is $\sqrt{3}$, then the area is $\sqrt{3} \cdot \sqrt{3}$, or $(\sqrt{3})^2$, or 3. The answer checks.

*State.* The length of a side of the square is $\sqrt{3}$.

## Exercise Set 8.2

**1.** $\sqrt{12} = \sqrt{4 \cdot 3}$    4 is a perfect square

$= \sqrt{4}\,\sqrt{3}$    Factoring into a product of radicals

$= 2\sqrt{3}$    Taking the square root

**3.** $\sqrt{75} = \sqrt{25 \cdot 3}$    25 is a perfect square

$= \sqrt{25}\,\sqrt{3}$    Factoring into a product of radicals

$= 5\sqrt{3}$    Taking the square root

**5.** $\sqrt{20} = \sqrt{4 \cdot 5}$    4 is a perfect square

$= \sqrt{4}\,\sqrt{5}$    Factoring into a product of radicals

$= 2\sqrt{5}$    Taking the square root

**7.** $\sqrt{600} = \sqrt{100 \cdot 6}$    100 is a perfect square

$= \sqrt{100} \cdot \sqrt{6}$    Factoring into a product of radicals

$= 10\sqrt{6}$    Taking the square root

**9.** $\sqrt{9x} = \sqrt{9 \cdot x} = \sqrt{9}\,\sqrt{x} = 3\sqrt{x}$

**11.** $\sqrt{48x} = \sqrt{16 \cdot 3x} = \sqrt{16}\,\sqrt{3x} = 4\sqrt{3x}$

**13.** $\sqrt{16a} = \sqrt{16 \cdot a} = \sqrt{16}\,\sqrt{a} = 4\sqrt{a}$

**15.** $\sqrt{64y^2} = \sqrt{64}\,\sqrt{y^2} = 8y$, or

$\sqrt{64y^2} = \sqrt{(8y)^2} = 8y$

**17.** $\sqrt{13x^2} = \sqrt{13}\,\sqrt{x^2} = \sqrt{13} \cdot x$, or $x\sqrt{13}$

**19.** $\sqrt{8t^2} = \sqrt{4 \cdot t^2 \cdot 2} = \sqrt{4}\,\sqrt{t^2}\,\sqrt{2} = 2t\sqrt{2}$

**21.** $\sqrt{180} = \sqrt{36 \cdot 5} = 6\sqrt{5}$

**23.** $\sqrt{288y} = \sqrt{144 \cdot 2y} = \sqrt{144}\,\sqrt{2y} = 12\sqrt{2y}$

**25.** $\sqrt{28x^2} = \sqrt{4 \cdot x^2 \cdot 7} = \sqrt{4}\,\sqrt{x^2}\,\sqrt{7} = 2x\sqrt{7}$

**27.** $\sqrt{8x^2+8x+2} = \sqrt{2(4x^2+4x+1)} =$

$\sqrt{2(2x+1)^2} = \sqrt{2}\,\sqrt{(2x+1)^2} = \sqrt{2}\,(2x+1)$

**29.** $\sqrt{36y + 12y^2 + y^3} = \sqrt{y(36 + 12y + y^2)} =$

$\sqrt{y(6+y)^2} = \sqrt{y}\,\sqrt{(6+y)^2} = \sqrt{y}\,(6+y)$

**31.** $\sqrt{x^6} = \sqrt{(x^3)^2} = x^3$

**33.** $\sqrt{x^{12}} = \sqrt{(x^6)^2} = x^6$

**35.** $\sqrt{x^5} = \sqrt{x^4 \cdot x}$    One factor is a perfect square

$= \sqrt{x^4}\,\sqrt{x}$

$= \sqrt{(x^2)^2}\,\sqrt{x}$

$= x^2\sqrt{x}$

**37.** $\sqrt{t^{19}} = \sqrt{t^{18} \cdot t} = \sqrt{t^{18}}\,\sqrt{t} = \sqrt{(t^9)^2}\,\sqrt{t} = t^9\sqrt{t}$

**39.** $\sqrt{(y-2)^8} = \sqrt{[(y-2)^4]^2} = (y-2)^4$

**41.** $\sqrt{4(x+5)^{10}} = \sqrt{4[(x+5)^5]^2} = \sqrt{4}\,\sqrt{[(x+5)^5]^2} = 2(x+5)^5$

**43.** $\sqrt{36m^3} = \sqrt{36 \cdot m^2 \cdot m} = \sqrt{36}\,\sqrt{m^2}\,\sqrt{m} = 6m\sqrt{m}$

**45.** $\sqrt{8a^5} = \sqrt{4a^4(2a)} = \sqrt{4(a^2)^2(2a)} = \sqrt{4}\,\sqrt{(a^2)^2}\,\sqrt{2a} = 2a^2\sqrt{2a}$

**47.** $\sqrt{104p^{17}} = \sqrt{4p^{16}(26p)} = \sqrt{4(p^8)^2(26p)} = \sqrt{4}\,\sqrt{(p^8)^2}\,\sqrt{26p} = 2p^8\sqrt{26p}$

**49.** $\sqrt{448x^6y^3} = \sqrt{64x^6y^2(7y)} = \sqrt{64(x^3)^2y^2(7y)} = \sqrt{64}\,\sqrt{(x^3)^2}\,\sqrt{y^2}\,\sqrt{7y} = 8x^3y\,\sqrt{7y}$

**51.** $\sqrt{3}\sqrt{18} = \sqrt{3\cdot 18}$     Multiplying

$= \sqrt{3\cdot 3\cdot 6}$     Looking for perfect-square factors or pairs of factors

$= \sqrt{3\cdot 3}\sqrt{6}$

$= 3\sqrt{6}$

**53.** $\sqrt{15}\sqrt{6} = \sqrt{15\cdot 6}$     Multiplying

$= \sqrt{5\cdot 3\cdot 3\cdot 2}$     Looking for perfect-square factors or pairs of factors

$= \sqrt{3\cdot 3}\sqrt{5\cdot 2}$

$= 3\sqrt{10}$

**55.** $\sqrt{18}\sqrt{14x} = \sqrt{18\cdot 14x} = \sqrt{3\cdot 3\cdot 2\cdot 2\cdot 7\cdot x} =$
$\sqrt{3\cdot 3}\sqrt{2\cdot 2}\sqrt{7x} = 3\cdot 2\sqrt{7x} = 6\sqrt{7x}$

**57.** $\sqrt{3x}\sqrt{12y} = \sqrt{3x\cdot 12y} = \sqrt{3\cdot x\cdot 3\cdot 4\cdot y} =$
$\sqrt{3\cdot 3\cdot 4\cdot x\cdot y} = \sqrt{3\cdot 3}\sqrt{4}\sqrt{x\cdot y} = 3\cdot 2\sqrt{xy} = 6\sqrt{xy}$

**59.** $\sqrt{13}\sqrt{13} = \sqrt{13\cdot 13} = 13$

**61.** $\sqrt{5b}\sqrt{15b} = \sqrt{5b\cdot 15b} = \sqrt{5\cdot b\cdot 5\cdot 3\cdot b} =$
$\sqrt{5\cdot 5\cdot b\cdot b\cdot 3} = \sqrt{5\cdot 5}\sqrt{b\cdot b}\sqrt{3} = 5b\sqrt{3}$

**63.** $\sqrt{2t}\sqrt{2t} = \sqrt{2t\cdot 2t} = 2t$

**65.** $\sqrt{ab}\sqrt{ac} = \sqrt{ab\cdot ac} = \sqrt{a\cdot a\cdot b\cdot c} = \sqrt{a\cdot a}\sqrt{b\cdot c} =$
$a\sqrt{bc}$

**67.** $\sqrt{2x^2y}\sqrt{4xy^2} = \sqrt{2x^2y\cdot 4xy^2} = \sqrt{4\cdot x^2\cdot y^2\cdot 2\cdot x\cdot y} =$
$\sqrt{4}\sqrt{x^2}\sqrt{y^2}\sqrt{2xy} = 2xy\sqrt{2xy}$

**69.** $\sqrt{18}\sqrt{18} = \sqrt{18\cdot 18} = 18$

**71.** $\sqrt{5}\sqrt{2x-1} = \sqrt{5(2x-1)} = \sqrt{10x-5}$

**73.** $\sqrt{x+2}\sqrt{x+2} = \sqrt{(x+2)^2} = x+2$

**75.** $\sqrt{18x^2y^3}\sqrt{6xy^4} = \sqrt{18x^2y^3\cdot 6xy^4} =$
$\sqrt{3\cdot 6\cdot x^2\cdot y^2\cdot y\cdot 6\cdot x\cdot y^4} = \sqrt{6\cdot 6\cdot x^2\cdot y^6\cdot 3\cdot x\cdot y} =$
$\sqrt{6\cdot 6}\sqrt{x^2}\sqrt{y^6}\sqrt{3xy} = 6xy^3\sqrt{3xy}$

**77.** $\sqrt{50x^4y^6}\sqrt{10xy} = \sqrt{50x^4y^6\cdot 10xy} =$
$\sqrt{5\cdot 10\cdot x^4\cdot y^6\cdot 10\cdot x\cdot y} = \sqrt{10\cdot 10\cdot x^4\cdot y^6\cdot 5\cdot x\cdot y} =$
$\sqrt{10\cdot 10}\sqrt{x^4}\sqrt{y^6}\sqrt{5xy} = 10x^2y^3\sqrt{5xy}$

**79.** $\sqrt{125} = \sqrt{25\cdot 5}$     Factoring the radicand

$= \sqrt{25}\sqrt{5}$     Factoring the radical expression

$= 5\sqrt{5}$

$\approx 5(2.236)$     From Table 2, $\sqrt{5}\approx 2.236$

$\approx 11.180$

**81.** $\sqrt{360} = \sqrt{36\cdot 10}$

$= \sqrt{36}\sqrt{10}$

$= 6\sqrt{10}$

$\approx 6(3.162)$     From Table 2

$\approx 18.972$

**83.** $\sqrt{700} = \sqrt{100\cdot 7}$

$= \sqrt{100}\sqrt{7}$

$= 10\sqrt{7}$

$\approx 10(2.646)$     From Table 2

$\approx 26.460$

**85.** $\sqrt{122} = \sqrt{2\cdot 61}$     Factoring using numbers shown in Table 2

$= \sqrt{2}\sqrt{61}$

$\approx 1.414 \times 7.810$     From Table 2

$\approx 11.043$     Rounding to 3 decimal places

**87.** First we substitute 20 for $L$ in the formula:
$r = 2\sqrt{5L} = 2\sqrt{5\cdot 20} = 2\sqrt{100} = 2\cdot 10 = 20$ mph
Then we substitute 150 for $L$:
$r = 2\sqrt{5\cdot 150} = 2\sqrt{750} = 2\sqrt{25\cdot 30} = 2\sqrt{25}\sqrt{30} =$
$2\cdot 5\sqrt{30} = 10\sqrt{30} \approx 10(5.477) \approx 54.77$ mph, or 54.8 mph (rounded to the nearest tenth)

**89.** $\quad x - y = -6 \quad (1)$

$\quad \underline{x + y = \phantom{-}2} \quad (2)$

$\quad 2x \phantom{aaaa} = -4 \quad$ Adding

$\quad\quad x = -2$

Now we substitute $-2$ for $x$ in one of the original equations and solve for $y$.

$\quad x + y = 2 \quad$ Equation (2)

$\quad -2 + y = 2 \quad$ Substituting

$\quad\quad y = 4$

Since $(-2, 4)$ checks in both equations, it is the solution.

**91.** $\quad 3x - 2y = 4, \quad (1)$

$\quad 2x + 5y = 9 \quad (2)$

We will us the elimination method. We multiply on both sides of Equation (1) by 5 and on both sides of Equation (2) by 2. Then we add

$\quad 15x - 10y = 20$

$\quad \underline{4x + 10y = 18}$

$\quad 19x \phantom{aaaa} = 38$

$\quad\quad x = \phantom{1}2$

Now we substitute 2 for $x$ in one of the original equations and solve for $y$.

$2x + 5y = 9$    Equation (2)

$2 \cdot 2 + 5y = 9$

$4 + 5y = 9$

$5y = 5$

$y = 1$

Since $(2, 1)$ checks, it is the solution.

**93.** *Familiarize*. We let $l$ = the length of the rectangle and $w$ = the width. Recall that the perimeter of a rectangle is $2l + 2w$, and the area is $lw$.

*Translate*. We translate the first statement.

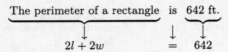

$$2l + 2w = 642$$

Now we translate the second statement.

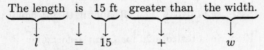

The resulting system is

$2l + 2w = 642$,    (1)

$l = 15 + w$.        (2)

*Solve*. We use the substitution method. We substitute $15 + w$ for $l$ in Equation (1) and solve for $w$.

$2l + 2w = 642$    Equation (1)

$2(15 + w) + 2w = 642$    Substituting

$30 + 2w + 2w = 642$

$30 + 4w = 642$

$4w = 612$

$w = 153$

Substitute 153 for $w$ in Equation (2) and compute $l$.

$l = 15 + w = 15 + 153 = 168$

When $l = 168$ ft and $w = 153$ ft, the area is $168$ ft $\cdot$ $153$ ft, or $25{,}704$ ft$^2$.

*Check*. When $l = 168$ ft and $w = 153$ ft, the perimeter is $2 \cdot 168$ ft $+ 2 \cdot 153$ ft, or $336$ ft $+ 306$ ft, or $642$ ft. The length, $168$ ft, is $15$ ft greater than $153$ ft, the width. We recheck the computation of the area.

*Check*. The area of the rectangle is $25{,}704$ ft$^2$.

**95.** $\sqrt{x^2 - x - 2} = \sqrt{(x-2)(x+1)} = \sqrt{x-2}\,\sqrt{x+1}$

**97.** $\sqrt{2x^2 - 5x - 12} = \sqrt{(2x+3)(x-4)} =$

$\sqrt{2x+3}\,\sqrt{x-4}$

**99.** $\sqrt{a^2 - b^2} = \sqrt{(a+b)(a-b)} = \sqrt{a+b}\,\sqrt{a-b}$

**101.** $\sqrt{0.25} = \sqrt{(0.5)^2} = 0.5$

**103.** $\sqrt{9a^6} = \sqrt{(3a^3)^2} = 3a^3$

**105.** $\sqrt{2y}\,\sqrt{3}\,\sqrt{8y} = \sqrt{2y \cdot 3 \cdot 8y} = \sqrt{2 \cdot y \cdot 3 \cdot 2 \cdot 4 \cdot y} =$

$\sqrt{2 \cdot 2 \cdot 4 \cdot y \cdot y \cdot 3} = \sqrt{2 \cdot 2}\,\sqrt{4}\,\sqrt{y \cdot y}\,\sqrt{3} =$

$2 \cdot 2 \cdot y\sqrt{3} = 4y\sqrt{3}$

**107.** $\sqrt{27(x+1)}\,\sqrt{12y(x+1)^2}$

$\sqrt{27(x+1) \cdot 12y(x+1)^2} =$

$\sqrt{9 \cdot 3 \cdot (x+1) \cdot 4 \cdot 3 \cdot y(x+1)^2} =$

$\sqrt{9 \cdot 3 \cdot 3 \cdot 4 \cdot (x+1)^2 \cdot (x+1)y} =$

$\sqrt{9}\,\sqrt{3 \cdot 3}\,\sqrt{4}\,\sqrt{(x+1)^2}\,\sqrt{(x+1)y} =$

$3 \cdot 3 \cdot 2(x+1)\sqrt{(x+1)y} = 18(x+1)\sqrt{(x+1)y}$

**109.** $\sqrt{x}\,\sqrt{2x}\,\sqrt{10x^5} = \sqrt{x \cdot 2x \cdot 10x^5} =$

$\sqrt{x \cdot 2 \cdot x \cdot 2 \cdot 5 \cdot x^4 \cdot x} = \sqrt{x \cdot x \cdot 2 \cdot 2 \cdot x^4 \cdot 5 \cdot x} =$

$\sqrt{x \cdot x}\,\sqrt{2 \cdot 2}\,\sqrt{x^4}\,\sqrt{5x} = x \cdot 2 \cdot x^2\sqrt{5x} = 2x^3\sqrt{5x}$

## Exercise Set 8.3

**1.** $\dfrac{\sqrt{18}}{\sqrt{2}} = \sqrt{\dfrac{18}{2}} = \sqrt{9} = 3$

**3.** $\dfrac{\sqrt{108}}{\sqrt{3}} = \sqrt{\dfrac{108}{3}} = \sqrt{36} = 6$

**5.** $\dfrac{\sqrt{65}}{\sqrt{13}} = \sqrt{\dfrac{65}{13}} = \sqrt{5}$

**7.** $\dfrac{\sqrt{3}}{\sqrt{75}} = \sqrt{\dfrac{3}{75}} = \sqrt{\dfrac{1}{25}} = \dfrac{1}{5}$

**9.** $\dfrac{\sqrt{12}}{\sqrt{75}} = \sqrt{\dfrac{12}{75}} = \sqrt{\dfrac{4}{25}} = \dfrac{2}{5}$

**11.** $\dfrac{\sqrt{8x}}{\sqrt{2x}} = \sqrt{\dfrac{8x}{2x}} = \sqrt{4} = 2$

**13.** $\dfrac{\sqrt{63y^3}}{\sqrt{7y}} = \sqrt{\dfrac{63y^3}{7y}} = \sqrt{9y^2} = 3y$

**15.** $\sqrt{\dfrac{16}{49}} = \dfrac{\sqrt{16}}{\sqrt{49}} = \dfrac{4}{7}$

**17.** $\sqrt{\dfrac{1}{36}} = \dfrac{\sqrt{1}}{\sqrt{36}} = \dfrac{1}{6}$

**19.** $-\sqrt{\dfrac{16}{81}} = -\dfrac{\sqrt{16}}{\sqrt{81}} = -\dfrac{4}{9}$

**21.** $\sqrt{\dfrac{64}{289}} = \dfrac{\sqrt{64}}{\sqrt{289}} = \dfrac{8}{17}$

**23.** $\sqrt{\dfrac{1690}{1960}} = \sqrt{\dfrac{169 \cdot 10}{196 \cdot 10}} = \sqrt{\dfrac{169}{196} \cdot \dfrac{10}{10}} = \sqrt{\dfrac{169}{196} \cdot 1} =$

$\sqrt{\dfrac{169}{196}} = \dfrac{\sqrt{169}}{\sqrt{196}} = \dfrac{13}{14}$

**25.** $\sqrt{\dfrac{25}{x^2}} = \dfrac{\sqrt{25}}{\sqrt{x^2}} = \dfrac{5}{x}$

**27.** $\sqrt{\dfrac{9a^2}{625}} = \dfrac{\sqrt{9a^2}}{\sqrt{625}} = \dfrac{3a}{25}$

**29.** $\sqrt{\dfrac{2}{5}} = \sqrt{\dfrac{2}{5} \cdot \dfrac{5}{5}} = \sqrt{\dfrac{10}{25}} = \dfrac{\sqrt{10}}{\sqrt{25}} = \dfrac{\sqrt{10}}{5}$

**31.** $\sqrt{\dfrac{7}{8}} = \sqrt{\dfrac{7}{8} \cdot \dfrac{2}{2}} = \sqrt{\dfrac{14}{16}} = \dfrac{\sqrt{14}}{\sqrt{16}} = \dfrac{\sqrt{14}}{4}$

**33.** $\sqrt{\dfrac{1}{12}} = \sqrt{\dfrac{1}{12} \cdot \dfrac{3}{3}} = \sqrt{\dfrac{3}{36}} = \dfrac{\sqrt{3}}{\sqrt{36}} = \dfrac{\sqrt{3}}{6}$

**35.** $\sqrt{\dfrac{5}{18}} = \sqrt{\dfrac{5}{18} \cdot \dfrac{2}{2}} = \sqrt{\dfrac{10}{36}} = \dfrac{\sqrt{10}}{\sqrt{36}} = \dfrac{\sqrt{10}}{6}$

**37.** $\dfrac{3}{\sqrt{5}} = \dfrac{3}{\sqrt{5}} \cdot \dfrac{\sqrt{5}}{\sqrt{5}} = \dfrac{3\sqrt{5}}{5}$

**39.** $\sqrt{\dfrac{8}{3}} = \sqrt{\dfrac{8}{3} \cdot \dfrac{3}{3}} = \sqrt{\dfrac{24}{9}} = \dfrac{\sqrt{4 \cdot 6}}{\sqrt{9}} = \dfrac{\sqrt{4}\,\sqrt{6}}{\sqrt{9}} = \dfrac{2\sqrt{6}}{3}$

**41.** $\sqrt{\dfrac{3}{x}} = \sqrt{\dfrac{3}{x} \cdot \dfrac{x}{x}} = \sqrt{\dfrac{3x}{x^2}} = \dfrac{\sqrt{3x}}{\sqrt{x^2}} = \dfrac{\sqrt{3x}}{x}$

**43.** $\sqrt{\dfrac{x}{y}} = \sqrt{\dfrac{x}{y} \cdot \dfrac{y}{y}} = \sqrt{\dfrac{xy}{y^2}} = \dfrac{\sqrt{xy}}{\sqrt{y^2}} = \dfrac{\sqrt{xy}}{y}$

**45.** $\sqrt{\dfrac{x^2}{20}} = \sqrt{\dfrac{x^2}{20} \cdot \dfrac{5}{5}} = \sqrt{\dfrac{5x^2}{100}} = \dfrac{\sqrt{x^2 \cdot 5}}{\sqrt{100}} = \dfrac{\sqrt{x^2}\,\sqrt{5}}{\sqrt{100}} = \dfrac{x\sqrt{5}}{10}$

**47.** $\dfrac{\sqrt{7}}{\sqrt{2}} = \dfrac{\sqrt{7}}{\sqrt{2}} \cdot \dfrac{\sqrt{2}}{\sqrt{2}} = \dfrac{\sqrt{14}}{2}$

**49.** $\dfrac{\sqrt{9}}{\sqrt{8}} = \dfrac{\sqrt{9}}{\sqrt{8}} \cdot \dfrac{\sqrt{2}}{\sqrt{2}} = \dfrac{\sqrt{9 \cdot 2}}{\sqrt{16}} = \dfrac{3\sqrt{2}}{4}$

**51.** $\dfrac{\sqrt{3}}{\sqrt{2}} = \dfrac{\sqrt{3}}{\sqrt{2}} \cdot \dfrac{\sqrt{2}}{\sqrt{2}} = \dfrac{\sqrt{6}}{2}$

**53.** $\dfrac{2}{\sqrt{2}} = \dfrac{2}{\sqrt{2}} \cdot \dfrac{\sqrt{2}}{\sqrt{2}} = \dfrac{2\sqrt{2}}{2} = \sqrt{2}$

**55.** $\dfrac{\sqrt{5}}{\sqrt{11}} = \dfrac{\sqrt{5}}{\sqrt{11}} \cdot \dfrac{\sqrt{11}}{\sqrt{11}} = \dfrac{\sqrt{55}}{11}$

**57.** $\dfrac{\sqrt{7}}{\sqrt{12}} = \dfrac{\sqrt{7}}{\sqrt{12}} \cdot \dfrac{\sqrt{3}}{\sqrt{3}} = \dfrac{\sqrt{21}}{\sqrt{36}} = \dfrac{\sqrt{21}}{6}$

**59.** $\dfrac{\sqrt{48}}{\sqrt{32}} = \sqrt{\dfrac{48}{32}} = \sqrt{\dfrac{3}{2}} = \sqrt{\dfrac{3}{2} \cdot \dfrac{2}{2}} = \sqrt{\dfrac{6}{4}} = \dfrac{\sqrt{6}}{\sqrt{4}} = \dfrac{\sqrt{6}}{2}$

**61.** $\dfrac{\sqrt{450}}{\sqrt{18}} = \sqrt{\dfrac{450}{18}} = \sqrt{25} = 5$

**63.** $\dfrac{\sqrt{3}}{\sqrt{x}} = \dfrac{\sqrt{3}}{\sqrt{x}} \cdot \dfrac{\sqrt{x}}{\sqrt{x}} = \dfrac{\sqrt{3x}}{x}$

**65.** $\dfrac{4y}{\sqrt{5}} = \dfrac{4y}{\sqrt{5}} \cdot \dfrac{\sqrt{5}}{\sqrt{5}} = \dfrac{4y\sqrt{5}}{5}$

**67.** $\dfrac{\sqrt{a^3}}{\sqrt{8}} = \dfrac{\sqrt{a^3}}{\sqrt{8}} \cdot \dfrac{\sqrt{2}}{\sqrt{2}} = \dfrac{\sqrt{2a^3}}{\sqrt{16}} = \dfrac{\sqrt{a^2 \cdot 2a}}{\sqrt{16}} = \dfrac{a\sqrt{2a}}{4}$

**69.** $\dfrac{\sqrt{56}}{\sqrt{12x}} = \sqrt{\dfrac{56}{12x}} = \sqrt{\dfrac{14}{3x}} = \sqrt{\dfrac{14}{3x} \cdot \dfrac{3x}{3x}} = \sqrt{\dfrac{42x}{3x \cdot 3x}} = \dfrac{\sqrt{42x}}{3x}$

**71.** $\dfrac{\sqrt{27c}}{\sqrt{32c^3}} = \sqrt{\dfrac{27c}{32c^3}} = \sqrt{\dfrac{27}{32c^2}} = \sqrt{\dfrac{27}{32c^2} \cdot \dfrac{2}{2}} = \sqrt{\dfrac{54}{64c^2}} = \sqrt{\dfrac{9 \cdot 6}{64c^2}} = \dfrac{3\sqrt{6}}{8c}$

**73.** $\dfrac{\sqrt{y^5}}{\sqrt{xy^2}} = \sqrt{\dfrac{y^5}{xy^2}} = \sqrt{\dfrac{y^3}{x}} = \sqrt{\dfrac{y^3}{x} \cdot \dfrac{x}{x}} = \sqrt{\dfrac{xy^3}{x^2}} = \sqrt{\dfrac{y^2 \cdot xy}{x^2}} = \dfrac{y\sqrt{xy}}{x}$

**75.** $\dfrac{\sqrt{45mn^2}}{\sqrt{32m}} = \sqrt{\dfrac{45mn^2}{32m}} = \sqrt{\dfrac{45n^2}{32}} = \sqrt{\dfrac{45n^2}{32} \cdot \dfrac{2}{2}} = \sqrt{\dfrac{90n^2}{64}} = \dfrac{\sqrt{90n^2}}{\sqrt{64}} = \dfrac{\sqrt{9 \cdot n^2 \cdot 10}}{8} = \dfrac{3n\sqrt{10}}{8}$

**77.** $\sqrt{\dfrac{1}{3}} \approx \sqrt{0.333333} \approx 0.577350269 \approx 0.577$

**79.** $\sqrt{\dfrac{7}{8}} = \sqrt{0.875} \approx 0.935414346 \approx 0.935$

**81.** $\sqrt{\dfrac{1}{12}} \approx \sqrt{0.0833333} \approx 0.288675134 \approx 0.289$

**83.** $\sqrt{\dfrac{1}{2}} = \sqrt{0.5} \approx 0.707106781 \approx 0.707$

**85.** $\dfrac{17}{\sqrt{20}} \approx \dfrac{17}{4.472135955} \approx 3.801315562 \approx 3.801$

**87.** $\dfrac{\sqrt{13}}{\sqrt{18}} \approx \dfrac{3.605551275}{4.242640687} \approx 0.849836585 \approx 0.850$

**89.** $x = y + 2, \quad (1)$

$x + y = 6 \quad (2)$

We substitute $y + 2$ for $x$ in Equation (2) and solve for $y$.

$(y + 2) + y = 6$

$2y + 2 = 6$

$2y = 4$

$y = 2$

Substitute 2 for $y$ in Equation (1) to find $x$.

$x = 2 + 2 = 4$

The ordered pair $(4, 2)$ checks in both equations. It is the solution.

**91.**  $2x - 3y = 7$   (1)
$2x - 3y = 9$   (2)

We multiply Eauation (2) by $-1$ and add.

$\begin{array}{r} 2x - 3y = \phantom{-}7 \\ -2x + 3y = -9 \\ \hline 0 = -2 \end{array}$

We get a false equation. The system of equations has no solution.

**93.**  $x + y = -7$   (1)
$\dfrac{x - y = \phantom{-}2}{2x \phantom{-+ y} = -5}$   (2)

$2x \phantom{{}+ y} = -5$   Adding

$x = -\dfrac{5}{2}$

Substitute $-\dfrac{5}{2}$ for $x$ in Equation (1) to find $y$.

$x + y = -7$   Equation (1)

$-\dfrac{5}{2} + y = -7$   Substituting

$y = -\dfrac{9}{2}$

The ordered pair $\left(-\dfrac{5}{2}, -\dfrac{9}{2}\right)$ checks in both equations. It is the solution.

**95.**  2 ft:  $T \approx 2(3.14)\sqrt{\dfrac{2}{32}} \approx 6.28\sqrt{\dfrac{1}{16}} \approx 6.28\left(\dfrac{1}{4}\right) \approx$
1.57 sec

8 ft:  $T \approx 2(3.14)\sqrt{\dfrac{8}{32}} \approx 6.28\sqrt{\dfrac{1}{4}} \approx 6.28\left(\dfrac{1}{2}\right) \approx$
3.14 sec

64 ft:  $T \approx 2(3.14)\sqrt{\dfrac{64}{32}} \approx 6.28\sqrt{2} \approx$
$(6.28)(1.414) \approx 8.88$ sec

100 ft:  $T \approx 2(3.14)\sqrt{\dfrac{100}{32}} \approx 6.28\sqrt{\dfrac{50}{16}} \approx \dfrac{6.28\sqrt{50}}{4} \approx$
$\dfrac{6.28(7.071)}{4} \approx 11.10$ sec

**97.**  $T = 2\pi\sqrt{\dfrac{\frac{32}{\pi^2}}{32}} = 2\pi\sqrt{\dfrac{32}{\pi^2} \cdot \dfrac{1}{32}} = 2\pi\sqrt{\dfrac{1}{\pi^2}} = 2\pi\left(\dfrac{1}{\pi}\right) = 2$ sec

The time it takes the pendulum to swing from one side to the other and back is 2 sec, so it takes 1 sec to swing from one side to the other.

**99.**  $\sqrt{\dfrac{5}{1600}} = \dfrac{\sqrt{5}}{\sqrt{1600}} = \dfrac{\sqrt{5}}{40}$

**101.**  $\sqrt{\dfrac{1}{5x^3}} = \sqrt{\dfrac{1}{5x^3} \cdot \dfrac{5x}{5x}} = \sqrt{\dfrac{5x}{25x^4}} = \dfrac{\sqrt{5x}}{\sqrt{25x^4}} = \dfrac{\sqrt{5x}}{5x^2}$

**103.**  $\sqrt{\dfrac{3a}{b}} = \sqrt{\dfrac{3a}{b} \cdot \dfrac{b}{b}} = \sqrt{\dfrac{3ab}{b^2}} = \dfrac{\sqrt{3ab}}{\sqrt{b^2}} = \dfrac{\sqrt{3ab}}{b}$

**105.**  $\sqrt{0.009} = \sqrt{\dfrac{9}{1000}} = \sqrt{\dfrac{9}{1000} \cdot \dfrac{10}{10}} = \sqrt{\dfrac{90}{10,000}} =$

$\dfrac{\sqrt{90}}{\sqrt{10,000}} = \dfrac{\sqrt{9 \cdot 10}}{100} = \dfrac{\sqrt{9}\sqrt{10}}{100} = \dfrac{3\sqrt{10}}{100}$

**107.**  $\sqrt{\dfrac{1}{x^2} - \dfrac{2}{xy} + \dfrac{1}{y^2}}$,  LCD is $x^2y^2$

$= \sqrt{\dfrac{1}{x^2} \cdot \dfrac{y^2}{y^2} - \dfrac{2}{xy} \cdot \dfrac{xy}{xy} + \dfrac{1}{y^2} \cdot \dfrac{x^2}{x^2}}$

$= \sqrt{\dfrac{y^2 - 2xy + x^2}{x^2y^2}}$

$= \sqrt{\dfrac{(y - x)^2}{x^2y^2}}$

$= \dfrac{\sqrt{(y - x)^2}}{\sqrt{x^2y^2}}$

$= \dfrac{y - x}{xy}$

## Exercise Set 8.4

**1.**  $7\sqrt{3} + 9\sqrt{3} = (7 + 9)\sqrt{3}$
$= 16\sqrt{3}$

**3.**  $7\sqrt{5} - 3\sqrt{5} = (7 - 3)\sqrt{5}$
$= 4\sqrt{5}$

**5.**  $6\sqrt{x} + 7\sqrt{x} = (6 + 7)\sqrt{x}$
$= 13\sqrt{x}$

**7.**  $4\sqrt{d} - 13\sqrt{d} = (4 - 13)\sqrt{d}$
$= -9\sqrt{d}$

**9.**  $5\sqrt{8} + 15\sqrt{2} = 5\sqrt{4 \cdot 2} + 15\sqrt{2}$
$= 5 \cdot 2\sqrt{2} + 15\sqrt{2}$
$= 10\sqrt{2} + 15\sqrt{2}$
$= 25\sqrt{2}$

**11.**  $\sqrt{27} - 2\sqrt{3} = \sqrt{9 \cdot 3} - 2\sqrt{3}$
$= 3\sqrt{3} - 2\sqrt{3}$
$= (3 - 2)\sqrt{3}$
$= 1\sqrt{3}$
$= \sqrt{3}$

**13.**  $\sqrt{45} - \sqrt{20} = \sqrt{9 \cdot 5} - \sqrt{4 \cdot 5}$
$= 3\sqrt{5} - 2\sqrt{5}$
$= (3 - 2)\sqrt{5}$
$= 1\sqrt{5}$
$= \sqrt{5}$

**15.** $\sqrt{72} + \sqrt{98} = \sqrt{36 \cdot 2} + \sqrt{49 \cdot 2}$
$\qquad = 6\sqrt{2} + 7\sqrt{2}$
$\qquad = (6 + 7)\sqrt{2}$
$\qquad = 13\sqrt{2}$

**17.** $2\sqrt{12} + \sqrt{27} - \sqrt{48} = 2\sqrt{4 \cdot 3} + \sqrt{9 \cdot 3} - \sqrt{16 \cdot 3}$
$\qquad = 2 \cdot 2\sqrt{3} + 3\sqrt{3} - 4\sqrt{3}$
$\qquad = 4\sqrt{3} + 3\sqrt{3} - 4\sqrt{3}$
$\qquad = (4 + 3 - 4)\sqrt{3}$
$\qquad = 3\sqrt{3}$

**19.** $\sqrt{18} - 3\sqrt{8} + \sqrt{50} = \sqrt{9 \cdot 2} - 3\sqrt{4 \cdot 2} + \sqrt{25 \cdot 2}$
$\qquad = 3\sqrt{2} - 3 \cdot 2\sqrt{2} + 5\sqrt{2}$
$\qquad = 3\sqrt{2} - 6\sqrt{2} + 5\sqrt{2}$
$\qquad = (3 - 6 + 5)\sqrt{2}$
$\qquad = 2\sqrt{2}$

**21.** $2\sqrt{27} - 3\sqrt{48} + 3\sqrt{12} = 2\sqrt{9 \cdot 3} - 3\sqrt{16 \cdot 3} + 3\sqrt{4 \cdot 3}$
$\qquad = 2 \cdot 3\sqrt{3} - 3 \cdot 4\sqrt{3} + 3 \cdot 2\sqrt{3}$
$\qquad = 6\sqrt{3} - 12\sqrt{3} + 6\sqrt{3}$
$\qquad = (6 - 12 + 6)\sqrt{3}$
$\qquad = 0\sqrt{3}$
$\qquad = 0$

**23.** $\sqrt{4x} + \sqrt{81x^3} = \sqrt{4 \cdot x} + \sqrt{81 \cdot x^2 \cdot x}$
$\qquad = 2\sqrt{x} + 9x\sqrt{x}$
$\qquad = (2 + 9x)\sqrt{x}$

**25.** $\sqrt{27} - \sqrt{12x^2} = \sqrt{9 \cdot 3} - \sqrt{4 \cdot 3 \cdot x^2}$
$\qquad = 3\sqrt{3} - 2x\sqrt{3}$
$\qquad = (3 - 2x)\sqrt{3}$

**27.** $\sqrt{8x + 8} + \sqrt{2x + 2} = \sqrt{4(2x + 2)} + \sqrt{2x + 2}$
$\qquad = 2\sqrt{2x + 2} + 1\sqrt{2x + 2}$
$\qquad = (2 + 1)\sqrt{2x + 2}$
$\qquad = 3\sqrt{2x + 2}$

**29.** $\sqrt{x^5 - x^2} + \sqrt{9x^3 - 9} = \sqrt{x^2(x^3 - 1)} + \sqrt{9(x^3 - 1)}$
$\qquad = x\sqrt{x^3 - 1} + 3\sqrt{x^3 - 1}$
$\qquad = (x + 3)\sqrt{x^3 - 1}$

**31.** $4a\sqrt{a^2 b} + a\sqrt{a^2 b^3} - 5\sqrt{b^3}$
$\qquad = 4a\sqrt{a^2 \cdot b} + a\sqrt{a^2 \cdot b^2 \cdot b} - 5\sqrt{b^2 \cdot b}$
$\qquad = 4a \cdot a\sqrt{b} + a \cdot a \cdot b\sqrt{b} - 5 \cdot b\sqrt{b}$
$\qquad = 4a^2\sqrt{b} + a^2 b\sqrt{b} - 5b\sqrt{b}$
$\qquad = (4a^2 + a^2 b - 5b)\sqrt{b}$

**33.** $\sqrt{3} - \sqrt{\dfrac{1}{3}} = \sqrt{3} - \sqrt{\dfrac{1}{3} \cdot \dfrac{3}{3}}$
$\qquad = \sqrt{3} - \dfrac{\sqrt{3}}{3}$
$\qquad = \left(1 - \dfrac{1}{3}\right)\sqrt{3}$
$\qquad = \dfrac{2}{3}\sqrt{3}, \text{ or } \dfrac{2\sqrt{3}}{3}$

**35.** $5\sqrt{2} + 3\sqrt{\dfrac{1}{2}} = 5\sqrt{2} + 3\sqrt{\dfrac{1}{2} \cdot \dfrac{2}{2}}$
$\qquad = 5\sqrt{2} + \dfrac{3}{2}\sqrt{2}$
$\qquad = \left(5 + \dfrac{3}{2}\right)\sqrt{2}$
$\qquad = \dfrac{13}{2}\sqrt{2}, \text{ or } \dfrac{13\sqrt{2}}{2}$

**37.** $\sqrt{\dfrac{2}{3}} - \sqrt{\dfrac{1}{6}} = \sqrt{\dfrac{2}{3} \cdot \dfrac{3}{3}} - \sqrt{\dfrac{1}{6} \cdot \dfrac{6}{6}}$
$\qquad = \dfrac{\sqrt{6}}{3} - \dfrac{\sqrt{6}}{6}$
$\qquad = \left(\dfrac{1}{3} - \dfrac{1}{6}\right)\sqrt{6}$
$\qquad = \dfrac{1}{6}\sqrt{6}, \text{ or } \dfrac{\sqrt{6}}{6}$

**39.** $\sqrt{3}(\sqrt{5} - 1) = \sqrt{3}\,\sqrt{5} - \sqrt{3} \cdot 1$
$\qquad = \sqrt{15} - \sqrt{3}$

**41.** $(\sqrt{2} + 8)(\sqrt{2} - 8)$
$= (\sqrt{2})^2 - 8^2 \quad \text{Using } (A + B)(A - B) = A^2 - B^2$
$= 2 - 64$
$= -62$

**43.** $(\sqrt{6} - \sqrt{5})(\sqrt{6} + \sqrt{5})$
$= (\sqrt{6})^2 - (\sqrt{5})^2 \quad \text{Using } (A + B)(A - B) = A^2 - B^2$
$= 6 - 5$
$= 1$

**45.** $(3\sqrt{5} - 2)(\sqrt{5} + 1)$
$= 3\sqrt{5}\,\sqrt{5} + 3\sqrt{5} - 2\sqrt{5} - 2 \qquad \text{Using FOIL}$
$= 3 \cdot 5 + 3\sqrt{5} - 2\sqrt{5} - 2$
$= 15 + \sqrt{5} - 2$
$= 13 + \sqrt{5}$

**47.** $(\sqrt{x} - \sqrt{y})^2 = (\sqrt{x})^2 - 2\sqrt{x}\,\sqrt{y} + (\sqrt{y})^2$
$\qquad\qquad \text{Using } (A - B)^2 = A^2 - 2AB + B^2$
$\qquad = x - 2\sqrt{xy} + y$

**49.** We multiply by 1 using the conjugate of $\sqrt{3} - \sqrt{5}$, which is $\sqrt{3} + \sqrt{5}$, as the numerator and denominator.

$$\frac{2}{\sqrt{3} - \sqrt{5}} = \frac{2}{\sqrt{3} - \sqrt{5}} \cdot \frac{\sqrt{3} + \sqrt{5}}{\sqrt{3} + \sqrt{5}} \quad \text{Multiplying by 1}$$

$$= \frac{2(\sqrt{3} + \sqrt{5})}{(\sqrt{3} - \sqrt{5})(\sqrt{3} + \sqrt{5})} \quad \text{Multiplying}$$

$$= \frac{2\sqrt{3} + 2\sqrt{5}}{(\sqrt{3})^2 - (\sqrt{5})^2} = \frac{2\sqrt{3} + 2\sqrt{5}}{3 - 5}$$

$$= \frac{2\sqrt{3} + 2\sqrt{5}}{-2} = \frac{2(\sqrt{3} + \sqrt{5})}{-2}$$

$$= -(\sqrt{3} + \sqrt{5}) = -\sqrt{3} - \sqrt{5}$$

**51.** We multiply by 1 using the conjugate of $\sqrt{3} + \sqrt{2}$, which is $\sqrt{3} - \sqrt{2}$, as the numerator and denominator.

$$\frac{\sqrt{3} - \sqrt{2}}{\sqrt{3} + \sqrt{2}} = \frac{\sqrt{3} - \sqrt{2}}{\sqrt{3} + \sqrt{2}} \cdot \frac{\sqrt{3} - \sqrt{2}}{\sqrt{3} - \sqrt{2}} \quad \text{Multiplying by 1}$$

$$= \frac{(\sqrt{3} - \sqrt{2})^2}{(\sqrt{3} + \sqrt{2})(\sqrt{3} - \sqrt{2})}$$

$$= \frac{(\sqrt{3})^2 - 2\sqrt{3}\sqrt{2} + (\sqrt{2})^2}{(\sqrt{3})^2 - (\sqrt{2})^2}$$

$$= \frac{3 - 2\sqrt{6} + 2}{3 - 2} = \frac{5 - 2\sqrt{6}}{1}$$

$$= 5 - 2\sqrt{6}$$

**53.** We multiply by 1 using the conjugate of $\sqrt{10} + 1$, which is $\sqrt{10} - 1$, as the numerator and denominator.

$$\frac{4}{\sqrt{10} + 1} = \frac{4}{\sqrt{10} + 1} \cdot \frac{\sqrt{10} - 1}{\sqrt{10} - 1}$$

$$= \frac{4(\sqrt{10} - 1)}{(\sqrt{10} + 1)(\sqrt{10} - 1)}$$

$$= \frac{4\sqrt{10} - 4}{(\sqrt{10})^2 - 1^2} = \frac{4\sqrt{10} - 4}{10 - 1}$$

$$= \frac{4\sqrt{10} - 4}{9}$$

**55.** We multiply by 1 using the conjugate of $3 + \sqrt{7}$, which is $3 - \sqrt{7}$, as the numerator and denominator.

$$\frac{1 - \sqrt{7}}{3 + \sqrt{7}} = \frac{1 - \sqrt{7}}{3 + \sqrt{7}} \cdot \frac{3 - \sqrt{7}}{3 - \sqrt{7}}$$

$$= \frac{(1 - \sqrt{7})(3 - \sqrt{7})}{(3 + \sqrt{7})(3 - \sqrt{7})}$$

$$= \frac{3 - \sqrt{7} - 3\sqrt{7} + \sqrt{7}\sqrt{7}}{3^2 - (\sqrt{7})^2}$$

$$= \frac{3 - \sqrt{7} - 3\sqrt{7} + 7}{9 - 7} = \frac{10 - 4\sqrt{7}}{2}$$

$$= \frac{2(5 - 2\sqrt{7})}{2} = 5 - 2\sqrt{7}$$

**57.** We multiply by 1 using the conjugate of $4 + \sqrt{x}$, which is $4 - \sqrt{x}$, as the numerator and denominator.

$$\frac{3}{4 + \sqrt{x}} = \frac{3}{4 + \sqrt{x}} \cdot \frac{4 - \sqrt{x}}{4 - \sqrt{x}}$$

$$= \frac{3(4 - \sqrt{x})}{(4 + \sqrt{x})(4 - \sqrt{x})}$$

$$= \frac{12 - 3\sqrt{x}}{4^2 - (\sqrt{x})^2}$$

$$= \frac{12 - 3\sqrt{x}}{16 - x}$$

**59.** We multiply by 1 using the conjugate of $8 - \sqrt{x}$, which is $8 + \sqrt{x}$, as the numerator and denominator.

$$\frac{3 + \sqrt{2}}{8 - \sqrt{x}} = \frac{3 + \sqrt{2}}{8 - \sqrt{x}} \cdot \frac{8 + \sqrt{x}}{8 + \sqrt{x}}$$

$$= \frac{(3 + \sqrt{2})(8 + \sqrt{x})}{(8 - \sqrt{x})(8 + \sqrt{x})}$$

$$= \frac{3 \cdot 8 + 3 \cdot \sqrt{x} + \sqrt{2} \cdot 8 + \sqrt{2} \cdot \sqrt{x}}{8^2 - (\sqrt{x})^2}$$

$$= \frac{24 + 3\sqrt{x} + 8\sqrt{2} + \sqrt{2x}}{64 - x}$$

**61.** *Familiarize.* The problem states that we have inverse variation between $t$ and $r$. Thus, an equation $t = \dfrac{k}{r}$, $k > 0$, applies. As the speed increases, the time decreases.

*Translate.* We write an equation of variation. Time varies inversely as speed. This translates to $t = \dfrac{k}{r}$.

*Solve.*

a) First find an equation of variation.

$$t = \frac{k}{r}$$

$$\frac{1}{2} = \frac{k}{40} \quad \text{Substituting } \frac{1}{2} \text{ for } t \text{ and } 40 \text{ for } r$$

$$20 = k \quad \text{Multiplying by 40}$$

The equation of variation is $t = \dfrac{20}{r}$.

b) We use the equation of variation to find how long it will take to travel the same distance at 60 mph.

$$t = \frac{20}{r}$$

$$t = \frac{20}{60} \quad \text{Substituting 60 for } r$$

$$t = \frac{1}{3}$$

*Check.* Let us do some reasoning about the answer. The time decreased from $\frac{1}{2}$ hr to $\frac{1}{3}$ hr when the speed increased from 40 mph to 60 mph. This is what we would expect with inverse variation.

*State*. It will take $\frac{1}{3}$ hr to travel the fixed distance at 60 mph.

Since $t = \frac{k}{r}$, or $rt = k$, the variation constant is the fixed distance. (Speed $\cdot$ Time = Distance)

**63.** $\sqrt{10} + \sqrt{50} = \sqrt{10} + \sqrt{10}\,\sqrt{5} = \sqrt{10}(1 + \sqrt{5})$

$\sqrt{10} + \sqrt{50} = \sqrt{10} + \sqrt{25 \cdot 2} = \sqrt{10} + 5\sqrt{2}$

$\sqrt{10} + \sqrt{50} = \sqrt{2}\,\sqrt{5} + \sqrt{2}\,\sqrt{25} =$

$\sqrt{2}(\sqrt{5} + \sqrt{25}) = \sqrt{2}(\sqrt{5} + 5)$, or $\sqrt{2}(5 + \sqrt{5})$

All three are correct.

**65.** $\frac{1}{3}\sqrt{27} + \sqrt{8} + \sqrt{300} - \sqrt{18} - \sqrt{162}$

$= \frac{1}{3}\sqrt{9 \cdot 3} + \sqrt{4 \cdot 2} + \sqrt{100 \cdot 3} - \sqrt{9 \cdot 2} - \sqrt{81 \cdot 2}$

$= \frac{1}{3} \cdot 3\sqrt{3} + 2\sqrt{2} + 10\sqrt{3} - 3\sqrt{2} - 9\sqrt{2}$

$= \sqrt{3} + 2\sqrt{2} + 10\sqrt{3} - 3\sqrt{2} - 9\sqrt{2}$

$= (1 + 10)\sqrt{3} + (2 - 3 - 9)\sqrt{2}$

$= 11\sqrt{3} - 10\sqrt{2}$

**67.** Since $\sqrt{a^2 + b^2} \neq \sqrt{a^2} + \sqrt{b^2}$ for $a = 2$ and $b = 3$, the two expressions are not equivalent.

**69.** $(3\sqrt{x+2})^2 = 3^2(\sqrt{x^2+2})^2$

$\qquad = 9(x + 2)$

The statement is true.

## Exercise Set 8.5

**1.** $\sqrt{x} = 6$

$(\sqrt{x})^2 = 6^2$   Squaring both sides

$x = 36$   Simplifying

Check: $\dfrac{\sqrt{x} = 6}{\begin{array}{c|c} \sqrt{36} & 6 \\ 6 & \text{TRUE} \end{array}}$

The solution is 36.

**3.** $\sqrt{x} = 4.3$

$(\sqrt{x})^2 = (4.3)^2$   Squaring both sides

$x = 18.49$   Simplifying

Check: $\dfrac{\sqrt{x} = 4.3}{\begin{array}{c|c} \sqrt{18.49} & 4.3 \\ 4.3 & \text{TRUE} \end{array}}$

The solution is 18.49.

**5.** $\sqrt{y+4} = 13$

$(\sqrt{y+4})^2 = 13^2$   Squaring both sides

$y + 4 = 169$   Simplifying

$y = 165$   Subtracting 4

Check: $\dfrac{\sqrt{y+4} = 13}{\begin{array}{c|c} \sqrt{165+4} & 13 \\ \sqrt{169} & \\ 13 & \text{TRUE} \end{array}}$

The solution is 165.

**7.** $\sqrt{2x+4} = 25$

$(\sqrt{2x+4})^2 = 25^2$   Squaring both sides

$2x + 4 = 625$   Simplifying

$2x = 621$   Subtracting 4

$x = \frac{621}{2}$   Dividing by 2

Check: $\dfrac{\sqrt{2x+4} = 25}{\begin{array}{c|c} \sqrt{2 \cdot \frac{621}{2} + 4} & 25 \\ \sqrt{621 + 4} & \\ \sqrt{625} & \\ 25 & \text{TRUE} \end{array}}$

The solution is $\frac{621}{2}$.

**9.** $3 + \sqrt{x-1} = 5$

$\sqrt{x-1} = 2$   Subtracting 3

$(\sqrt{x-1})^2 = 2^2$   Squaring both sides

$x - 1 = 4$

$x = 5$

Check: $\dfrac{3 + \sqrt{x-1} = 5}{\begin{array}{c|c} 3 + \sqrt{5-1} & 5 \\ 3 + \sqrt{4} & \\ 3 + 2 & \\ 5 & \text{TRUE} \end{array}}$

The solution is 5.

**11.** $6 - 2\sqrt{3n} = 0$

$6 = 2\sqrt{3n}$   Adding $2\sqrt{3n}$

$6^2 = (2\sqrt{3n})^2$   Squaring both sides

$36 = 4 \cdot 3n$

$36 = 12n$

$3 = n$

Check: $\dfrac{6 - 2\sqrt{3n} = 0}{\begin{array}{c|c} 6 - 2\sqrt{3 \cdot 3} & 0 \\ 6 - 2 \cdot 3 & \\ 6 - 6 & \\ 0 & \text{TRUE} \end{array}}$

The solution is 3.

**13.**
$$\sqrt{5x-7} = \sqrt{x+10}$$
$$(\sqrt{5x-7})^2 = (\sqrt{x+10})^2 \qquad \text{Squaring both sides}$$
$$5x - 7 = x + 10$$
$$4x = 17$$
$$x = \frac{17}{4}$$

Check: 

$$\sqrt{5x-7} = \sqrt{x+10}$$

| $\sqrt{5 \cdot \dfrac{17}{4} - 7}$ | $\sqrt{\dfrac{17}{4} + 10}$ |
|---|---|
| $\sqrt{\dfrac{85}{4} - \dfrac{28}{4}}$ | $\sqrt{\dfrac{57}{4}}$ |
| $\sqrt{\dfrac{57}{4}}$ | |
| | TRUE |

The solution is $\dfrac{17}{4}$.

**15.** $\sqrt{x} = -7$

There is no solution. The principal square root of $x$ cannot be negative.

**17.**
$$\sqrt{2y+6} = \sqrt{2y-5}$$
$$(\sqrt{2y+6})^2 = (\sqrt{2y-5})^2$$
$$2y + 6 = 2y - 5$$
$$6 = -5$$

The equation $6 = -5$ is false; there is no solution.

**19.**
$$x - 7 = \sqrt{x-5}$$
$$(x-7)^2 = (\sqrt{x-5})^2$$
$$x^2 - 14x + 49 = x - 5$$
$$x^2 - 15 + 54 = 0$$
$$(x-9)(x-6) = 0$$
$$x - 9 = 0 \quad \text{or} \quad x - 6 = 0$$
$$x = 9 \quad \text{or} \qquad x = 6$$

Check: 

$$x - 7 = \sqrt{x-5}$$

| $9 - 7$ | $\sqrt{9-5}$ |
|---|---|
| $2$ | $\sqrt{4}$ |
| | $2$ |
| | TRUE |

$$x - 7 = \sqrt{x-5}$$

| $6 - 7$ | $\sqrt{6-5}$ |
|---|---|
| $-1$ | $\sqrt{1}$ |
| | $1$ |
| | FALSE |

The number 9 checks, but 6 does not. The solution is 9.

**21.**
$$x - 9 = \sqrt{x-3}$$
$$(x-9)^2 = (\sqrt{x-3})^2$$
$$x^2 - 18x + 81 = x - 3$$
$$x^2 - 19x + 84 = 0$$
$$(x-12)(x-7) = 0$$
$$x - 12 = 0 \quad \text{or} \quad x - 7 = 0$$
$$x = 12 \quad \text{or} \qquad x = 7$$

Check: 

$$x - 9 = \sqrt{x-3}$$

| $12 - 9$ | $\sqrt{12-3}$ |
|---|---|
| $3$ | $\sqrt{9}$ |
| | $3$ |
| | TRUE |

$$x - 9 = \sqrt{x-3}$$

| $7 - 9$ | $\sqrt{7-3}$ |
|---|---|
| $-2$ | $\sqrt{4}$ |
| | $2$ |
| | FALSE |

The number 12 checks, but 7 does not. The solution is 12.

**23.**
$$2\sqrt{x-1} = x - 1$$
$$(2\sqrt{x-1})^2 = (x-1)^2$$
$$4(x-1) = x^2 - 2x + 1$$
$$4x - 4 = x^2 - 2x + 1$$
$$0 = x^2 - 6x + 5$$
$$0 = (x-5)(x-1)$$
$$x - 5 = 0 \quad \text{or} \quad x - 1 = 0$$
$$x = 5 \quad \text{or} \qquad x = 1$$

Both numbers check. The solutions are 5 and 1.

**25.**
$$\sqrt{5x+21} = x + 3$$
$$(\sqrt{5x+21})^2 = (x+3)^2$$
$$5x + 21 = x^2 + 6x + 9$$
$$0 = x^2 + x - 12$$
$$0 = (x+4)(x-3)$$
$$x + 4 = 0 \quad \text{or} \quad x - 3 = 0$$
$$x = -4 \quad \text{or} \qquad x = 3$$

Check: 

$$\sqrt{5x+21} = x + 3$$

| $\sqrt{5(-4)+21}$ | $-4+3$ |
|---|---|
| $\sqrt{1}$ | $-1$ |
| $1$ | |
| | FALSE |

$$\sqrt{5x+21} = x + 3$$

| $\sqrt{5 \cdot 3 + 21}$ | $3+3$ |
|---|---|
| $\sqrt{36}$ | $6$ |
| $6$ | |
| | TRUE |

The number 3 checks, but $-4$ does not. The solution is 3.

**27.** $\sqrt{2x-1}+2=x$

$\qquad \sqrt{2x-1}=x-2 \qquad$ Isolating the radical

$\qquad (\sqrt{2x-1})^2=(x-2)^2$

$\qquad 2x-1=x^2-4x+4$

$\qquad 0=x^2-6x+5$

$\qquad 0=(x-5)(x-1)$

$x-5=0 \quad \text{or} \quad x-1=0$

$\quad x=5 \quad \text{or} \qquad x=1$

Check: $\dfrac{\sqrt{2x-1}+2=x}{}$

$\qquad \sqrt{2\cdot 5-1}+2 \ \bigg| \ 5$

$\qquad \sqrt{10-1}+2 \ \bigg|$

$\qquad \sqrt{9}+2 \ \bigg|$

$\qquad 3+2 \ \bigg|$

$\qquad\qquad 5 \ \bigg| \quad$ TRUE

$\dfrac{\sqrt{2x-1}+2=x}{}$

$\sqrt{2\cdot 1-1}+2 \ \bigg| \ 1$

$\sqrt{2-1}+2 \ \bigg|$

$\sqrt{1}+2 \ \bigg|$

$1+2 \ \bigg|$

$\qquad 3 \ \bigg| \quad$ FALSE

The number 5 checks, but 1 does not. The solution is 5.

**29.** $\sqrt{x^2+6}-x+3=0$

$\qquad \sqrt{x^2+6}=x-3 \qquad$ Isolating the radical

$\qquad (\sqrt{x^2+6})^2=(x-3)^2$

$\qquad x^2+6=x^2-6x+9$

$\qquad -3=-6x \qquad$ Adding $-x^2$ and $-9$

$\qquad \dfrac{1}{2}=x$

Check: $\dfrac{\sqrt{x^2+6}-x+3=0}{}$

$\sqrt{\left(\dfrac{1}{2}\right)^2+6}-\dfrac{1}{2}+3 \ \bigg| \ 0$

$\sqrt{\dfrac{25}{4}}-\dfrac{1}{2}+3 \ \bigg|$

$\dfrac{5}{2}-\dfrac{1}{2}+3 \ \bigg|$

$\qquad\qquad 5 \ \bigg| \quad$ FALSE

The number $\dfrac{1}{2}$ does not check. There is no solution.

**31.** $\sqrt{(p+6)(p+1)}-2=p+1$

$\qquad \sqrt{(p+6)(p+1)}=p+3 \quad$ Isolating the radical

$\qquad \left(\sqrt{(p+6)(p+1)}\right)^2=(p+3)^2$

$\qquad (p+6)(p+1)=p^2+6p+9$

$\qquad p^2+7p+6=p^2+6p+9$

$\qquad\qquad p=3$

The number 3 checks. It is the solution.

**33.** $\sqrt{4x-10}=\sqrt{2-x}$

$\qquad (\sqrt{4x-10})^2=(\sqrt{2-x})^2$

$\qquad 4x-10=2-x$

$\qquad 5x=12 \qquad$ Adding 10 and $x$

$\qquad x=\dfrac{12}{5}$

Check: $\dfrac{\sqrt{4x-10}=\sqrt{2-x}}{}$

$\sqrt{4\cdot\dfrac{12}{5}-10} \ \bigg| \ \sqrt{2-\dfrac{12}{5}}$

$\sqrt{\dfrac{48}{5}-10} \ \bigg| \ \sqrt{-\dfrac{2}{5}}$

Since $\sqrt{-\dfrac{2}{5}}$ does not represent a real number, there is no solution.

**35.** $\sqrt{x-5}=5-\sqrt{x}$

$\qquad (\sqrt{x-5})^2=(5-\sqrt{x})^2 \qquad$ Squaring both sides

$\qquad x-5=25-10\sqrt{x}+x$

$\qquad -30=-10\sqrt{x} \qquad$ Isolating the radical

$\qquad 3=\sqrt{x} \qquad$ Dividing by $-10$

$\qquad 3^2=(\sqrt{x})^2 \qquad$ Squaring both sides

$\qquad 9=x$

The number 9 checks. It is the solution.

**37.** $\sqrt{y+8}-\sqrt{y}=2$

$\qquad \sqrt{y+8}=\sqrt{y}+2 \qquad$ Isolating one radical

$\qquad (\sqrt{y+8})^2=(\sqrt{y}+2)^2 \quad$ Squaring both sides

$\qquad y+8=y+4\sqrt{y}+4$

$\qquad 4=4\sqrt{x} \qquad$ Isolating the radical

$\qquad 1=\sqrt{y} \qquad$ Dividing by 4

$\qquad 1^2=(\sqrt{y})^2$

$\qquad 1=y$

The number 1 checks. It is the solution.

**39.** $V=3.5\sqrt{h}$

$\qquad V=3.5\sqrt{24} \qquad$ Substituting 24 for $h$

$\qquad V\approx 3.5(4.898979486)$

$\qquad V\approx 17$

The sailor can see approximately 17 km to the horizon.

**41.**
$$V = 3.5\sqrt{h}$$
$$21.4 = 3.5\sqrt{h} \quad \text{Substituting 21.4 for } V$$
$$\frac{21.4}{3.5} = \sqrt{h}$$
$$\left(\frac{21.4}{3.5}\right)^2 = (\sqrt{h})^2$$
$$37 \approx h$$

The mast is about 37 m high.

**43.**
$$r = 2\sqrt{5L}$$
$$60 = 2\sqrt{5L} \quad \text{Substituting 60 for } r$$
$$30 = \sqrt{5L}$$
$$(30)^2 = (\sqrt{5L})^2$$
$$900 = 5L$$
$$180 = L$$

A car will skid 180 ft at 60 mph.

$$100 = 2\sqrt{5L} \quad \text{Substituting 100 for } r$$
$$50 = \sqrt{5L}$$
$$(50)^2 = (\sqrt{5L})^2$$
$$2500 = 5L$$
$$500 = L$$

A car will skid 500 ft at 100 mph.

**45. *Familiarize*.** Let $n$ = the number.

***Translate*.**

$$\underbrace{\text{The square root of 4 more than 5 times a number}}_{\sqrt{5n+4}} \quad \underbrace{\text{is}}_{=} \quad \underbrace{8.}_{8}$$

***Solve*.**
$$\sqrt{5n+4} = 8$$
$$(\sqrt{5n+4})^2 = 8^2$$
$$5n + 4 = 64$$
$$5n = 60$$
$$n = 12$$

***Check*.** Four more than 5 times 12 is $60 + 4$, or 64, and $\sqrt{64} = 8$. The result checks.

***State*.** The number is 12.

**47.**
$$\frac{x^2 - 49}{x + 8} \div \frac{x^2 - 14x + 49}{x^2 + 15x + 56}$$
$$= \frac{x^2 - 49}{x + 8} \cdot \frac{x^2 + 15x + 56}{x^2 - 14x + 49}$$
$$= \frac{(x^2 - 49)(x^2 + 15x + 56)}{(x + 8)(x^2 - 14x + 49)}$$
$$= \frac{(x + 7)(x - 7)(x + 7)(x + 8)}{(x + 8)(x - 7)(x - 7)}$$
$$= \frac{(x + 7)(x - 7)(x + 7)(x + 8)}{(x + 8)(x - 7)(x - 7)}$$
$$= \frac{(x + 7)(x + 7)}{x - 7}, \text{ or } \frac{(x + 7)^2}{x - 7}$$

**47.**
$$\frac{a^2 - 25}{6} \div \frac{a + 5}{3} = \frac{a^2 - 25}{6} \cdot \frac{3}{a + 5}$$
$$= \frac{(a^2 - 25) \cdot 3}{6(a + 5)}$$
$$= \frac{(a + 5)(a - 5) \cdot 3}{2 \cdot 3 \cdot (a + 5)}$$
$$= \frac{(a + 5)(a - 5) \cdot 3}{2 \cdot 3 \cdot (a + 5)}$$
$$= \frac{a - 5}{2}$$

**51. *Familiarize*.** Let $x$ and $y$ represent the angles. Recall that supplementary angles are angles whose sum is 180°.

***Translate*.** We reword the problem.

$$\underbrace{\text{The sum of two angles}}_{x + y} \quad \underbrace{\text{is}}_{=} \quad \underbrace{180°.}_{180}$$

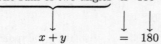

$$\underbrace{\text{One angle}}_{x} \quad \underbrace{\text{is}}_{=} \quad \underbrace{\text{two}}_{2} \quad \underbrace{\text{times}}_{\cdot} \quad \underbrace{\text{the other}}_{y} \quad \underbrace{\text{less}}_{-} \quad \underbrace{3°.}_{3}$$

The resulting system is
$$x + y = 180, \quad (1)$$
$$x = 2y - 3. \quad (2)$$

***Solve*.** We use substitution. We substitute $2y - 3$ for $x$ in Equation (1) and solve for $y$.

$$(2y - 3) + y = 180$$
$$3y - 3 = 180$$
$$3y = 183$$
$$y = 61$$

We substitute 61 for $y$ in Equation (2) to find $x$.

$$x = 2(61) - 3 = 122 - 3 = 119$$

***Check*.** $61° + 119° = 180°$, so the angles are supplementary. Also, 3° less than twice 61° is $2 \cdot 61° - 3°$, or $122° - 3°$, or 119°. The numbers check.

***State*.** The angles are 61° and 119°.

**53.** 
$$\sqrt{5x^2 + 5} = 5$$
$$(\sqrt{5x^2 + 5})^2 = 5^2$$
$$5x^2 + 5 = 25$$
$$5x^2 - 20 = 0$$
$$5(x^2 - 4) = 0$$
$$5(x + 2)(x - 2) = 0$$

$$x + 2 = 0 \quad \text{or} \quad x - 2 = 0$$
$$x = -2 \quad \text{or} \quad x = 2$$

Both numbers check, so the solutions are 2 and $-2$.

**55.** 
$$4 + \sqrt{19 - x} = 6 + \sqrt{4 - x}$$
$$\sqrt{19 - x} = 2 + \sqrt{4 - x} \quad \text{Isolating one radical}$$
$$(\sqrt{19 - x})^2 = (2 + \sqrt{4 - x})^2$$
$$19 - x = 4 + 4\sqrt{4 - x} + (4 - x)$$
$$19 - x = 4\sqrt{4 - x} + 8 - x$$
$$11 = 4\sqrt{4 - x}$$
$$11^2 = (4\sqrt{4 - x})^2$$
$$121 = 16(4 - x)$$
$$121 = 64 - 16x$$
$$57 = -16x$$
$$-\frac{57}{16} = x$$

$-\frac{57}{16}$ checks, so it is the solution.

**57.** 
$$\sqrt{x + 3} = \frac{8}{\sqrt{x - 9}}$$
$$(\sqrt{x + 3})^2 = \left(\frac{8}{\sqrt{x - 9}}\right)^2$$
$$x + 3 = \frac{64}{x - 9}$$
$$(x - 9)(x + 3) = 64 \quad \text{Multiplying by } x - 9$$
$$x^2 - 6x - 27 = 64$$
$$x^2 - 6x - 91 = 0$$
$$(x - 13)(x + 7) = 0$$
$$x - 13 = 0 \quad \text{or} \quad x + 7 = 0$$
$$x = 13 \quad \text{or} \quad x = -7$$

The number 13 checks, but $-7$ does not. The solution is 13.

---

## Exercise Set 8.6

**1.** 
$$a^2 + b^2 = c^2$$
$$8^2 + 15^2 = c^2 \quad \text{Substituting}$$
$$64 + 225 = c^2$$
$$289 = c^2$$
$$\sqrt{289} = c$$
$$17 = c$$

**3.** 
$$a^2 + b^2 = c^2$$
$$4^2 + 4^2 = c^2 \quad \text{Substituting}$$
$$16 + 16 = c^2$$
$$32 = c^2$$
$$\sqrt{32} = c \quad \text{Exact answer}$$
$$5.657 \approx c \quad \text{Approximation}$$

**5.** 
$$a^2 + b^2 = c^2$$
$$5^2 + b^2 = 13^2$$
$$25 + b^2 = 169$$
$$b^2 = 144$$
$$b = 12$$

**7.** 
$$a^2 + b^2 = c^2$$
$$(4\sqrt{3})^2 + b^2 = 8^2$$
$$16 \cdot 3 + b^2 = 64$$
$$48 + b^2 = 64$$
$$b^2 = 16$$
$$b = 4$$

**9.** 
$$a^2 + b^2 = c^2$$
$$10^2 + 24^2 = c^2$$
$$100 + 576 = c^2$$
$$676 = c^2$$
$$26 = c$$

**11.** 
$$a^2 + b^2 = c^2$$
$$9^2 + b^2 = 15^2$$
$$81 + b^2 = 225$$
$$b^2 = 144$$
$$b = 12$$

**13.** 
$$a^2 + b^2 = c^2$$
$$a^2 + 1^2 = (\sqrt{5})^2$$
$$a^2 + 1 = 5$$
$$a^2 = 4$$
$$a = 2$$

**15.** 
$$a^2 + b^2 = c^2$$
$$1^2 + b^2 = (\sqrt{3})^2$$
$$1 + b^2 = 3$$
$$b^2 = 2$$
$$b = \sqrt{2} \quad \text{Exact answer}$$
$$b \approx 1.414 \quad \text{Approximation}$$

**17.**
$$a^2 + b^2 = c^2$$
$$a^2 + (5\sqrt{3})^2 = 10^2$$
$$a^2 + 25 \cdot 3 = 100$$
$$a^2 + 75 = 100$$
$$a^2 = 25$$
$$a = 5$$

**19.** We first make a drawing. We label the diagonal $d$.

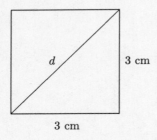

We know that $3^2 + 3^2 = d^2$. We solve this equation.
$$3^2 + 3^2 = d^2$$
$$9 + 9 = d^2$$
$$18 = d^2$$
$$\sqrt{18} \text{ cm} = d \quad \text{Exact answer}$$
$$4.243 \text{ cm} \approx d \quad \text{Approximation}$$

**21.** We first make a drawing. We label the diagonal $d$.

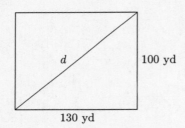

We know that $130^2 + 100^2 = d^2$. We solve this equation.
$$130^2 + 100^2 = d^2$$
$$16,900 + 10,000 = d^2$$
$$26,900 = d^2$$
$$\sqrt{26,900} \text{ yd} = d \quad \text{Exact answer}$$
$$164.012 \text{ yd} \approx d \quad \text{Approximation}$$

**23.** We use the drawing in the text, labeling the horizontal distance $h$.

We know that $4100^2 + h^2 = 15,100^2$. We solve this equation.
$$16,810,000 + h^2 = 228,010,000$$
$$h^2 = 211,200,000$$
$$h = \sqrt{211,200,000} \text{ ft} \quad \text{Exact answer}$$
$$h \approx 14,533 \text{ ft} \quad \text{Approximation}$$

**25.**
$$5x + 7 = 8y,$$
$$3x = 8y - 4$$
$$5x - 8y = -7 \quad (1) \quad \text{Rewriting}$$
$$3x - 8y = -4 \quad (2) \quad \text{the equations}$$
We multiply Equation (2) by $-1$ and add.
$$5x - 8y = -7$$
$$\underline{-3x + 8y = \phantom{-}4}$$
$$2x \phantom{- 8y} = -3$$
$$x = -\frac{3}{2}$$
Substitute $-\frac{3}{2}$ for $x$ in Equation (1) and solve for $y$.
$$5x - 8y = -7$$
$$5\left(-\frac{3}{2}\right) - 8y = -7$$
$$-\frac{15}{2} - 8y = -7$$
$$-8y = \frac{1}{2}$$
$$y = -\frac{1}{16}$$
The ordered pair $\left(-\frac{3}{2}, -\frac{1}{16}\right)$ checks. It is the solution.

**27.**
$$3x - 4y = -11 \quad (1)$$
$$5x + 6y = 12 \quad (2)$$
We multiply Equation (1) by 3 and Equation (2) by 2, and then we add.
$$9x - 12y = -33$$
$$\underline{10x + 12y = \phantom{-}24}$$
$$19x \phantom{+ 12y} = -9$$
$$x = -\frac{9}{19}$$
Substitute $-\frac{9}{19}$ for $x$ in Equation (2) and solve for $y$.

$$5x + 6y = 12$$

$$5\left(-\frac{9}{19}\right) + 6y = 12$$

$$-\frac{45}{19} + 6y = 12$$

$$6y = \frac{273}{19} \quad \text{Adding } \frac{45}{19}$$

$$y = \frac{273}{6 \cdot 19} \quad \text{Dividing by 6}$$

$$y = \frac{91}{38} \quad \text{Simplifying}$$

The ordered pair $\left(-\frac{9}{19}, \frac{91}{38}\right)$ checks. It is the solution.

**29.** After one-half hour, the car traveling east has gone $\frac{1}{2} \cdot 50$, or 25 mi, and the car traveling south has gone $\frac{1}{2} \cdot 60$, or 30 mi. We make a drawing. We label the distance between the cards $d$.

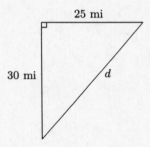

We know that $30^2 + 25^2 = d^2$. We solve this equation.

$$30^2 + 25^2 = d^2$$
$$900 + 625 = d^2$$
$$1525 = d^2$$
$$\sqrt{1525} \text{ mi} = d \quad \text{Exact answer}$$
$$39.1 \text{ mi} \approx d \quad \text{Approximation}$$

**31.**

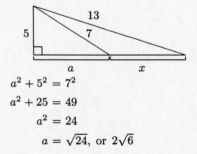

$$a^2 + 5^2 = 7^2$$
$$a^2 + 25 = 49$$
$$a^2 = 24$$
$$a = \sqrt{24}, \text{ or } 2\sqrt{6}$$

$$(a + x)^2 + 5^2 = 13^2$$
$$(2\sqrt{6} + x)^2 + 5^2 = 13^2 \quad \text{Substituting } 2\sqrt{6} \text{ for } a$$
$$(2\sqrt{6} + x)^2 + 25 = 169$$
$$(2\sqrt{6} + x)^2 = 144$$
$$2\sqrt{6} + x = 12 \quad \text{Taking the principal square root}$$
$$x = 12 - 2\sqrt{6}$$
$$x \approx 7.101$$

**33.** Using the Pythagorean equation we can label the figure with additional information.

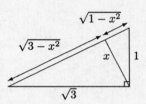

Next we use the Pythagorean equation with the largest right triangle and solve for $x$.

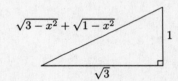

$$(\sqrt{3})^2 + 1^2 = (\sqrt{3 - x^2} + \sqrt{1 - x^2})^2$$
$$3 + 1 = (3 - x^2) + 2\sqrt{(3 - x^2)(1 - x^2)} + (1 - x^2)$$
$$4 = 4 - 2x^2 + 2\sqrt{3 - 4x^2 + x^4}$$
$$\text{Collecting like terms}$$
$$2x^2 = 2\sqrt{3 - 4x^2 + x^4}$$
$$\text{Subtracting 4 and adding } 2x^2$$
$$x^2 = \sqrt{3 - 4x^2 + x^4} \quad \text{Dividing by 2}$$
$$(x^2)^2 = (\sqrt{3 - 4x^2 + x^4})^2$$
$$x^4 = 3 - 4x^2 + x^4$$
$$4x^2 = 3 \quad \text{Subtracting } x^4 \text{ and adding } 4x^2$$
$$x^2 = \frac{3}{4}$$
$$x = \sqrt{\frac{3}{4}}$$
$$x = \frac{\sqrt{3}}{2} \quad \text{Exact answer}$$
$$x \approx 0.866 \quad \text{Approximation}$$

# Chapter 9

# Quadratic Equations

---

**Exercise Set 9.1**

---

**1.** $x^2 - 3x + 2 = 0$

This equation is already in standard form.

$a = 1,\ b = -3,\ c = 2$

**3.** $\qquad 7x^2 = 4x - 3$

$7x^2 - 4x + 3 = 0 \qquad$ Standard form

$a = 7,\ b = -4,\ c = 3$

**5.** $\qquad 5 = -2x^2 + 3x$

$2x^2 - 3x + 5 = 0 \qquad$ Standard form

$a = 2,\ b = -3,\ c = 5$

**7.** $\quad x^2 + 5x = 0$

$x(x + 5) = 0$

$x = 0 \qquad$ or $\qquad x + 5 = 0$

$x = 0 \qquad$ or $\qquad x = -5$

The solutions are 0 and $-5$.

**9.** $\quad 3x^2 + 6x = 0$

$3x(x + 2) = 0$

$3x = 0 \qquad$ or $\qquad x + 2 = 0$

$x = 0 \qquad$ or $\qquad x = -2$

The solutions are 0 and $-2$.

**11.** $\qquad 5x^2 = 2x$

$5x^2 - 2x = 0$

$x(5x - 2) = 0$

$x = 0 \qquad$ or $\qquad 5x - 2 = 0$

$x = 0 \qquad$ or $\qquad 5x = 2$

$x = 0 \qquad$ or $\qquad x = \dfrac{2}{5}$

The solutions are 0 and $\dfrac{2}{5}$.

**13.** $\quad 4x^2 + 4x = 0$

$4x(x + 1) = 0$

$4x = 0 \qquad$ or $\qquad x + 1 = 0$

$x = 0 \qquad$ or $\qquad x = -1$

The solutions are 0 and $-1$.

**15.** $\quad 0 = 10x^2 - 30x$

$0 = 10x(x - 3)$

$10x = 0 \qquad$ or $\qquad x - 3 = 0$

$x = 0 \qquad$ or $\qquad x = 3$

The solutions are 0 and 3.

**17.** $\qquad 11x = 55x^2$

$0 = 55x^2 - 11x$

$0 = 11x(5x - 1)$

$11x = 0 \qquad$ or $\qquad 5x - 1 = 0$

$x = 0 \qquad$ or $\qquad 5x = 1$

$x = 0 \qquad$ or $\qquad x = \dfrac{1}{5}$

The solutions are 0 and $\dfrac{1}{5}$.

**19.** $\qquad 14t^2 = 3t$

$14t^2 - 3t = 0$

$t(14t - 3) = 0$

$t = 0 \qquad$ or $\qquad 14t - 3 = 0$

$t = 0 \qquad$ or $\qquad 14t = 3$

$t = 0 \qquad$ or $\qquad t = \dfrac{3}{14}$

The solutions are 0 and $\dfrac{3}{14}$.

**21.** $\quad 5y^2 - 3y^2 = 72y + 9y$

$\qquad\qquad 2y^2 = 81y$

$\quad 2y^2 - 81y = 0$

$\quad y(2y - 81) = 0$

$y = 0 \qquad$ or $\qquad 2y - 81 = 0$

$y = 0 \qquad$ or $\qquad 2y = 81$

$y = 0 \qquad$ or $\qquad y = \dfrac{81}{2}$

The solutions are 0 and $\dfrac{81}{2}$.

**23.** $\quad x^2 + 8x - 48 = 0$

$(x + 12)(x - 4) = 0$

$x + 12 = 0 \qquad$ or $\qquad x - 4 = 0$

$\qquad x = -12 \qquad$ or $\qquad\qquad x = 4$

The solutions are $-12$ and 4.

**25.** $\quad 5 + 6x + x^2 = 0$

$(5 + x)(1 + x) = 0$

$5 + x = 0 \qquad$ or $\qquad 1 + x = 0$

$\quad x = -5 \qquad$ or $\qquad\quad x = -1$

The solutions are $-5$ and $-1$.

**27.** $\quad 18 = 7p + p^2$

$0 = p^2 + 7p - 18$

$0 = (p + 9)(p - 2)$

$p + 9 = 0 \qquad$ or $\qquad p - 2 = 0$

$\quad p = -9 \qquad$ or $\qquad\quad p = 2$

The solutions are $-9$ and 2.

**29.**  $-15 = -8y + y^2$
$0 = y^2 - 8y + 15$
$0 = (y - 5)(y - 3)$

$y - 5 = 0$    or    $y - 3 = 0$
$y = 5$    or    $y = 3$

The solutions are 5 and 3.

**31.**  $x^2 + 10x + 25 = 0$
$(x + 5)(x + 5) = 0$

$x + 5 = 0$    or    $x + 5 = 0$
$x = -5$    or    $x = -5$

The solution is $-5$.

**33.**  $r^2 = 8r - 16$
$r^2 - 8r + 16 = 0$
$(r - 4)(r - 4) = 0$

$r - 4 = 0$    or    $r - 4 = 0$
$r = 4$    or    $r = 4$

The solution is 4.

**35.**  $6x^2 + x - 2 = 0$
$(3x + 2)(2x - 1) = 0$

$3x + 2 = 0$    or    $2x - 1 = 0$
$3x = -2$    or    $2x = 1$
$x = -\dfrac{2}{3}$    or    $x = \dfrac{1}{2}$

The solutions are $-\dfrac{2}{3}$ and $\dfrac{1}{2}$.

**37.**  $3a^2 = 10a + 8$
$3a^2 - 10a - 8 = 0$
$(3a + 2)(a - 4) = 0$

$3a + 2 = 0$    or    $a - 4 = 0$
$3a = -2$    or    $a = 4$
$a = -\dfrac{2}{3}$    or    $a = 4$

The solutions are $-\dfrac{2}{3}$ and 4.

**39.**  $6x^2 - 4x = 10$
$6x^2 - 4x - 10 = 0$
$2(3x^2 - 2x - 5) = 0$
$2(3x - 5)(x + 1) = 0$

$3x - 5 = 0$    or    $x + 1 = 0$
$3x = 5$    or    $x = -1$
$x = \dfrac{5}{3}$    or    $x = -1$

The solutions are $\dfrac{5}{3}$ and $-1$.

**41.**  $2t^2 + 12t = -10$
$2t^2 + 12t + 10 = 0$
$2(t^2 + 6t + 5) = 0$
$2(t + 5)(t + 1) = 0$

$t + 5 = 0$    or    $t + 1 = 0$
$t = -5$    or    $t = -1$

The solutions are $-5$ and $-1$.

**43.**  $t(t - 5) = 14$
$t^2 - 5t = 14$
$t^2 - 5t - 14 = 0$
$(t + 2)(t - 7) = 0$

$t + 2 = 0$    or    $t - 7 = 0$
$t = -2$    or    $t = 7$

The solutions are $-2$ and 7.

**45.**  $t(9 + t) = 4(2t + 5)$
$9t + t^2 = 8t + 20$
$t^2 + t - 20 = 0$
$(t + 5)(t - 4) = 0$

$t + 5 = 0$    or    $t - 4 = 0$
$t = -5$    or    $t = 4$

The solutions are $-5$ and 4.

**47.**  $16(p - 1) = p(p + 8)$
$16p - 16 = p^2 + 8p$
$0 = p^2 - 8p + 16$
$0 = (p - 4)(p - 4)$

$p - 4 = 0$    or    $p - 4 = 0$
$p = 4$    or    $p = 4$

The solution is 4.

**49.**  $(t - 1)(t + 3) = t - 1$
$t^2 + 2t - 3 = t - 1$
$t^2 + t - 2 = 0$
$(t + 2)(t - 1) = 0$

$t + 2 = 0$    or    $t - 1 = 0$
$t = -2$    or    $t = 1$

The solutions are $-2$ and 1.

**51.**  $\dfrac{24}{x - 2} + \dfrac{24}{x + 2} = 5$

The LCM is $(x - 2)(x + 2)$.

$$(x - 2)(x + 2)\left(\dfrac{24}{x - 2} + \dfrac{24}{x + 2}\right) =$$
$$(x - 2)(x + 2) \cdot 5$$

$$(x - 2)(x + 2) \cdot \dfrac{24}{x - 2} + (x - 2)(x + 2) \cdot \dfrac{24}{x + 2} =$$
$$5(x - 2)(x + 2)$$
$$24(x + 2) + 24(x - 2) =$$
$$5(x^2 - 4)$$
$$24x + 48 + 24x - 48 = 5x^2 - 20$$
$$48x = 5x^2 - 20$$
$$0 = 5x^2 - 48x - 20$$
$$0 = (5x + 2)(x - 10)$$

$5x + 2 = 0$    or    $x - 10 = 0$
$5x = -2$    or    $x = 10$
$x = -\dfrac{2}{5}$    or    $x = 10$

Both numbers check. The solutions are $-\dfrac{2}{5}$ and 10.

**53.**
$$\frac{1}{x} + \frac{1}{x+6} = \frac{1}{4}$$
The LCM is $4x(x+6)$.

$$4x(x+6)\left(\frac{1}{x} + \frac{1}{x+6}\right) = 4x(x+6) \cdot \frac{1}{4}$$

$$4x(x+6) \cdot \frac{1}{x} + 4x(x+6) \cdot \frac{1}{x+6} = x(x+6)$$

$$4(x+6) + 4x = x(x+6)$$
$$4x + 24 + 4x = x^2 + 6x$$
$$8x + 24 = x^2 + 6x$$
$$0 = x^2 - 2x - 24$$
$$0 = (x-6)(x+4)$$

$$x - 6 = 0 \quad \text{or} \quad x + 4 = 0$$
$$x = 6 \quad \text{or} \quad x = -4$$

Both numbers check. The solutions are 6 and $-4$.

**55.**
$$1 + \frac{12}{x^2 - 4} = \frac{3}{x - 2}$$
The LCM is $(x+2)(x-2)$.

$$(x+2)(x-2)\left(1 + \frac{12}{(x+2)(x-2)}\right) =$$
$$(x+2)(x-2) \cdot \frac{3}{x-2}$$

$$(x+2)(x-2) \cdot 1 + (x+2)(x-2) \cdot \frac{12}{(x+2)(x-2)} =$$
$$3(x+2)$$

$$x^2 - 4 + 12 = 3x + 6$$
$$x^2 + 8 = 3x + 6$$
$$x^2 - 3x + 2 = 0$$
$$(x-2)(x-1) = 0$$

$$x - 2 = 0 \quad \text{or} \quad x - 1 = 0$$
$$x = 2 \quad \text{or} \quad x = 1$$

The number 1 checks, but 2 does not. (It makes the denominators $x^2 - 4$ and $x - 2$ zero.) The solution is 1.

**57.**
$$\frac{r}{r-1} + \frac{2}{r^2 - 1} = \frac{8}{r+1}$$
The LCM is $(r-1)(r+1)$.

$$(r-1)(r+1)\left(\frac{r}{r-1} + \frac{2}{(r-1)(r+1)}\right) =$$
$$(r-1)(r+1) \cdot \frac{8}{r+1}$$

$$(r-1)(r+1) \cdot \frac{r}{r-1} + (r-1)(r+1) \cdot \frac{2}{(r-1)(r+1)} =$$
$$8(r-1)$$

$$r(r+1) + 2 = 8(r-1)$$
$$r^2 + r + 2 = 8r - 8$$
$$r^2 - 7r + 10 = 0$$
$$(r-5)(r-2) = 0$$

$$r - 5 = 0 \quad \text{or} \quad r - 2 = 0$$
$$r = 5 \quad \text{or} \quad r = 2$$

Both numbers check. The solutions are 5 and 2.

**59.**
$$\frac{x-1}{1-x} = -\frac{x+8}{x-8}$$
The LCM is $(1-x)(x-8)$.

$$(1-x)(x-8) \cdot \frac{x-1}{1-x} = (1-x)(x-8)\left(-\frac{x+8}{x-8}\right)$$
$$(x-8)(x-1) = -(1-x)(x+8)$$
$$x^2 - 9x + 8 = -(x + 8 - x^2 - 8x)$$
$$x^2 - 9x + 8 = -(-x^2 - 7x + 8)$$
$$x^2 - 9x + 8 = x^2 + 7x - 8$$
$$16 = 16x$$
$$1 = x$$

The number 1 does not check. (It makes the denominator $1 - x$ zero.) There is no solution.

**61.** *Familiarize.* We will use the formula
$$d = \frac{n^2 - 3n}{2},$$
where $d$ is the number of diagonals and $n$ is the number of sides.

*Translate.* We substitute 10 for $n$.
$$d = \frac{10^2 - 3 \cdot 10}{2}$$

*Solve.* We do the computation.
$$d = \frac{10^2 - 3 \cdot 10}{2} = \frac{100 - 30}{2} = \frac{70}{2} = 35$$

*Check.* We can recheck our computation. We can also substitute 35 for $d$ in the original formula and determine whether this yields $n = 10$. Our result checks.

*State.* A decagon has 35 diagonals.

**63.** *Familiarize.* We will use the formula
$$d = \frac{n^2 - 3n}{2},$$
where $d$ is the number of diagonals and $n$ is the number of sides.

*Translate.* We substitute 14 for $d$.
$$14 = \frac{n^2 - 3n}{2}$$

*Solve.* We solve the equation.
$$\frac{n^2 - 3n}{2} = 14$$
$$n^2 - 3n = 28 \quad \text{Multiplying by 2}$$
$$n^2 - 3n - 28 = 0$$
$$(n-7)(n+4) = 0$$

$$n - 7 = 0 \quad \text{or} \quad n + 4 = 0$$
$$n = 7 \quad \text{or} \quad n = -4$$

*Check.* Since the number of sides cannot be negative, $-4$ cannot be a solution. To check 7, we substitute 7 for $n$ in the original formula and determine if this yields $d = 14$. Our result checks.

*State.* The polygon has 7 sides.

**65.** $\sqrt{20} = \sqrt{4 \cdot 5} = \sqrt{4}\sqrt{5} = 2\sqrt{5}$

**67.** $\sqrt{\dfrac{3240}{2560}} = \sqrt{\dfrac{81 \cdot 40}{64 \cdot 40}} = \sqrt{\dfrac{81}{64}} = \dfrac{\sqrt{81}}{\sqrt{64}} = \dfrac{9}{8}$

**69.** $\sqrt{405} = \sqrt{81 \cdot 5} = \sqrt{81}\sqrt{5} = 9\sqrt{5}$

**71.**
$$4m^2 - (m+1)^2 = 0$$
$$4m^2 - (m^2 + 2m + 1) = 0$$
$$4m^2 - m^2 - 2m - 1 = 0$$
$$3m^2 - 2m - 1 = 0$$
$$(3m+1)(m-1) = 0$$

$$\begin{array}{rcl}
3m+1 = 0 & \text{or} & m-1 = 0 \\
3m = -1 & \text{or} & m = 1 \\
m = -\dfrac{1}{3} & \text{or} & m = 1
\end{array}$$

The solutions are $-\dfrac{1}{3}$ and 1.

**73.**
$$\sqrt{5}x^2 - x = 0$$
$$x(\sqrt{5}x - 1) = 0$$

$$\begin{array}{rcl}
x = 0 & \text{or} & \sqrt{5}x - 1 = 0 \\
x = 0 & \text{or} & \sqrt{5}x = 1 \\
x = 0 & \text{or} & x = \dfrac{1}{\sqrt{5}}, \text{ or } \dfrac{\sqrt{5}}{5}
\end{array}$$

The solutions are 0 and $\dfrac{\sqrt{5}}{5}$.

**75.**
$$\frac{5}{y+4} - \frac{3}{y-2} = 4$$

The LCM is $(y+4)(y-2)$.

$$(y+4)(y-2)\left(\frac{5}{y+4} - \frac{3}{y-2}\right) = (y+4)(y-2) \cdot 4$$
$$5(y-2) - 3(y+4) = 4(y^2 + 2y - 8)$$
$$5y - 10 - 3y - 12 = 4y^2 + 8y - 32$$
$$2y - 22 = 4y^2 + 8y - 32$$
$$0 = 4y^2 + 6y - 10$$
$$0 = 2(2y^2 + 3y - 5)$$
$$0 = 2(2y+5)(y-1)$$

$$\begin{array}{rcl}
2y + 5 = 0 & \text{or} & y - 1 = 0 \\
2y = -5 & \text{or} & y = 1 \\
y = -\dfrac{5}{2} & \text{or} & y = 1
\end{array}$$

The solutions are $-\dfrac{5}{2}$ and 1.

**77.**
$$ax^2 + bx = 0$$
$$x(ax + b) = 0$$

$$\begin{array}{rcl}
x = 0 & \text{or} & ax + b = 0 \\
x = 0 & \text{or} & ax = -b \\
x = 0 & \text{or} & x = -\dfrac{b}{a}
\end{array}$$

The solutions are 0 and $-\dfrac{b}{a}$.

**79.** $z - 10\sqrt{z} + 9 = 0$

Let $x = \sqrt{z}$. The $x^2 = z$. We substitute $x$ for $\sqrt{z}$ and $x^2$ for $z$.

$$x^2 - 10x + 9 = 0$$
$$(x-9)(x-1) = 0$$

$$\begin{array}{rcl}
x - 9 = 0 & \text{or} & x - 1 = 0 \\
x = 9 & \text{or} & x = 1 \\
\sqrt{z} = 9 & \text{or} & \sqrt{z} = 1 \quad \text{Substituting } \sqrt{z} \text{ for } x \\
(\sqrt{z})^2 = 9^2 & \text{or} & (\sqrt{z})^2 = 1^2 \\
z = 81 & \text{or} & z = 1
\end{array}$$

The solutions are 81 and 1.

## Exercise Set 9.2

**1.** $x^2 = 121$

$x = 11$ or $x = -11$     Principle of square roots

The solutions are 11 and $-11$.

**3.** $5x^2 = 35$

$\quad x^2 = 7$     Dividing by 5

$x = \sqrt{7}$ or $x = -\sqrt{7}$     Principle of square roots

The solutions are $\sqrt{7}$ and $-\sqrt{7}$.

**5.** $5x^2 = 3$

$\quad x^2 = \dfrac{3}{5}$

$$x = \sqrt{\frac{3}{5}} \quad \text{or} \quad x = -\sqrt{\frac{3}{5}} \qquad \begin{array}{l}\text{Principle of}\\ \text{square roots}\end{array}$$

$$x = \sqrt{\frac{3}{5} \cdot \frac{5}{5}} \quad \text{or} \quad x = -\sqrt{\frac{3}{5} \cdot \frac{5}{5}} \qquad \begin{array}{l}\text{Rationalizing}\\ \text{denominators}\end{array}$$

$$x = \frac{\sqrt{15}}{5} \quad \text{or} \quad x = -\frac{\sqrt{15}}{5}$$

The solutions are $\dfrac{\sqrt{15}}{5}$ and $-\dfrac{\sqrt{15}}{5}$.

**7.** $4x^2 - 25 = 0$

$\quad 4x^2 = 25$

$\quad\quad x^2 = \dfrac{25}{4}$

$x = \dfrac{5}{2}$ or $x = -\dfrac{5}{2}$

The solutions are $\dfrac{5}{2}$ and $-\dfrac{5}{2}$.

**9.** $3x^2 - 49 = 0$

$\quad 3x^2 = 49$

$\quad\quad x^2 = \dfrac{49}{3}$

$$x = \frac{7}{\sqrt{3}} \quad \text{or} \quad x = -\frac{7}{\sqrt{3}}$$

$$x = \frac{7}{\sqrt{3}} \cdot \frac{\sqrt{3}}{\sqrt{3}} \quad \text{or} \quad x = -\frac{7}{\sqrt{3}} \cdot \frac{\sqrt{3}}{\sqrt{3}}$$

$$x = \frac{7\sqrt{3}}{3} \quad \text{or} \quad x = -\frac{7\sqrt{3}}{3}$$

The solutions are $\dfrac{7\sqrt{3}}{3}$ and $-\dfrac{7\sqrt{3}}{3}$.

**11.** $4y^2 - 3 = 9$

$\quad 4y^2 = 12$

$\quad\quad y^2 = 3$

$y = \sqrt{3}$ or $y = -\sqrt{3}$

The solutions are $\sqrt{3}$ and $-\sqrt{3}$.

**13.** $49y^2 - 64 = 0$

$\quad 49y^2 = 64$

$\quad\quad y^2 = \dfrac{64}{49}$

$y = \dfrac{8}{7}$ or $y = -\dfrac{8}{7}$

The solutions are $\dfrac{8}{7}$ and $-\dfrac{8}{7}$.

**15.** $(x+3)^2 = 16$

$x + 3 = 4$ or $x + 3 = -4$    Principle of square roots
$x = 1$ or $x = -7$

The solutions are 1 and $-7$.

**17.** $(x+3)^2 = 21$

$x + 3 = \sqrt{21}$    or    $x + 3 = -\sqrt{21}$    Principle of square roots

$x = -3 + \sqrt{21}$ or    $x = -3 - \sqrt{21}$

The solutions are $-3 + \sqrt{21}$ and $-3 - \sqrt{21}$, or $-3 \pm \sqrt{21}$.

**19.** $(x+13)^2 = 8$

$x + 13 = \sqrt{8}$    or    $x + 13 = -\sqrt{8}$
$x + 13 = 2\sqrt{2}$    or    $x + 13 = -2\sqrt{2}$
$x = -13 + 2\sqrt{2}$ or    $x = -13 - 2\sqrt{2}$

The solutions are $-13 + 2\sqrt{2}$ and $-13 - 2\sqrt{2}$, or $-13 \pm 2\sqrt{2}$.

**21.** $(x-7)^2 = 12$

$x - 7 = \sqrt{12}$    or    $x - 7 = -\sqrt{12}$
$x - 7 = 2\sqrt{3}$    or    $x - 7 = -2\sqrt{3}$
$x = 7 + 2\sqrt{3}$ or    $x = 7 - 2\sqrt{3}$

The solutions are $7 + 2\sqrt{3}$ and $7 - 2\sqrt{3}$, or $7 \pm 2\sqrt{3}$.

**23.** $(x+9)^2 = 34$

$x + 9 = \sqrt{34}$    or    $x + 9 = -\sqrt{34}$
$x = -9 + \sqrt{34}$ or    $x = -9 - \sqrt{34}$

The solutions are $-9 + \sqrt{34}$ and $-9 - \sqrt{34}$, or $-9 \pm \sqrt{34}$.

**25.** $\left(x + \dfrac{3}{2}\right)^2 = \dfrac{7}{2}$

$x + \dfrac{3}{2} = \sqrt{\dfrac{7}{2}}$    or    $x + \dfrac{3}{2} = -\sqrt{\dfrac{7}{2}}$

$x = -\dfrac{3}{2} + \sqrt{\dfrac{7}{2}}$    or    $x = -\dfrac{3}{2} - \sqrt{\dfrac{7}{2}}$

$x = -\dfrac{3}{2} + \sqrt{\dfrac{7}{2} \cdot \dfrac{2}{2}}$    or    $x = -\dfrac{3}{2} - \sqrt{\dfrac{7}{2} \cdot \dfrac{2}{2}}$

$x = -\dfrac{3}{2} + \dfrac{\sqrt{14}}{2}$    or    $x = -\dfrac{3}{2} - \dfrac{\sqrt{14}}{2}$

$x = \dfrac{-3 + \sqrt{14}}{2}$    or    $x = \dfrac{-3 - \sqrt{14}}{2}$

The solutions are $\dfrac{-3 \pm \sqrt{14}}{2}$.

**27.** $x^2 - 6x + 9 = 64$
$(x - 3)^2 = 64$    Factoring the left side

$x - 3 = 8$    or    $x - 3 = -8$    Principle of square roots
$x = 11$    or    $x = -5$

The solutions are 11 and $-5$.

**29.** $x^2 + 14x + 49 = 64$
$(x + 7)^2 = 64$    Factoring the left side

$x + 7 = 8$    or    $x + 7 = -8$    Principle of square roots

$x = 1$    or    $x = -15$

The solutions are 1 and $-15$.

**31.** $x^2 - 6x - 16 = 0$
$x^2 - 6x = 16$    Adding 16
$x^2 - 6x + 9 = 16 + 9$    Adding 9: $\left(\dfrac{-6}{2}\right)^2 = (-3)^2 = 9$

$(x - 3)^2 = 25$

$x - 3 = 5$ or    $x - 3 = -5$    Principle of square roots

$x = 8$ or    $x = -2$

The solutions are 8 and $-2$.

**33.** $x^2 + 22x + 21 = 0$
$x^2 + 22x = -21$    Subtracting 21
$x^2 + 22x + 121 = -21 + 121$    Adding 121: $\left(\dfrac{22}{2}\right)^2 = 11^2 = 121$

$(x + 11)^2 = 100$

$x + 11 = 10$    or    $x + 11 = -10$    Principle of square roots

$x = -1$ or    $x = -21$

The solutions are $-1$ and $-21$.

**35.** $x^2 - 2x - 5 = 0$
$x^2 - 2x = 5$
$x^2 - 2x + 1 = 5 + 1$    Adding 1: $\left(\dfrac{-2}{2}\right)^2 = (-1)^2 = 1$

$(x - 1)^2 = 6$

$x - 1 = \sqrt{6}$    or    $x - 1 = -\sqrt{6}$
$x = 1 + \sqrt{6}$ or    $x = 1 - \sqrt{6}$

The solutions are $1 \pm \sqrt{6}$.

**37.** $x^2 - 22x + 102 = 0$
$x^2 - 22x = -102$
$x^2 - 22x + 121 = -102 + 121$    Adding 121:
$\left(\dfrac{-22}{2}\right)^2 = (-11)^2 = 121$

$(x - 11)^2 = 19$

$x - 11 = \sqrt{19}$    or    $x - 11 = -\sqrt{19}$
$x = 11 + \sqrt{19}$ or    $x = 11 - \sqrt{19}$

The solutions are $11 \pm \sqrt{19}$.

**39.** $x^2 + 10x - 4 = 0$

$x^2 + 10x \quad\quad = 4$

$x^2 + 10x + 25 = 4 + 25$ Adding 25: $\left(\dfrac{10}{2}\right)^2 =$
$$5^2 = 25$$

$(x + 5)^2 = 29$

$x + 5 = \sqrt{29}$ or $x + 5 = -\sqrt{29}$

$x = -5 + \sqrt{29}$ or $x = -5 - \sqrt{29}$

The solutions are $-5 \pm \sqrt{29}$.

**41.** $x^2 - 7x - 2 = 0$

$x^2 - 7x \quad\quad = 2$

$x^2 - 7x + \dfrac{49}{4} = 2 + \dfrac{49}{4}$ Adding $\dfrac{49}{4}$:
$$\left(\dfrac{-7}{2}\right)^2 = \dfrac{49}{4}$$

$\left(x - \dfrac{7}{2}\right)^2 = \dfrac{8}{4} + \dfrac{49}{4} = \dfrac{57}{4}$

$x - \dfrac{7}{2} = \dfrac{\sqrt{57}}{2}$ or $x - \dfrac{7}{2} = -\dfrac{\sqrt{57}}{2}$

$x = \dfrac{7}{2} + \dfrac{\sqrt{57}}{2}$ or $x = \dfrac{7}{2} - \dfrac{\sqrt{57}}{2}$

$x = \dfrac{7 + \sqrt{57}}{2}$ or $x = \dfrac{7 - \sqrt{57}}{2}$

The solutions are $\dfrac{7 \pm \sqrt{57}}{2}$.

**43.** $x^2 + 3x - 28 = 0$

$x^2 + 3x \quad\quad = 28$

$x^2 + 3x + \dfrac{9}{4} = 28 + \dfrac{9}{4}$ Adding $\dfrac{9}{4}$: $\left(\dfrac{3}{2}\right)^2 = \dfrac{9}{4}$

$\left(x + \dfrac{3}{2}\right)^2 = \dfrac{121}{4}$

$x + \dfrac{3}{2} = \dfrac{11}{2}$ or $x + \dfrac{3}{2} = -\dfrac{11}{2}$

$x = \dfrac{8}{2}$ or $x = -\dfrac{14}{2}$

$x = 4$ or $x = -7$

The solutions are 4 and $-7$.

**45.** $x^2 + \dfrac{3}{2}x - \dfrac{1}{2} = 0$

$x^2 + \dfrac{3}{2}x \quad\quad = \dfrac{1}{2}$

$x^2 + \dfrac{3}{2}x + \dfrac{9}{16} = \dfrac{1}{2} + \dfrac{9}{16}$ Adding $\dfrac{9}{16}$: $\left(\dfrac{3/2}{2}\right)^2 =$
$$\left(\dfrac{3}{4}\right)^2 = \dfrac{9}{16}$$

$\left(x + \dfrac{3}{4}\right)^2 = \dfrac{17}{16}$

$x + \dfrac{3}{4} = \dfrac{\sqrt{17}}{4}$ or $x + \dfrac{3}{4} = -\dfrac{\sqrt{17}}{4}$

$x = -\dfrac{3}{4} + \dfrac{\sqrt{17}}{4}$ or $x = -\dfrac{3}{4} - \dfrac{\sqrt{17}}{4}$

$x = \dfrac{-3 + \sqrt{17}}{4}$ or $x = \dfrac{-3 - \sqrt{17}}{4}$

The solutions are $\dfrac{-3 \pm \sqrt{17}}{4}$.

**47.** $2x^2 + 3x - 17 = 0$

$\dfrac{1}{2}(2x^2 + 3x - 17) = \dfrac{1}{2} \cdot 0$ Multiplying by $\dfrac{1}{2}$ to make the $x^2$-coefficient 1

$x^2 + \dfrac{3}{2}x - \dfrac{17}{2} = 0$

$x^2 + \dfrac{3}{2}x \quad\quad = \dfrac{17}{2}$

$x^2 + \dfrac{3}{2}x + \dfrac{9}{16} = \dfrac{17}{2} + \dfrac{9}{16}$ Adding $\dfrac{9}{16}$: $\left(\dfrac{3/2}{2}\right)^2 =$
$$\left(\dfrac{3}{4}\right)^2 = \dfrac{9}{16}$$

$\left(x + \dfrac{3}{4}\right)^2 = \dfrac{145}{16}$

$x + \dfrac{3}{4} = \dfrac{\sqrt{145}}{4}$ or $x + \dfrac{3}{4} = -\dfrac{\sqrt{145}}{4}$

$x = \dfrac{-3 + \sqrt{145}}{4}$ or $x = \dfrac{-3 - \sqrt{145}}{4}$

The solutions are $\dfrac{-3 \pm \sqrt{145}}{4}$.

**49.** $3x^2 + 4x - 1 = 0$

$\dfrac{1}{3}(3x^2 + 4x - 1) = \dfrac{1}{3} \cdot 0$

$x^2 + \dfrac{4}{3}x - \dfrac{1}{3} = 0$

$x^2 + \dfrac{4}{3}x \quad\quad = \dfrac{1}{3}$

$x^2 + \dfrac{4}{3}x + \dfrac{4}{9} = \dfrac{1}{3} + \dfrac{4}{9}$

$\left(x + \dfrac{2}{3}\right)^2 = \dfrac{7}{9}$

$x + \dfrac{2}{3} = \dfrac{\sqrt{7}}{3}$ or $x + \dfrac{2}{3} = -\dfrac{\sqrt{7}}{3}$

$x = \dfrac{-2 + \sqrt{7}}{3}$ or $x = -\dfrac{-2 - \sqrt{7}}{3}$

The solutions are $\dfrac{-2 \pm \sqrt{7}}{3}$.

**51.** $2x^2 = 9x + 5$

$2x^2 - 9x - 5 = 0$ Standard form

$\dfrac{1}{2}(2x^2 - 9x - 5) = \dfrac{1}{2} \cdot 0$

$x^2 - \dfrac{9}{2}x - \dfrac{5}{2} = 0$

$x^2 - \dfrac{9}{2}x \quad\quad = \dfrac{5}{2}$

$x^2 - \dfrac{9}{2}x + \dfrac{81}{16} = \dfrac{5}{2} + \dfrac{81}{16}$

$\left(x - \dfrac{9}{4}\right)^2 = \dfrac{121}{16}$

$$x - \frac{9}{4} = \frac{11}{4} \quad \text{or} \quad x - \frac{9}{4} = -\frac{11}{4}$$
$$x = \frac{20}{4} \quad \text{or} \quad x = -\frac{2}{4}$$
$$x = 5 \quad \text{or} \quad x = -\frac{1}{2}$$

The solutions are 5 and $-\frac{1}{2}$.

**53.**
$$6x^2 + 11x = 10$$
$$6x^2 + 11x - 10 = 0 \qquad \text{Standard form}$$
$$\frac{1}{6}(6x^2 + 11x - 10) = \frac{1}{6} \cdot 0$$
$$x^2 + \frac{11}{6}x - \frac{5}{3} = 0$$
$$x^2 + \frac{11}{6}x = \frac{5}{3}$$
$$x^2 + \frac{11}{6}x + \frac{121}{144} = \frac{5}{3} + \frac{121}{144}$$
$$\left(x + \frac{11}{12}\right)^2 = \frac{361}{144}$$
$$x + \frac{11}{12} = \frac{19}{12} \quad \text{or} \quad x + \frac{11}{12} = -\frac{19}{12}$$
$$x = \frac{8}{12} \quad \text{or} \quad x = -\frac{30}{12}$$
$$x = \frac{2}{3} \quad \text{or} \quad x = -\frac{5}{2}$$

The solutions are $\frac{2}{3}$ and $-\frac{5}{2}$.

**55. *Familiarize.*** We will use the formula $s = 16t^2$.

***Translate.*** We substitute 1451 for $s$.
$$1451 = 16t^2$$

***Solve.*** We solve the equation.
$$1451 = 16t^2$$
$$\frac{1451}{16} = t^2 \qquad \text{Solving for } t^2$$
$$90.6875 = t^2 \qquad \text{Dividing}$$
$$\sqrt{90.6875} = t \quad \text{or} \quad -\sqrt{90.6875} = t \qquad \text{Principle of square roots}$$
$$9.5 \approx t \quad \text{or} \qquad -9.5 \approx t \qquad \text{Using a calculator and rounding to the nearest tenth}$$

***Check.*** The number $-9.5$ cannot be a solution, because time cannot be negative in this situation. We substitute 9.5 in the original equation.
$$s = 16(9.5)^2 = 16(90.25) = 1444$$

This is close. Remember that we approximated a solution. Thus we have a check.

***State.*** It takes about 9.5 sec for an object to fall to the ground from the top of the Sears Tower.

**57. *Familiarize.*** We will use the formula $s = 16t^2$.

***Translate.*** We substitute 311 for $s$.
$$311 = 16t^2$$

***Solve.*** We solve the equation.
$$311 = 16t^2$$
$$\frac{311}{16} = t^2 \qquad \text{Solving for } t^2$$
$$19.4375 = t^2 \qquad \text{Dividing}$$
$$\sqrt{19.4375} = t \quad \text{or} \quad -\sqrt{19.4375} = t \qquad \text{Principle of square roots}$$
$$4.4 \approx t \quad \text{or} \qquad -4.4 \approx t \qquad \text{Using a calculator and rounding to the nearest tenth}$$

***Check.*** The number $-4.4$ cannot be a solution, because time cannot be negative in this situation. We substitute 4.4 in the original equation.
$$s = 16(4.4)^2 = 16(19.36) = 309.76$$

This is close. Remember that we approximated a solution. Thus we have a check.

***State.*** The fall took approximately 4.4 sec.

**59.**
$$y = \frac{k}{x} \qquad \text{Inverse variation}$$
$$235 = \frac{k}{0.6} \qquad \text{Substituting 0.6 for } x \text{ and 235 for } y$$
$$141 = k \qquad \text{Constant of variation}$$
$$y = \frac{141}{x} \qquad \text{Equation of variation}$$

**61.** $\sqrt{3x} \cdot \sqrt{6x} = \sqrt{18x^2} = \sqrt{9 \cdot x^2 \cdot 2} = \sqrt{9}\sqrt{x^2}\sqrt{2} = 3x\sqrt{2}$

**63.** $3\sqrt{t} \cdot \sqrt{t} = 3\sqrt{t^2} = 3t$

**65.** $x^2 + bx + 36$

The trinomial is a square if the square of one-half the $x$-coefficient is equal to 36. Thus we have:
$$\left(\frac{b}{2}\right)^2 = 36$$
$$\frac{b^2}{4} = 36$$
$$b^2 = 144$$
$$b = 12 \text{ or } b = -12 \qquad \text{Principle of square roots}$$

**67.** $x^2 + bx + 128$

The trinomial is a square if the square of one-half the $x$-coefficient is equal to 128. Thus we have:
$$\left(\frac{b}{2}\right)^2 = 128$$
$$\frac{b^2}{4} = 128$$
$$b^2 = 512$$
$$b = \sqrt{512} \quad \text{or} \quad b = -\sqrt{512}$$
$$b = 16\sqrt{2} \quad \text{or} \quad b = -16\sqrt{2}$$

**69.** $x^2 + bx + c$

The trinomial is a square if the square of one-half the $x$-coefficient is equal to $c$. Thus we have:

$$\left(\frac{b}{2}\right)^2 = c$$

$$\frac{b^2}{4} = c$$

$$b^2 = 4c$$

$b = \sqrt{4c}$  or  $b = -\sqrt{4c}$
$b = 2\sqrt{c}$  or  $b = -2\sqrt{c}$

**71.**  $4.82x^2 = 12{,}000$

$$x^2 = \frac{12{,}000}{4.82}$$

$x = \sqrt{\dfrac{12{,}000}{4.82}}$  or  $x = -\sqrt{\dfrac{12{,}000}{4.82}}$  Principle of square roots

$x \approx 4.896$  or  $x \approx -4.896$  Using a calculator and rounding

The solutions are approximately 4.896 and −4.896.

**73.**    $\dfrac{x}{2} = \dfrac{32}{x}$,    LCM is $2x$

$$2x \cdot \frac{x}{2} = 2x \cdot \frac{32}{x}$$

$$x^2 = 64$$

$x = 8$ or $x = -8$    Principle of square roots

Both numbers check. The solutions are 8 and −8.

**75.**    $\dfrac{4}{m^2 - 7} = 1$

$4 = m^2 - 7$  Multiplying by $m^2 - 7$

$11 = m^2$

$\sqrt{11} = m$ or $-\sqrt{11} = m$

Both numbers check. The solutions are $\sqrt{11}$ and $-\sqrt{11}$.

## Exercise Set 9.3

**1.**     $x^2 - 4x = 21$
$x^2 - 4x - 21 = 0$    Standard form
$a = 1,\ b = -4,\ c = -21$

We compute the discriminant:

$b^2 - 4ac = (-4)^2 - 4 \cdot 1 \cdot (-21) = 16 + 84 = 100$

The discriminant is a perfect square, so we can factor.

$x^2 - 4x - 21 = 0$
$(x - 7)(x + 3) = 0$

$x - 7 = 0$  or  $x + 3 = 0$
$x = 7$  or     $x = -3$

The solutions are 7 and −3.

**3.**        $x^2 = 6x - 9$
$x^2 - 6x + 9 = 0$    Standard form
$a = 1,\ b = -6,\ c = 9$

We compute the discriminant:

$b^2 - 4ac = (-6)^2 - 4 \cdot 1 \cdot 9 = 36 - 36 = 0$

The discriminant is a perfect square, so we can factor.

$x^2 - 6x + 9 = 0$
$(x - 3)(x - 3) = 0$

$x - 3 = 0$  or  $x - 3 = 0$
$x = 3$  or     $x = 3$

The solution is 3.

**5.**  $3y^2 - 2y - 8 = 0$

$a = 3,\ b = -2,\ c = -8$

We compute the discriminant:

$b^2 - 4ac = (-2)^2 - 4 \cdot 3 \cdot (-8) = 4 + 96 = 100$

The discriminant is a perfect square, so we can factor.

$3y^2 - 2y - 8 = 0$
$(3y + 4)(y - 2) = 0$

$3y + 4 = 0$    or   $y - 2 = 0$
$3y = -4$    or       $y = 2$

$y = -\dfrac{4}{3}$  or       $y = 2$

The solutions are $-\dfrac{4}{3}$ and 2.

**7.**        $4x^2 + 4x = 15$
$4x^2 + 4x - 15 = 0$      Standard form
$a = 4,\ b = 4,\ c = -15$

We compute the discriminant:

$b^2 - 4ac = 4^2 - 4 \cdot 4 \cdot (-15) = 16 + 240 = 256$

The discriminant is a perfect square, so we can factor.

$4x^2 + 4x - 15 = 0$
$(2x - 3)(2x + 5) = 0$

$2x - 3 = 0$  or  $2x + 5 = 0$
$2x = 3$  or     $2x = -5$

$x = \dfrac{3}{2}$  or     $x = -\dfrac{5}{2}$

The solutions are $\dfrac{3}{2}$ and $-\dfrac{5}{2}$.

**9.**        $x^2 - 9 = 0$    Difference of squares
$(x + 3)(x - 3) = 0$

$x + 3 = 0$  or  $x - 3 = 0$
$x = -3$  or     $x = 3$

The solutions are −3 and 3.

**11.** $x^2 - 2x - 2 = 0$

$a = 1,\ b = -2,\ c = -2$

We compute the discriminant:

$b^2 - 4ac = (-2)^2 - 4 \cdot 1 \cdot (-2) = 4 + 8 = 12$

The discriminant is positive, so there are real-number solutions. They are given by

$$x = \frac{-(-2) \pm \sqrt{12}}{2 \cdot 1}$$

$$= \frac{2 \pm \sqrt{12}}{2} = \frac{2 \pm \sqrt{4 \cdot 3}}{2}$$

$$= \frac{2 \pm 2\sqrt{3}}{2} = \frac{2(1 \pm \sqrt{3})}{2}$$

$$= 1 \pm \sqrt{3}.$$

**13.** $y^2 - 10y + 22 = 0$

$a = 1, \ b = -10, \ c = 22$

We compute the discriminant:

$b^2 - 4ac = (-10)^2 - 4 \cdot 1 \cdot 22 = 100 - 88 = 12$

The discriminant is positive, so there are real-number solutions. They are given by

$$y = \frac{-(-10) \pm \sqrt{12}}{2 \cdot 1}$$

$$= \frac{10 \pm \sqrt{12}}{2} = \frac{10 \pm \sqrt{4 \cdot 3}}{2}$$

$$= \frac{10 \pm 2\sqrt{3}}{2} = \frac{2(5 \pm \sqrt{3})}{2}$$

$$= 5 \pm \sqrt{3}.$$

**15.** $x^2 + 4x + 4 = 7$

$x^2 + 4x - 3 = 0$     Adding $-7$ to get standard form

$a = 1, \ b = 4, \ c = -3$

We compute the discriminant:

$b^2 - 4ac = 4^2 - 4 \cdot 1 \cdot (-3) = 16 + 12 = 28$

The discriminant is positive, so there are real-number solutions. They are given by

$$x = \frac{-4 \pm \sqrt{28}}{2 \cdot 1}$$

$$= \frac{-4 \pm \sqrt{28}}{2} = \frac{-4 \pm \sqrt{4 \cdot 7}}{2}$$

$$= \frac{-4 \pm 2\sqrt{7}}{2} = \frac{2(-2 \pm \sqrt{7})}{2}$$

$$= -2 \pm \sqrt{7}.$$

**17.** $3x^2 + 8x + 2 = 0$

$a = 3, \ b = 8, \ c = 2$

We compute the discriminant:

$b^2 - 4ac = 8^2 - 4 \cdot 3 \cdot 2 = 64 - 24 = 40$

The discriminant is positive, so there are real-number solutions. They are given by

$$x = \frac{-8 \pm \sqrt{40}}{2 \cdot 3}$$

$$= \frac{-8 \pm \sqrt{40}}{6} = \frac{-8 \pm \sqrt{4 \cdot 10}}{6}$$

$$= \frac{-8 \pm 2\sqrt{10}}{6} = \frac{2(-4 \pm \sqrt{10})}{2 \cdot 3}$$

$$= \frac{-4 \pm \sqrt{10}}{3}.$$

**19.** $2x^2 - 5x = 1$

$2x^2 - 5x - 1 = 0$     Adding $-1$ to get standard form

$a = 2, \ b = -5, \ c = -1$

We compute the discriminant:

$b^2 - 4ac = (-5)^2 - 4 \cdot 2 \cdot (-1) = 25 + 8 = 33$

The discriminant is positive, so there are real-number solutions. They are given by

$$x = \frac{-(-5) \pm \sqrt{33}}{2 \cdot 2}$$

$$= \frac{5 \pm \sqrt{33}}{4}$$

**21.** $2y^2 - 2y - 1 = 0$

$a = 2, \ b = -2, \ c = -1$

We compute the discriminant:

$b^2 - 4ac = (-2)^2 - 4 \cdot 2 \cdot (-1) = 4 + 8 = 12$

The discriminant is positive, so there are real-number solutions. They are given by

$$x = \frac{-(-2) \pm \sqrt{12}}{2 \cdot 2}$$

$$= \frac{2 \pm \sqrt{12}}{4} = \frac{2 \pm \sqrt{4 \cdot 3}}{4}$$

$$= \frac{2 \pm 2\sqrt{3}}{4} = \frac{2(1 \pm \sqrt{3})}{2 \cdot 2}$$

$$= \frac{1 \pm \sqrt{3}}{2}.$$

**23.** $2t^2 + 6t + 5 = 0$

$a = 2, \ b = 6, \ c = 5$

We compute the discriminant:

$b^2 - 4ac = 6^2 - 4 \cdot 2 \cdot 5 = 36 - 40 = -4$

The discriminant is negative, so there are no real-number solutions.

**25.** $3x^2 = 5x + 4$

$3x^2 - 5x - 4 = 0$

$a = 3, \ b = -5, \ c = -4$

We compute the discriminant:

$b^2 - 4ac = (-5)^2 - 4 \cdot 3 \cdot (-4) = 25 + 48 = 73$

The discriminant is positive, so there are real-number solutions. They are given by

$$x = \frac{-(-5) \pm \sqrt{73}}{2 \cdot 3}$$

$$= \frac{5 \pm \sqrt{73}}{6}.$$

**27.** $2y^2 - 6y = 10$

$2y^2 - 6y - 10 = 0$

$y^2 - 3y - 5 = 0$     Multiplying by $\frac{1}{2}$ to simplify

$a = 1, \ b = -3, \ c = -5$

We compute the discriminant:

$b^2 - 4ac = (-3)^2 - 4 \cdot 1 \cdot (-5) = 9 + 20 = 29$

The discriminant is positive, so there are real-number solutions. They are given by

$$y = \frac{-(-3) \pm \sqrt{29}}{2 \cdot 1}$$
$$= \frac{3 \pm \sqrt{29}}{2}.$$

**29.**
$$\frac{x^2}{x+3} - \frac{5}{x+3} = 0, \quad \text{LCM is } x+3$$
$$(x+3)\left(\frac{x^2}{x+3} - \frac{5}{x+3}\right) = (x+3) \cdot 0$$
$$x^2 - 5 = 0$$
$$x^2 = 5$$

$x = \sqrt{5}$ or $x = -\sqrt{5}$    Principle of square roots

Both numbers check. The solutions are $\sqrt{5}$ and $-\sqrt{5}$.

**31.**
$$x + 2 = \frac{3}{x+2}$$
$$(x+2)(x+2) = (x+2) \cdot \frac{3}{x+2} \quad \text{Clearing the fraction}$$
$$x^2 + 4x + 4 = 3$$
$$x^2 + 4x + 1 = 0$$

$a = 1, \ b = 4, \ c = 1$

We compute the discriminant:

$b^2 - 4ac = 4^2 - 4 \cdot 1 \cdot 1 = 16 - 4 = 12$

The discriminant is positive, so there are real-number solutions. They are given by

$$x = \frac{-4 \pm \sqrt{12}}{2 \cdot 1}$$
$$= \frac{-4 \pm \sqrt{12}}{2} = \frac{-4 \pm \sqrt{4 \cdot 3}}{2}$$
$$= \frac{-4 \pm 2\sqrt{3}}{2} = \frac{2(-2 \pm \sqrt{3})}{2}$$
$$= -2 \pm \sqrt{3}.$$

Both numbers check. The solutions are $-2 + \sqrt{3}$ and $-2 - \sqrt{3}$.

**33.**
$$\frac{1}{x} + \frac{1}{x+1} = \frac{1}{3}, \quad \text{LCM is } 3x(x+1)$$
$$3x(x+1)\left(\frac{1}{x} + \frac{1}{x+1}\right) = 3x(x+1) \cdot \frac{1}{3}$$
$$3(x+1) + 3x = x(x+1)$$
$$3x + 3 + 3x = x^2 + x$$
$$6x + 3 = x^2 + x$$
$$0 = x^2 - 5x - 3$$

$a = 1, \ b = -5, \ c = -3$

We compute the discriminant:

$b^2 - 4ac = (-5)^2 - 4 \cdot 1 \cdot (-3) = 25 + 12 = 37$

The discriminant is positive, so there are real-number solutions. They are given by

$$x = \frac{-(-5) \pm \sqrt{37}}{2 \cdot 1}$$
$$= \frac{5 \pm \sqrt{37}}{2}.$$

**35.** $x^2 - 4x - 7 = 0$

$a = 1, \ b = -4, \ c = -7$

$$x = \frac{-(-4) \pm \sqrt{(-4)^2 - 4 \cdot 1 \cdot (-7)}}{2 \cdot 1}$$
$$= \frac{4 \pm \sqrt{16 + 28}}{2} = \frac{4 \pm \sqrt{44}}{2}$$
$$= \frac{4 \pm \sqrt{4 \cdot 11}}{2} = \frac{4 \pm 2\sqrt{11}}{2}$$
$$= \frac{2(2 \pm \sqrt{11})}{2} = 2 \pm \sqrt{11}$$

Using a calculator or Table 2, we see that $\sqrt{11} \approx 3.317$:

$$2 + \sqrt{11} \approx 2 + 3.317 \quad \text{or} \quad 2 - \sqrt{11} \approx 2 - 3.317$$
$$\approx 5.3 \qquad\qquad \text{or} \qquad\qquad \approx -1.3$$

The approximate solutions, to the nearest tenth, are 5.3 and $-1.3$.

**37.** $y^2 - 6y - 1 = 0$

$a = 1, \ b = -6, \ c = -1$

$$y = \frac{-(-6) \pm \sqrt{(-6)^2 - 4 \cdot 1 \cdot (-1)}}{2 \cdot 1}$$
$$= \frac{6 \pm \sqrt{36 + 4}}{2} = \frac{6 \pm \sqrt{40}}{2}$$
$$= \frac{6 \pm \sqrt{4 \cdot 10}}{2} = \frac{6 \pm 2\sqrt{10}}{2}$$
$$= \frac{2(3 \pm \sqrt{10})}{2} = 3 \pm \sqrt{10}$$

Using a calculator or Table 2, we see that $\sqrt{10} \approx 3.162$:

$$3 + \sqrt{10} \approx 3 + 3.162 \quad \text{or} \quad 3 - \sqrt{10} \approx 3 - 3.162$$
$$\approx 6.2 \qquad\qquad \text{or} \qquad\qquad \approx -0.2$$

The approximate solutions, to the nearest tenth, are 6.2 and $-0.2$.

**39.**
$$4x^2 + 4x = 1$$
$$4x^2 + 4x - 1 = 0 \quad \text{Standard form}$$

$a = 4, \ b = 4, \ c = -1$

$$x = \frac{-4 \pm \sqrt{4^2 - 4 \cdot 4 \cdot (-1)}}{2 \cdot 4}$$
$$= \frac{-4 \pm \sqrt{16 + 16}}{8} = \frac{-4 \pm \sqrt{32}}{8}$$
$$= \frac{-4 \pm \sqrt{16 \cdot 2}}{8} = \frac{-4 \pm 4\sqrt{2}}{8}$$
$$= \frac{4(-1 \pm \sqrt{2})}{4 \cdot 2} = \frac{-1 \pm \sqrt{2}}{2}$$

Using a calculator or Table 2, we see that $\sqrt{2} \approx 1.414$:

$$\frac{-1 + \sqrt{2}}{2} \approx \frac{-1 + 1.414}{2} \quad \text{or} \quad \frac{-1 - \sqrt{2}}{2} \approx \frac{-1 - 1.414}{2}$$
$$\approx \frac{0.414}{2} \qquad \text{or} \qquad \approx \frac{-2.414}{2}$$
$$\approx 0.2 \qquad\quad \text{or} \qquad\quad \approx -1.2$$

The approximate solutions, to the nearest tenth, are 0.2 and $-1.2$.

**41.** $3x^2 - 8x + 2 = 0$

$a = 3, \ b = -8, \ c = 2$

$$x = \frac{-(-8) \pm \sqrt{(-8)^2 - 4 \cdot 3 \cdot 2}}{2 \cdot 3}$$

$$= \frac{8 \pm \sqrt{64 - 24}}{6} = \frac{8 \pm \sqrt{40}}{6}$$

$$= \frac{8 \pm \sqrt{4 \cdot 10}}{6} = \frac{8 \pm 2\sqrt{10}}{6}$$

$$= \frac{2(4 \pm \sqrt{10})}{2 \cdot 3} = \frac{4 \pm \sqrt{10}}{3}$$

Using a calculator or Table 2, we see that $\sqrt{10} \approx 3.162$:

$$\frac{4 + \sqrt{10}}{3} \approx \frac{4 + 3.162}{3} \quad \text{or} \quad \frac{4 - \sqrt{10}}{3} \approx \frac{4 - 3.162}{3}$$

$$\approx \frac{7.162}{3} \quad \text{or} \quad \approx \frac{0.838}{3}$$

$$\approx 2.4 \quad \text{or} \quad \approx 0.3$$

The approximate solutions, to the nearest tenth, are 2.4 and 0.3.

**43.** $\sqrt{3x^2}\sqrt{9x^3} = \sqrt{27x^5} = \sqrt{9x^4 \cdot 3x} = \sqrt{9x^4}\sqrt{3x} = 3x^2\sqrt{3x}$

**45.** $\sqrt{80} = \sqrt{16 \cdot 5} = \sqrt{16}\sqrt{5} = 4\sqrt{5}$

**47.** $\sqrt{18} + \sqrt{50} - 3\sqrt{8} = \sqrt{9 \cdot 2} + \sqrt{25 \cdot 2} - 3\sqrt{4 \cdot 2}$

$$= \sqrt{9}\sqrt{2} + \sqrt{25}\sqrt{2} - 3\sqrt{4}\sqrt{2}$$

$$= 3\sqrt{2} + 5\sqrt{2} - 3 \cdot 2\sqrt{2}$$

$$= 3\sqrt{2} + 5\sqrt{2} - 6\sqrt{2}$$

$$= (3 + 5 - 6)\sqrt{2}$$

$$= 2\sqrt{2}$$

**49.** $5x + x(x - 7) = 0$

$5x + x^2 - 7x = 0$

$x^2 - 2x = 0$    We can factor.

$x(x - 2) = 0$

$x = 0 \quad \text{or} \quad x - 2 = 0$

$x = 0 \quad \text{or} \quad x = 2$

The solutions are 0 and 2.

**51.** $3 - x(x - 3) = 4$

$3 - x^2 + 3x = 4$

$0 = x^2 - 3x + 1$    Standard form

$a = 1, \ b = -3, \ c = 1$

We compute the discriminant:

$b^2 - 4ac = (-3)^2 - 4 \cdot 1 \cdot 1 = 9 - 4 = 5$

The discriminant is positive, so there are real-number solutions. They are given by

$$x = \frac{-(-3) \pm \sqrt{5}}{2 \cdot 1}$$

$$= \frac{3 \pm \sqrt{5}}{2}.$$

**53.** $(y + 4)(y + 3) = 15$

$y^2 + 7y + 12 = 15$

$y^2 + 7y - 3 = 0$    Standard form

$a = 1, \ b = 7, \ c = -3$

We compute the discriminant:

$b^2 - 4ac = 7^2 - 4 \cdot 1 \cdot (-3) = 49 + 12 = 61$

The discriminant is positive, so there are real-number solutions. They are given by

$$y = \frac{-7 \pm \sqrt{61}}{2 \cdot 1}$$

$$= \frac{-7 \pm \sqrt{61}}{2}.$$

**55.** $x^2 + (x + 2)^2 = 7$

$x^2 + x^2 + 4x + 4 = 7$

$2x^2 + 4x + 4 = 7$

$2x^2 + 4x - 3 = 0$    Standard form

$a = 2, \ b = 4, \ c = -3$

We compute the discriminant:

$b^2 - 4ac = 4^2 - 4 \cdot 2 \cdot (-3) = 16 + 24 = 40$

The discriminant is positive, so there are real-number solutions. They are given by

$$x = \frac{-4 \pm \sqrt{40}}{2 \cdot 2} = \frac{-4 \pm \sqrt{4 \cdot 10}}{4}$$

$$= \frac{-4 \pm 2\sqrt{10}}{4} = \frac{2(-2 \pm \sqrt{10})}{2 \cdot 2}$$

$$= \frac{-2 \pm \sqrt{10}}{2}.$$

## Exercise Set 9.4

**1.** $P = 17\sqrt{Q}$

$\dfrac{P}{17} = \sqrt{Q}$    Isolating the radical

$\left(\dfrac{P}{17}\right)^2 = (\sqrt{Q})^2$    Principle of squaring

$\dfrac{P^2}{289} = Q$    Simplifying

**3.** $v = \sqrt{\dfrac{2gE}{m}}$

$v^2 = \left(\sqrt{\dfrac{2gE}{m}}\right)^2$    Principle of squaring

$v^2 = \dfrac{2gE}{m}$

$mv^2 = 2gE$    Multipying by $m$

$\dfrac{mv^2}{2g} = E$    Dividing by $2g$

**5.** $S = 4\pi r^2$

$\dfrac{S}{4\pi} = r^2$    Dividing by $4\pi$

$\sqrt{\dfrac{S}{4\pi}} = r$    Principle of square roots. Assume $r$ is nonnegative.

$\sqrt{\dfrac{1}{4} \cdot \dfrac{S}{\pi}} = r$

$\dfrac{1}{2}\sqrt{\dfrac{S}{\pi}} = r$

7.  $P = kA^2 + mA$

$0 = kA^2 + mA - P$     Standard form

$a = k,\ b = m,\ c = -P$

$A = \dfrac{-b \pm \sqrt{b^2 - 4ac}}{2a}$     Quadratic formula

$= \dfrac{-m \pm \sqrt{m^2 - 4 \cdot k \cdot (-P)}}{2 \cdot k}$     Substituting

$= \dfrac{-m + \sqrt{m^2 + 4kP}}{2k}$     Using the positive root

9.  $c^2 = a^2 + b^2$

$c^2 - b^2 = a^2$

$\sqrt{c^2 - b^2} = a$          Principle of square roots.
Assume $a$ is nonnegative.

11.  $s = 16t^2$

$\dfrac{s}{16} = t^2$

$\sqrt{\dfrac{s}{16}} = t$          Principle of square roots.
Assume $t$ is nonnegative.

$\dfrac{\sqrt{s}}{4} = t$

13.  $A = \pi r^2 + 2\pi r h$

$0 = \pi r^2 + 2\pi h r - A$

$a = \pi,\ b = 2\pi h,\ c = -A$

$r = \dfrac{-b \pm \sqrt{b^2 - 4ac}}{2a}$

$= \dfrac{-2\pi h \pm \sqrt{(2\pi h)^2 - 4 \cdot \pi \cdot (-A)}}{2 \cdot \pi}$

$= \dfrac{-2\pi h + \sqrt{4\pi^2 h^2 + 4\pi A}}{2\pi}$     Using the positive root

$= \dfrac{-2\pi h + \sqrt{4(\pi^2 h^2 + \pi A)}}{2\pi}$

$= \dfrac{-2\pi h + 2\sqrt{\pi^2 h^2 + \pi A}}{2\pi}$

$= \dfrac{2\left(-\pi h + \sqrt{\pi^2 h^2 + \pi A}\right)}{2\pi}$

$= \dfrac{-\pi h + \sqrt{\pi^2 h^2 + \pi A}}{\pi}$

15.
$F = \dfrac{Av^2}{400}$

$400F = Av^2$     Multiplying by 400

$\dfrac{400F}{A} = v^2$     Dividing by $A$

$\sqrt{\dfrac{400F}{A}} = v$     Principle of square roots.
Assume $v$ is nonnegative.

$\sqrt{400 \cdot \dfrac{F}{A}} = v$

$20\sqrt{\dfrac{F}{a}} = v$

17.          $c = \sqrt{a^2 + b^2}$

$c^2 = (\sqrt{a^2 + b^2})^2$     Principle of squaring

$c^2 = a^2 + b^2$

$c^2 - b^2 = a^2$

$\sqrt{c^2 - b^2} = a$          Principle of square roots.
Assume $a$ is nonnegative.

19.          $h = \dfrac{a}{2}\sqrt{3}$

$2h = a\sqrt{3}$

$\dfrac{2h}{\sqrt{3}} = a$

$\dfrac{2h\sqrt{3}}{3} = a$     Rationalizing the denominator

21.  $n = aT^2 - 4T + m$

$0 = aT^2 - 4T + m - n$

$a = a,\ b = -4,\ c = m - n$

$T = \dfrac{-b \pm \sqrt{b^2 - 4ac}}{2a}$

$= \dfrac{-(-4) \pm \sqrt{(-4)^2 - 4 \cdot a \cdot (m - n)}}{2 \cdot a}$

$= \dfrac{4 + \sqrt{16 - 4a(m - n)}}{2a}$     Using the positive root

$= \dfrac{4 + \sqrt{4[4 - a(m - n)]}}{2a}$

$= \dfrac{4 + 2\sqrt{4 - a(m - n)}}{2a}$

$= \dfrac{2\left(2 + \sqrt{4 - a(m - n)}\right)}{2 \cdot a}$

$= \dfrac{2 + \sqrt{4 - a(m - n)}}{a}$

**23.**
$$v = 2\sqrt{\frac{2kT}{\pi m}}$$

$$\frac{v}{2} = \sqrt{\frac{2kT}{\pi m}} \qquad \text{Isolating the radical}$$

$$\left(\frac{v}{2}\right)^2 = \left(\sqrt{\frac{2kT}{\pi m}}\right)^2 \qquad \text{Principle of squaring}$$

$$\frac{v^2}{4} = \frac{2kT}{\pi m}$$

$$\frac{v^2}{4} \cdot \frac{\pi m}{2k} = \frac{2kT}{\pi m} \cdot \frac{\pi m}{2k} \qquad \text{Multiplying by } \frac{\pi m}{2k}$$

$$\frac{v^2 \pi m}{8k} = T$$

**25.** $3x^2 = d^2$

$$x^2 = \frac{d^2}{3} \qquad \text{Dividing by 3}$$

$$x = \frac{d}{\sqrt{3}} \qquad \text{Principle of square roots. Assume } x \text{ is nonnegative.}$$

$$x = \frac{d}{\sqrt{3}} \cdot \frac{\sqrt{3}}{\sqrt{3}} \qquad \text{Rationalizing the denominator}$$

$$x = \frac{d\sqrt{3}}{3}$$

**27.** $N = \dfrac{n^2 - n}{2}$

$$2N = n^2 - n \qquad \text{Multiplying by 2}$$
$$0 = n^2 - n - 2N \qquad \text{Finding standard form}$$
$$a = 1, \ b = -1, \ c = -2N$$

$$n = \frac{-b \pm \sqrt{b^2 - 4ac}}{2a}$$

$$= \frac{-(-1) \pm \sqrt{(-1)^2 - 4 \cdot 1 \cdot (-2N)}}{2 \cdot 1} \qquad \text{Substituting}$$

$$= \frac{1 + \sqrt{1 + 8N}}{2} \qquad \text{Using the positive root}$$

**29.** $a^2 + b^2 = c^2$    Pythagorean equation
$4^2 + 7^2 = c^2$    Substituting
$16 + 49 = c^2$
$65 = c^2$
$\sqrt{65} = c$    Exact answer
$8.062 \approx c$    Approximate answer

**31.** $a^2 + b^2 = c^2$    Pythagorean equation
$4^2 + 5^2 = c^2$    Substituting
$16 + 25 = c^2$
$41 = c^2$
$\sqrt{41} = c$    Exact answer
$6.403 \approx c$    Approximate answer

**33.** $a^2 + b^2 = c^2$    Pythagorean equation
$2^2 + b^2 = (8\sqrt{17})^2$    Substituting
$4 + b^2 = 64 \cdot 17$
$4 + b^2 = 1088$
$b^2 = 1084$
$b = \sqrt{1084}$    Exact answer
$b \approx 32.924$    Approximate answer

**35.** a) $C = 2\pi r$

$$\frac{C}{2\pi} = r$$

b) $A = \pi r^2$

$$A = \pi \cdot \left(\frac{C}{2\pi}\right)^2 \qquad \text{Substituting } \frac{C}{2\pi} \text{ for } r$$

$$A = \pi \cdot \frac{C^2}{4\pi^2}$$

$$A = \frac{C^2}{4\pi}$$

**37.** $3ax^2 - x - 3ax + 1 = 0$
$3ax^2 + (-1 - 3a)x + 1 = 0$
$a = 3a, \ b = -1 - 3a, \ c = 1$

$$x = \frac{-b \pm \sqrt{b^2 - 4ac}}{2a}$$

$$= \frac{-(-1 - 3a) \pm \sqrt{(-1 - 3a)^2 - 4 \cdot 3a \cdot 1}}{2 \cdot 3a}$$

$$= \frac{1 + 3a \pm \sqrt{1 + 6a + 9a^2 - 12a}}{6a}$$

$$= \frac{1 + 3a \pm \sqrt{9a^2 - 6a + 1}}{6a}$$

$$= \frac{1 + 3a \pm \sqrt{(3a - 1)^2}}{6a}$$

$$= \frac{1 + 3a \pm (3a - 1)}{6a}$$

$$x = \frac{1 + 3a + 3a - 1}{6a} \quad \text{or} \quad x = \frac{1 + 3a - 3a + 1}{6a}$$

$$x = \frac{6a}{6a} \quad \text{or} \quad x = \frac{2}{6a}$$

$$x = 1 \quad \text{or} \quad x = \frac{1}{3a}$$

The solutions are 1 and $\dfrac{1}{3a}$.

---

## Exercise Set 9.5

**1.** *Familiarize.* Using the labels on the drawing in the text we have $w$ = the width of the rectangle and $w + 3$ = the length.

*Translate.* Recall that area is length × width. Then we have

$$(w + 3)(w) = 70.$$

*Solve.* We solve the equation.
$$w^2 + 3w = 70$$
$$w^2 + 3w - 70 = 0$$
$$(w + 10)(w - 7) = 0$$

$$w + 10 = 0 \quad \text{or} \quad w - 7 = 0$$
$$w = -10 \quad \text{or} \quad w = 7$$

*Check.* We know that $-10$ is not a solution of the original problem, because width cannot be negative. When $w = 7$, then $w + 3 = 10$, and the area is $10 \cdot 7$, or 70. This checks.

**State.** The width of the rectangle is 7 cm, and the length is 10 cm.

3. **Familiarize.** From the drawing in the text we have $s =$ the length of the shorter leg and $s + 14 =$ the length of the other leg.

**Translate.** We use the Pythagorean equation.
$$s^2 + (s + 14)^2 = 26^2.$$

**Solve.** We solve the equation.
$$
\begin{aligned}
s^2 + s^2 + 28s + 196 &= 676 \\
2s^2 + 28s - 480 &= 0 \\
s^2 + 14s - 240 &= 0 \qquad \text{Dividing by 2} \\
(s + 24)(s - 10) &= 0
\end{aligned}
$$

$$
\begin{aligned}
s + 24 &= 0 \quad \text{or} \quad s - 10 = 0 \\
s &= -24 \quad \text{or} \qquad s = 10
\end{aligned}
$$

**Check.** Since the length of a leg cannot be negative, $-24$ does not check. When $s = 10$, then $s + 14 = 24$, and $10^2 + 24^2 = 100 + 576 = 676 = 24^2$. This checks.

**State.** The legs measure 10 yd and 24 yd.

5. **Familiarize.** We first make a drawing. We let $x$ represent the length. Then $x - 4$ represents the width.

$$
\boxed{320 \text{ cm}^2} \quad x - 4
$$
$$
x
$$

**Translate.** The area is length × width. Thus, we have two expressions for the area of the rectangle: $x(x - 4)$ and 320. This gives us a translation.
$$x(x - 4) = 320.$$

**Solve.** We solve the equation.
$$
\begin{aligned}
x^2 - 4x &= 320 \\
x^2 - 4x - 320 &= 0 \\
(x - 20)(x + 16) &= 0
\end{aligned}
$$

$$
\begin{aligned}
x - 20 &= 0 \quad \text{or} \quad x + 16 = 0 \\
x &= 20 \quad \text{or} \qquad x = -16
\end{aligned}
$$

**Check.** Since the length of a side cannot be negative, $-16$ does not check. But 20 does check. If the length is 20, then the width is $20 - 4$, or 16. The area is $20 \times 16$, or 320. This checks.

**State.** The length is 20 cm, and the width is 16 cm.

7. **Familiarize.** We first make a drawing. We let $x$ represent the width. Then $2x$ represents the length.

$$
\boxed{50 \text{ m}^2} \quad x
$$
$$
2x
$$

**Translate.** The area is length × width. Thus, we have two expressions for the area of the rectangle: $2x \cdot x$ and 50. This gives us a translation.

$2x \cdot x = 50.$

**Solve.** We solve the equation.
$$
\begin{aligned}
2x^2 &= 50 \\
x^2 &= 25
\end{aligned}
$$

$x = 5$ or $x = -5$      Principle of square roots

**Check.** Since the length of a side cannot be negative, $-5$ does not check. But 5 does check. If the width is 5, then the length is $2 \cdot 5$, or 10. The area is $10 \times 5$, or 50. This checks.

**State.** The length is 10 m, and the width is 5 m.

9. **Familiarize.** We first make a drawing. We let $x$ represent the length of one leg. Then $x + 2$ represents the length of the other leg.

**Translate.** We use the Pythagorean equation.
$$x^2 + (x + 2)^2 = 8^2.$$

**Solve.** We solve the equation.
$$
\begin{aligned}
x^2 + x^2 + 4x + 4 &= 64 \\
2x^2 + 4x + 4 &= 64 \\
2x^2 + 4x - 60 &= 0 \\
x^2 + 2x - 30 &= 0 \qquad \text{Dividing by 2}
\end{aligned}
$$
$a = 1$, $b = 2$, $c = -30$
$$
\begin{aligned}
x &= \frac{-2 \pm \sqrt{2^2 - 4 \cdot 1 \cdot (-30)}}{2 \cdot 1} \\
&= \frac{-2 \pm \sqrt{4 + 120}}{2} = \frac{-2 \pm \sqrt{124}}{2} \\
&= \frac{-2 \pm \sqrt{4 \cdot 31}}{2} = \frac{-2 \pm 2\sqrt{31}}{2} \\
&= \frac{2(-1 \pm \sqrt{31})}{2} = -1 \pm \sqrt{31}
\end{aligned}
$$

Using a calculator or Table 2 we find that $\sqrt{31} \approx 5.568$:

$$
\begin{aligned}
-1 + \sqrt{31} &\approx -1 + 5.568 \quad \text{or} \quad -1 - \sqrt{31} \approx -1 - 5.568 \\
&\approx 4.6 \qquad\qquad\qquad \text{or} \qquad\qquad\quad \approx -6.6
\end{aligned}
$$

**Check.** Since the length of a leg cannot be negative, $-6.6$ does not check. But 4.6 does check. If the shorter leg is 4.6, then the other leg is $4.6 + 2$, or 6.6. Then $4.6^2 + 6.6^2 = 21.16 + 43.56 = 64.72$ and using a calculator, $\sqrt{64.72} \approx 8.04 \approx 8$. Note that our check is not exact since we are using an approximation.

**State.** One leg is about 4.6 m, and the other is about 6.6 m long.

11. **Familiarize.** We first make a drawing. We let $x$ represent the width and $x + 2$ the length.

**Translate.** The area is length × width. We have two expressions for the area of the rectangle: $(x+2)x$ and 20. This gives us a translation.

$$(x+2)x = 20.$$

**Solve.** We solve the equation.

$$x^2 + 2x = 20$$
$$x^2 + 2x - 20 = 0$$

$a = 1,\ b = 2,\ c = -20$

$$x = \frac{-2 \pm \sqrt{2^2 - 4 \cdot 1 \cdot (-20)}}{2 \cdot 1}$$

$$= \frac{-2 \pm \sqrt{4 + 80}}{2} = \frac{-2 \pm \sqrt{84}}{2}$$

$$= \frac{-2 \pm \sqrt{4 \cdot 21}}{2} = \frac{-2 \pm 2\sqrt{21}}{2}$$

$$= \frac{2(-1 \pm \sqrt{21})}{2} = -1 \pm \sqrt{21}$$

Using a calculator or Table 2 we find that $\sqrt{21} \approx 4.583$:

$$-1 + \sqrt{21} \approx -1 + 4.583 \quad \text{or} \quad -1 - \sqrt{21} \approx -1 - 4.583$$
$$\approx 3.6 \qquad\qquad \text{or} \qquad\qquad \approx -5.6$$

**Check.** Since the length of a side cannot be negative, $-5.6$ does not check. But 3.6 does check. If the width is 3.6, then the length is $3.6 + 2$, or 5.6. The area is $5.6(3.6)$, or $20.16 \approx 20$. This checks.

**State.** The length is about 5.6 in., and the width is about 3.6 in.

13. **Familiarize.** We make a drawing and label it. We let $w$ = the width of the rectangle and $2w$ = the length.

**Translate.** Recall that area = length × width. Then we have

$$2w \cdot w = 20.$$

**Solve.** We solve the equation.

$$2w^2 = 20$$
$$w^2 = 10 \qquad \text{Dividing by 2}$$

$$w = \sqrt{10} \quad \text{or} \quad w = -\sqrt{10} \qquad \text{Principle of square roots}$$
$$w \approx 3.2 \quad \text{or} \quad w \approx -3.2$$

**Check.** We know that $-3.2$ is not a solution of the original problem, because width cannot be negative. When $w \approx 3.2$, then $2w \approx 6.4$ and the area is about $(6.4)(3.2)$, or $20.48$. This checks, although the check is not exact since we used an approximation for $\sqrt{10}$.

**State.** The length is about 6.4 cm, and the width is about 3.2 cm.

15. **Familiarize.** Using the drawing in the text, we have $x$ = the thickness of the frame, $20 - 2x$ = the width of the picture showing, and $25 - 2x$ = the length of the picture showing.

**Translate.** Recall that area = length × width. Then we have

$$(25 - 2x)(20 - 2x) = 266.$$

**Solve.** We solve the equation.

$$500 - 90x + 4x^2 = 266$$
$$4x^2 - 90x + 234 = 0$$
$$2x^2 - 45x + 117 = 0 \qquad \text{Dividing by 2}$$
$$(2x - 39)(x - 3) = 0$$

$$2x - 39 = 0 \qquad \text{or} \quad x - 3 = 0$$
$$2x = 39 \qquad \text{or} \qquad\quad x = 3$$
$$x = 19.5 \quad \text{or} \qquad\quad x = 3$$

**Check.** The number 19.5 cannot be a solution, because when $x = 19.5$ then $20 - 2x = -19$, and the width cannot be negative. When $x = 3$, then $20 - 2x = 20 - 2 \cdot 3$, or 14 and $25 - 2x = 25 - 2 \cdot 3$, or 19 and $19 \cdot 14 = 266$. This checks.

**State.** The thickness of the frame is 3 cm.

17. **Familiarize.** Referring to the drawing in the text, we complete the table.

|            | $d$ | $r$    | $t$   |
|------------|-----|--------|-------|
| Upstream   | 40  | $r - 3$ | $t_1$ |
| Downstream | 40  | $r + 3$ | $t_2$ |

**Translate.** Using $t = d/r$ and the rows of the table, we have

$$t_1 = \frac{40}{r - 3} \text{ and } t_2 = \frac{40}{r + 3}.$$

Since the total time is 14 hr, $t_1 + t_2 = 14$, and we have

$$\frac{40}{r - 3} + \frac{40}{r + 3} = 14.$$

**Solve.** We solve the equation. We multiply by $(r - 3)(r + 3)$, the LCM of the denominators.

$$(r - 3)(r + 3)\left(\frac{40}{r - 3} + \frac{40}{r + 3}\right) = (r - 3)(r + 3) \cdot 14$$
$$40(r + 3) + 40(r - 3) = 14(r^2 - 9)$$
$$40r + 120 + 40r - 120 = 14r^2 - 126$$
$$80r = 14r^2 - 126$$
$$0 = 14r^2 - 80r - 126$$
$$0 = 7r^2 - 40r - 63$$
$$0 = (7r + 9)(r - 7)$$

$$7r + 9 = 0 \qquad \text{or} \quad r - 7 = 0$$
$$7r = -9 \qquad \text{or} \qquad\quad r = 7$$
$$r = -\frac{9}{7} \quad \text{or} \qquad\quad r = 7$$

**Check.** Since speed cannot be negative, $-\frac{9}{7}$ cannot be a solution. If the speed of the boat is 7 km/h, the speed upstream is $7 - 3$, or 4 km/h, and the speed downstream is $7 + 3$, or 10 km/h. The time upstream is $\frac{40}{4}$, or 10 hr. The time downstream is $\frac{40}{10}$, or 4 hr. The total time is 14 hr. This checks.

**State.** The speed of the boat in still water is 7 km/h.

**19. Familiarize.** We first make a drawing. We let $r$ represent the speed of the boat in still water. Then $r - 4$ is the speed of the boat traveling upstream and $r + 4$ is the speed of the boat traveling downstream.

Upstream
$r - 4$ mph

4 mi

Downstream
$r + 4$ mph

12 mi

We summarize the information in a table.

|            | $d$ | $r$     | $t$   |
|------------|-----|---------|-------|
| Upstream   | 4   | $r - 4$ | $t_1$ |
| Downstream | 12  | $r + 4$ | $t_2$ |

**Translate.** Using $t = d/r$ and the rows of the table, we have

$$t_1 = \frac{4}{r - 4} \text{ and } t_2 = \frac{12}{r + 4}.$$

Since the total time is 2 hr, $t_1 + t_2 = 2$, and we have

$$\frac{4}{r - 4} + \frac{12}{r + 4} = 2.$$

**Solve.** We solve the equation. We multiply by $(r - 4)(r + 4)$, the LCM of the denominators.

$$(r - 4)(r + 4)\left(\frac{4}{r - 4} + \frac{12}{r + 4}\right) = (r - 4)(r + 4) \cdot 2$$
$$4(r + 4) + 12(r - 4) = 2(r^2 - 16)$$
$$4r + 16 + 12r - 48 = 2r^2 - 32$$
$$16r - 32 = 2r^2 - 32$$
$$0 = 2r^2 - 16r$$
$$0 = 2r(r - 8)$$

$$2r = 0 \quad \text{or} \quad r - 8 = 0$$
$$r = 0 \quad \text{or} \qquad r = 8$$

**Check.** If $r = 0$, then the speed upstream, $0 - 4$, would be negative. Since speed cannot be negative, 0 cannot be a solution. If the speed of the boat is 8 mph, the speed upstream is $8 - 4$, or 4 mph, and the speed downstream is $8 + 4$, or 12 mph. The time upstream is $\frac{4}{4}$, or 1 hr. The time downstream is $\frac{12}{12}$, or 1 hr. The total time is 2 hr. This checks.

**State.** The speed of the boat in still water is 8 mph.

**21. Familiarize.** We first make a drawing. We let $r$ represent the speed of the current. Then $10 - r$ is the speed of the boat traveling upstream and $10 + r$ is the speed of the boat traveling downstream.

Upstream
$10 - r$ km/h

12 km

Downstream
$10 + r$ km/h

28 km

We summarize the information in a table.

|            | $d$ | $r$      | $t$   |
|------------|-----|----------|-------|
| Upstream   | 12  | $10 - r$ | $t_1$ |
| Downstream | 28  | $10 + r$ | $t_2$ |

**Translate.** Using $t = d/r$ and the rows of the table, we have

$$t_1 = \frac{12}{10 - r} \text{ and } t_2 = \frac{28}{10 + r}.$$

Since the total time is 4 hr, $t_1 + t_2 = 4$, and we have

$$\frac{12}{10 - r} + \frac{28}{10 + r} = 4.$$

**Solve.** We solve the equation. We multiply by $(10 - r)(10 + r)$, the LCM of the denominators.

$$(10 - r)(10 + r)\left(\frac{12}{10 - r} + \frac{28}{10 + r}\right) =$$
$$(10 - r)(10 + r) \cdot 4$$
$$12(10 + r) + 28(10 - r) = 4(100 - r^2)$$
$$120 + 12r + 280 - 28r = 400 - 4r^2$$
$$400 - 16r = 400 - 4r^2$$
$$4r^2 - 16r = 0$$
$$4r(r - 4) = 0$$

$$4r = 0 \quad \text{or} \quad r - 4 = 0$$
$$r = 0 \quad \text{or} \qquad r = 4$$

**Check.** Since a stream is defined to be a flow of running water, its rate must be greater than 0. Thus, 0 cannot be a solution. If the speed of the current is 4 km/h, the speed upstream is $10 - 4$, or 6 km/h, and the speed downstream is $10 + 4$, or 14 km/h. The time upstream is $\frac{12}{6}$, or 2 hr. The time downstream is $\frac{28}{14}$, or 2 hr. The total time is 4 hr. This checks.

**State.** The speed of the stream is 4 km/h.

**23. Familiarize.** We first make a drawing. We let $r$ represent the speed of the wind. Then the speed of the plane flying against the wind is $200 - r$ and the speed of the plane flying with the wind is $200 + r$.

Against the wind
$200 - r$ mph

738 miles

With the wind
$200 + r$ mph

1062 miles

We summarize the information in a table.

|             | $d$  | $r$       | $t$   |
|-------------|------|-----------|-------|
| Against wind | 738  | $200 - r$ | $t_1$ |
| With wind    | 1062 | $200 + r$ | $t_2$ |

**Translate.** Using $t = d/r$ and the rows of the table, we have

$t_1 = \dfrac{738}{200 - r}$ and $t_2 = \dfrac{1062}{200 + r}$.

Since the total time is 9 hr, $t_1 + t_2 = 9$, and we have

$$\dfrac{738}{200 - r} + \dfrac{1062}{200 + r} = 9.$$

**Solve.** We solve the equation. We multiply by $(200 - r)(200 + r)$, the LCM of the denominators.

$$(200 - r)(200 + r)\left(\dfrac{738}{200 - r} + \dfrac{1062}{200 + r}\right) =$$
$$(200 - r)(200 + r) \cdot 9$$
$$738(200 + r) + 1062(200 - r) =$$
$$9(40{,}000 - r^2)$$
$$147{,}600 + 738r + 212{,}400 - 1062r = 360{,}000 - 9r^2$$
$$360{,}000 - 324r = 360{,}000 - 9r^2$$
$$9r^2 - 324r = 0$$
$$9r(r - 36) = 0$$

$$9r = 0 \quad \text{or} \quad r - 36 = 0$$
$$r = 0 \quad \text{or} \quad r = 36$$

**Check.** In this problem we assume there is a wind. Thus, the speed of the wind must be greater than 0 and the number 0 cannot be a solution. If the speed of the wind is 36 mph, the speed of the airplane against the wind is $200 - 36$, or 164 mph, and the speed with the wind is $200 + 36$, or 236 mph. The time against the wind is $\dfrac{738}{164}$, or $4\dfrac{1}{2}$ hr. The time with the wind is $\dfrac{1062}{236}$, or $4\dfrac{1}{2}$ hr. The total time is 9 hr. The value checks.

**State.** The speed of the wind is 36 mph.

**25. Familiarize.** We first make a drawing. We let $r$ represent the speed of the stream. Then $9 - r$ represents the speed of the boat traveling upstream and $9 + r$ represents the speed of the boat traveling downstream.

Upstream
$9 - r$ km/h

80 km

Downstream
$9 + r$ km/h

80 km

We summarize the information in a table.

|            | $d$ | $r$     | $t$   |
|------------|-----|---------|-------|
| Upstream   | 80  | $9 - r$ | $t_1$ |
| Downstream | 80  | $9 + r$ | $t_2$ |

**Translate.** Using $t = d/r$ and the rows of the table, we have

$$t_1 = \dfrac{80}{9 - r} \text{ and } t_2 = \dfrac{80}{9 + r}.$$

Since the total time is 18 hr, $t_1 + t_2 = 18$, and we have

$$\dfrac{80}{9 - r} + \dfrac{80}{9 + r} = 18.$$

**Solve.** We solve the equation. We multiply by $(9 - r)(9 + r)$, the LCM of the denominators.

$$(9 - r)(9 + r)\left(\dfrac{80}{9 - r} + \dfrac{80}{9 + r}\right) = (9 - r)(9 + r) \cdot 18$$
$$80(9 + r) + 80(9 - r) = 18(81 - r^2)$$
$$720 + 80r + 720 - 80r = 1458 - 18r^2$$
$$1440 = 1458 - 18r^2$$
$$18r^2 = 18$$
$$r^2 = 1$$

$$r = 1 \text{ or } r = -1 \qquad \text{Principle of square roots}$$

**Check.** Since speed cannot be negative, $-1$ cannot be a solution. If the speed of the stream is 1 km/h, the speed upstream is $9 - 1$, or 8 km/h, and the speed downstream is $9 + 1$, or 10 km/h. The time upstream is $\dfrac{80}{8}$, or 10 hr. The time downstream is $\dfrac{80}{10}$, or 8 hr. The total time is 18 hr. This checks.

**State.** The speed of the stream is 1 km/h.

**27.**
$$5\sqrt{2} + \sqrt{18} = 5\sqrt{2} + \sqrt{9 \cdot 2}$$
$$= 5\sqrt{2} + \sqrt{9}\sqrt{2}$$
$$= 5\sqrt{2} + 3\sqrt{2}$$
$$= (5 + 3)\sqrt{2}$$
$$= 8\sqrt{2}$$

**29.**
$$\sqrt{4x^3} - 7\sqrt{x} = \sqrt{4 \cdot x^2 \cdot x} - 7\sqrt{x}$$
$$= \sqrt{4}\sqrt{x^2}\sqrt{x} - 7\sqrt{x}$$
$$= 2x\sqrt{x} - 7\sqrt{x}$$
$$= (2x - 7)\sqrt{x}$$

**31.**
$$\sqrt{2} + \sqrt{\dfrac{1}{2}} = \sqrt{2} + \sqrt{\dfrac{1}{2} \cdot \dfrac{2}{2}}$$
$$= \sqrt{2} + \sqrt{\dfrac{2}{4}}$$
$$= \sqrt{2} + \dfrac{\sqrt{2}}{\sqrt{4}}$$
$$= \sqrt{2} + \dfrac{\sqrt{2}}{2}$$
$$= \left(1 + \dfrac{1}{2}\right)\sqrt{2}$$
$$= \dfrac{3}{2}\sqrt{2}, \text{ or } \dfrac{3\sqrt{2}}{2}$$

**33. Familiarize.** Let $x =$ one integer and $x + 1 =$ the other. Then the squares of the numbers are $x^2$ and $(x + 1)^2$, respectively. Note that the difference of the squares of the numbers is 25, a positive number. Now when $x$ is nonnegative, then $x + 1$ is greater than $x$, so $(x + 1)^2 - x^2$ will be a positive number. On the other hand, when $x$ is negative, then $x^2$ is greater than $(x + 1)^2$, so $x^2 - (x + 1)^2$ will be a positive number. We must translate to two equations in order to allow for both possibilities.

**Translate.** When $x$ is nonnegative, we have

$$(x + 1)^2 - x^2 = 25.$$

When $x$ is negative, we have

$x^2 - (x+1)^2 = 25$.

**Solve.** We solve the first equation.

$$x^2 + 2x + 1 - x^2 = 25$$
$$2x + 1 = 25$$
$$2x = 24$$
$$x = 12$$

Next we solve the second equation.

$$x^2 - (x^2 + 2x + 1) = 25$$
$$x^2 - x^2 - 2x - 1 = 25$$
$$-2x - 1 = 25$$
$$-2x = 26$$
$$x = -13$$

**Check.** When $x = 12$, then $x + 1 = 13$ and $13^2 - 12^2 = 169 - 144$, or 25. When $x = -13$, then $x + 1 = -12$ and $(-13)^2 - (-12)^2 = 169 - 144$, or 25. Both pairs of integers check.

**State.** The integers and 12 and 13 or −13 and −12.

**35. Familiarize.** From the drawing in the text, we see that we have a right triangle where $r$ = the length of each leg and $r + 1$ = the length of the hypotenuse.

**Translate.** We use the Pythagorean equation.

$$r^2 + r^2 = (r+1)^2.$$

**Solve.** We solve the equation.

$$2r^2 = r^2 + 2r + 1$$
$$r^2 - 2r - 1 = 0$$

$a = 1, \; b = -2, \; c = -1$

$$r = \frac{-(-2) \pm \sqrt{(-2)^2 - 4 \cdot 1 \cdot (-1)}}{2 \cdot 1}$$

$$= \frac{2 \pm \sqrt{4 + 4}}{2} = \frac{2 \pm \sqrt{8}}{2}$$

$$= \frac{2 \pm \sqrt{4 \cdot 2}}{2} = \frac{2 \pm 2\sqrt{2}}{2}$$

$$= \frac{2(1 \pm \sqrt{2})}{2 \cdot 1} = 1 \pm \sqrt{2}$$

$x = 1 - \sqrt{2}$    or    $x = 1 + \sqrt{2}$
$x \approx 1 - 1.414$   or   $x \approx 1 + 1.414$
$x \approx -0.414$    or    $x \approx 2.414$
$x \approx -0.41$    or    $x \approx 2.41$    Rounding to the nearest hundredth

**Check.** Since the length of a leg cannot be negative, −0.41 cannot be a solution of the original equation. When $x \approx 2.41$, then $x + 1 \approx 3.41$ and $(2.41)^2 + (2, 41)^2 = 5.8081 + 5.8081 = 11.6162 \approx (3.41)^2$. This checks.

**State.** In the figure, $r = 1 + \sqrt{2} \approx 2.41$ cm.

**37. Familiarize.** The radius of a 10-in. pizza is $\frac{10}{2}$, or 5 in.

The radius of a $d$-in. pizza is $\frac{d}{2}$ in. The area of a circle is $\pi r^2$.

**Translate.**

| Area of $d$-in. pizza | is | Area of 10-in. pizza | plus | Area of 10-in. pizza |
|:---:|:---:|:---:|:---:|:---:|
| ↓ | ↓ | ↓ | ↓ | ↓ |
| $\pi\left(\dfrac{d}{2}\right)^2$ | $=$ | $\pi \cdot 5^2$ | $+$ | $\pi \cdot 5^2$ |

**Solve.** We solve the equation.

$$\frac{d^2}{4}\pi = 25\pi + 25\pi$$

$$\frac{d^2}{4}\pi = 50\pi$$

$$\frac{d^2}{4} = 50 \qquad \text{Dividing by } \pi$$

$$d^2 = 200$$

$d = \sqrt{200}$   or   $d = -\sqrt{200}$
$d = 10\sqrt{2}$   or   $d = -10\sqrt{2}$
$d \approx 14.14$   or   $d \approx -14.14$    Using a calculator or Table 2

**Check.** Since the diameter cannot be negative, −14.14 is not a solution. If $d = 10\sqrt{2}$, or 14.14, then $r = 5\sqrt{2}$ and the area is $\pi(5\sqrt{2})^2$, or $50\pi$. The area of the two 10-in. pizzas is $2 \cdot \pi \cdot 5^2$, or $50\pi$. The value checks.

**State.** The diameter of the pizza should be $10\sqrt{2} \approx 14.14$ in.

The area of two 10-in. pizzas is approximately the same as a 14-in. pizza. Thus, you get more to eat with two 10-in. pizzas than with a 13-in. pizza.

## Exercise Set 9.6

**1.** $y = x^2 + 1$

We first find the vertex. The $x$-coordinate is

$$-\frac{b}{2a} = -\frac{0}{2 \cdot 1} = 0.$$

We substitute into the equation to find the second coordinate of the vertex.

$$y = x^2 + 1 = 0^2 + 1 = 1$$

The vertex is (0,1). The line of symmetry is $x = 0$, the $y$-axis.

We choose some $x$-values on both sides of the vertex and graph the parabola.

When $x = 1$, $y = 1^2 + 1 = 1 + 1 = 2$.

When $x = -1$, $y = (-1)^2 + 1 = 1 + 1 = 2$.

When $x = 2$, $y = 2^2 + 1 = 4 + 1 = 5$.

When $x = -2$, $y = (-2)^2 + 1 = 4 + 1 = 5$.

| $x$ | $y$ | |
|---|---|---|
| 0 | 1 | ←Vertex |
| 1 | 2 | |
| −1 | 2 | |
| 2 | 5 | |
| −2 | 5 | |

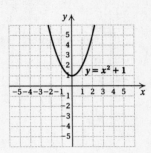

| $x$ | $y$ | |
|---|---|---|
| 1 | 1 | ←Vertex |
| 0 | 0 | ←$y$-intercept |
| −1 | −3 | |
| 2 | 0 | |
| 3 | −3 | |

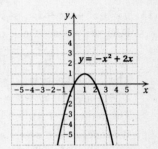

**3.** $y = -1 \cdot x^2$

Find the vertex. The $x$-coordinate is
$$-\frac{b}{2a} = -\frac{0}{2(-1)} = 0.$$
The $y$-coordinate is
$$y = -1 \cdot x^2 = -1 \cdot 0^2 = 0.$$
The vertex is $(0,0)$. The line of symmetry is $x = 0$, the $y$-axis.

Choose some $x$-values on both sides of the vertex and graph the parabola.

When $x = -2$, $y = -1 \cdot (-2)^2 = -1 \cdot 4 = -4$.

When $x = -1$, $y = -1 \cdot (-1)^2 = -1 \cdot 1 = -1$.

When $x = 1$, $y = -1 \cdot 1^2 = -1 \cdot 1 = -1$.

When $x = 2$, $y = -1 \cdot 2^2 = -1 \cdot 4 = -4$.

| $x$ | $y$ | |
|---|---|---|
| 0 | 0 | ←Vertex |
| −2 | −4 | |
| −1 | −1 | |
| 1 | −1 | |
| 2 | −4 | |

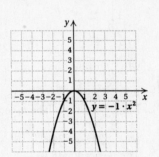

**5.** $y = -x^2 + 2x$

Find the vertex. The $x$-coordinate is
$$-\frac{b}{2a} = -\frac{2}{2(-1)} = -(-1) = 1.$$
The $y$-coordinate is
$$y = -x^2 + 2x = -(1)^2 + 2 \cdot 1 = -1 + 2 = 1.$$
The vertex is $(1,1)$.

We choose some $x$-values on both sides of the vertex and graph the parabola. We make sure we find $y$ when $x = 0$. This gives us the $y$-intercept.

**7.** $y = 5 - x - x^2$, or $y = -x^2 - x + 5$

Find the vertex. The $x$-coordinate is
$$-\frac{b}{2a} = -\frac{-1}{2(-1)} = -\frac{1}{2}.$$
The $y$-coordinate is
$$y = 5 - x - x^2 = 5 - \left(-\frac{1}{2}\right) - \left(-\frac{1}{2}\right)^2 = 5 + \frac{1}{2} - \frac{1}{4} = \frac{21}{4}.$$
The vertex is $\left(-\frac{1}{2}, \frac{21}{4}\right)$.

We choose some $x$-values on both sides of the vertex and graph the parabola.

| $x$ | $y$ | |
|---|---|---|
| $-\frac{1}{2}$ | $\frac{21}{4}$ | ←Vertex |
| 0 | 5 | ←$y$-intercept |
| −1 | 5 | |
| −2 | 3 | |
| 1 | 3 | |

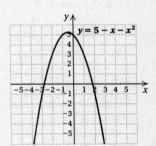

**9.** $y = x^2 - 2x + 1$

Find the vertex. The $x$-coordinate is
$$-\frac{b}{2a} = -\frac{-2}{2 \cdot 1} = -(-1) = 1.$$
The $y$-coordinate is
$$y = x^2 - 2x + 1 = 1^2 - 2 \cdot 1 + 1 = 1 - 2 + 1 = 0.$$
The vertex is $(1, 0)$.

We choose some $x$-values on both sides of the vertex and graph the parabola.

| $x$ | $y$ | |
|---|---|---|
| 1 | 0 | ←Vertex |
| 0 | 1 | ←$y$-intercept |
| −1 | 4 | |
| 2 | 1 | |
| 3 | 4 | |

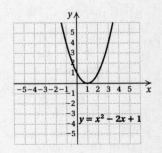

**11.** $y = -x^2 + 2x + 3$

Find the vertex. The $x$-coordinate is

$$-\frac{b}{2a} = -\frac{2}{2(-1)} = -(-1) = 1.$$

The $y$-coordinate is

$$y = -x^2 + 2x + 3 = -(1)^2 + 2 \cdot 1 + 3 = -1 + 2 + 3 = 4.$$

The vertex is $(1, 4)$.

We choose some $x$-values on both sides of the vertex and graph the parabola.

| $x$ | $y$ | |
|---|---|---|
| 1 | 4 | ←Vertex |
| 0 | 3 | ←$y$-intercept |
| −1 | 0 | |
| 2 | 3 | |
| 3 | 0 | |

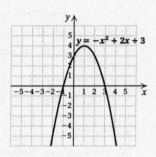

**13.** $y = -2x^2 - 4x + 1$

Find the vertex. The $x$-coordinate is

$$-\frac{b}{2a} = -\frac{-4}{2(-2)} = -1.$$

The $y$-coordinate is

$$y = -2x^2 - 4x + 1 = -2(-1)^2 - 4(-1) + 1 = -2 + 4 + 1 = 3.$$

The vertex is $(-1, 3)$.

We choose some $x$-values on both sides of the vertex and graph the parabola.

| $x$ | $y$ | |
|---|---|---|
| −1 | 3 | ←Vertex |
| 0 | 1 | ←$y$-intercept |
| 1 | −5 | |
| −2 | 1 | |
| −3 | −5 | |

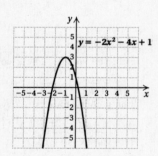

**15.** $y = \frac{1}{4}x^2$

Find the vertex. The $x$-coordinate is

$$-\frac{b}{2a} = -\frac{0}{2\left(\frac{1}{4}\right)} = 0.$$

The $y$-coordinate is

$$y = \frac{1}{4}x^2 = \frac{1}{4} \cdot 0^2 = 0.$$

The vertex is $(0, 0)$.

We choose some $x$-values on both sides of the vertex and graph the parabola.

| $x$ | $y$ | |
|---|---|---|
| 0 | 0 | ←Vertex |
| −2 | 1 | |
| −4 | 4 | |
| 2 | 1 | |
| 4 | 4 | |

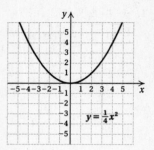

**17.** $y = 3 - x^2$, or $y = -x^2 + 3$

Find the vertex. The $x$-coordinate is

$$-\frac{b}{2a} = -\frac{0}{2(-1)} = 0.$$

The $y$-coordinate is

$$y = 3 - x^2 = 3 - 0^2 = 3.$$

The vertex is $(0, 3)$.

We choose some $x$-values on both sides of the vertex and graph the parabola.

| $x$ | $y$ | |
|---|---|---|
| 0 | 3 | ←Vertex |
| −1 | 2 | |
| −2 | −1 | |
| 1 | 2 | |
| 2 | −1 | |

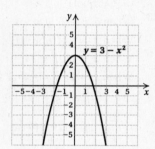

**19.** $y = -x^2 + x - 1$

Find the vertex. The $x$-coordinate is

$$-\frac{b}{2a} = -\frac{1}{2(-1)} = -\left(-\frac{1}{2}\right) = \frac{1}{2}.$$

The $y$-coordinate is

$$y = -x^2 + x - 1 = -\left(\frac{1}{2}\right)^2 + \frac{1}{2} - 1 = -\frac{1}{4} + \frac{1}{2} - 1 = -\frac{3}{4}.$$

The vertex is $\left(\frac{1}{2}, -\frac{3}{4}\right)$.

We choose some $x$-values on both sides of the vertex and graph the parabola.

| $x$ | $y$ | |
|---|---|---|
| $\frac{1}{2}$ | $-\frac{3}{4}$ | ←Vertex |
| 0 | −1 | ←$y$-intercept |
| −1 | −3 | |
| 1 | −1 | |
| 2 | −3 | |

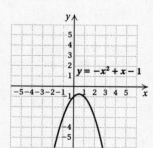

**21.** $y = -2x^2$

Find the vertex. The $x$-coordinate is
$$-\frac{b}{2a} = -\frac{0}{2(-2)} = 0.$$

The $y$-coordinate is
$$y = -2x^2 = -2 \cdot 0^2 = 0.$$

The vertex is $(0, 0)$.

We choose some $x$-values on both sides of the vertex and graph the parabola.

| $x$ | $y$ | |
|-----|-----|-----|
| 0 | 0 | ←Vertex |
| -1 | -2 | |
| -2 | -8 | |
| 1 | -2 | |
| 2 | -8 | |

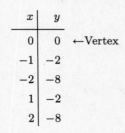

**23.** $y = x^2 - x - 6$

Find the vertex. The $x$-coordinate is
$$-\frac{b}{2a} = -\frac{-1}{2 \cdot 1} = -\left(-\frac{1}{2}\right) = \frac{1}{2}.$$

The $y$-coordinate is
$$y = x^2 - x - 6 = \left(\frac{1}{2}\right)^2 - \frac{1}{2} - 6 = \frac{1}{4} - \frac{1}{2} - 6 = -\frac{25}{4}.$$

The vertex is $\left(\frac{1}{2}, -\frac{25}{4}\right)$.

We choose some $x$-values on both sides of the vertex and graph the parabola.

| $x$ | $y$ | |
|-----|-----|-----|
| $\frac{1}{2}$ | $-\frac{25}{4}$ | ←Vertex |
| 0 | -6 | ←$y$-intercept |
| -1 | -4 | |
| 1 | -6 | |
| 2 | -4 | |

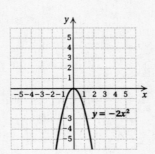

**25.** $y = x^2 - 2$

To find the $x$-intercepts we solve the equation $x^2 - 2 = 0$.
$$x^2 - 2 = 0$$
$$x^2 = 2$$
$x = \sqrt{2}$ or $x = -\sqrt{2}$    Principle of square roots

The $x$-intercepts are $(\sqrt{2}, 0)$ and $(-\sqrt{2}, 0)$.

**27.** $y = x^2 + 5x$

To find the $x$-intercepts we solve the equation $x^2 + 5x = 0$.
$$x^2 + 5x = 0$$
$$x(x + 5) = 0$$

$x = 0$  or  $x + 5 = 0$
$x = 0$  or     $x = -5$

The $x$-intercepts are $(0, 0)$ and $(-5, 0)$.

**29.** $y = 8 - x - x^2$

To find the $x$-intercepts we solve the equation $8 - x - x^2 = 0$.
$$8 - x - x^2 = 0$$
$$x^2 + x - 8 = 0 \qquad \text{Standard form}$$
$a = 1, \ b = 1, \ c = -8$

$$x = \frac{-1 \pm \sqrt{1^2 - 4 \cdot 1 \cdot (-8)}}{2 \cdot 1}$$
$$= \frac{-1 \pm \sqrt{33}}{2}$$

The $x$-intercepts are $\left(\frac{-1 + \sqrt{33}}{2}, 0\right)$ and $\left(\frac{-1 - \sqrt{33}}{2}, 0\right)$.

**31.** $y = x^2 - 6x + 9$

To find the $x$-intercepts we solve the equation $x^2 - 6x + 9 = 0$.
$$x^2 - 6x + 9 = 0$$
$$(x - 3)(x - 3) = 0$$

$x - 3 = 0$  or  $x - 3 = 0$
$x = 3$  or     $x = 3$

The $x$-intercept is $(3, 0)$.

**33.** $y = -x^2 - 4x + 1$

To find the $x$-intercepts we solve the equation $-x^2 - 4x + 1 = 0$.
$$-x^2 - 4x + 1 = 0$$
$$x^2 + 4x - 1 = 0 \qquad \text{Standard form}$$
$a = 1, \ b = 4, \ c = -1$

$$x = \frac{-4 \pm \sqrt{4^2 - 4 \cdot 1 \cdot (-1)}}{2 \cdot 1}$$
$$= \frac{-4 \pm \sqrt{20}}{2} = \frac{-4 \pm \sqrt{4 \cdot 5}}{2} = \frac{-4 \pm 2\sqrt{5}}{2}$$
$$= \frac{2(-2 \pm \sqrt{5})}{2} = -2 \pm \sqrt{5}$$

The $x$-intercepts are $(-2 + \sqrt{5}, 0)$ and $(-2 - \sqrt{5}, 0)$.

**35.** $y = x^2 + 9$

To find the $x$-intercepts we solve the equation $x^2 + 9 = 0$.
$$x^2 + 9 = 0$$
$$x^2 = -9$$

The negative number $-9$ has no real-number square roots. Thus there are no $x$-intercepts.

**37.** 
$$\sqrt{x^3 - x^2} + \sqrt{4x - 4} = \sqrt{x^2(x - 1)} + \sqrt{4(x - 1)}$$
$$= \sqrt{x^2}\sqrt{x - 1} + \sqrt{4}\sqrt{x - 1}$$
$$= x\sqrt{x - 1} + 2\sqrt{x - 1}$$
$$= (x + 2)\sqrt{x - 1}$$

**39.** $\sqrt{2}\sqrt{14} = \sqrt{2 \cdot 14} = \sqrt{2 \cdot 2 \cdot 7} = 2\sqrt{7}$

**41.** a) We substitute 128 for $H$ and solve for $t$:

$$128 = -16t^2 + 96t$$
$$16t^2 - 96t + 128 = 0$$
$$16(t^2 - 6t + 8) = 0$$
$$16(t - 2)(t - 4) = 0$$

$$t - 2 = 0 \quad \text{or} \quad t - 4 = 0$$
$$t = 2 \quad \text{or} \quad t = 4$$

The projectile is 128 ft from the ground 2 sec after launch and again 4 sec after launch. The graph confirms this.

b) We find the first coordinate of the vertex of the function $H = -16t^2 + 96t$:

$$-\frac{b}{2a} = -\frac{96}{2(-16)} = -\frac{96}{-32} = -(-3) = 3$$

The projectile reaches its maximum height 3 sec after launch. The graph confirms this.

c) We substitute 0 for $H$ and solve for $t$:

$$0 = -16t^2 + 96t$$
$$0 = -16t(t - 6)$$

$$-16t = 0 \quad \text{or} \quad t - 6 = 0$$
$$t = 0 \quad \text{or} \quad t = 6$$

At $t = 0$ sec the projectile has not yet been launched. Thus, we use $t = 6$. The projectile returns to the ground 6 sec after launch. The graph confirms this.